RAMAN SPECTROSCOPY FOR SOFT MATTER APPLICATIONS

RAMAN SPECTROSCOPY FOR SOFT MATTER APPLICATIONS

Edited by

Maher S. Amer

WILEY

A JOHN WILEY & SONS, INC., PUBLICATION

Published by John Wiley & Sons, Inc., Hoboken, New Jersey
Published simultaneously in Canada

Library of Congress Cataloging-in-Publication Data:

Amer, Maher S.
Raman spectroscopy for soft matter applications / Maher S. Amer.
p. cm.
Includes index.
ISBN 978-0-470-45383-4 (cloth)
1. Polymers–Spectra. 2. Raman spectroscopy. I. Title.
QC463.P5A44 2009
543′.57—dc22

2008049905

Printed in the United States of America

10 9 8 7 6 5 4 3 2 1

To my wife, Eman, and to my three best friends:
Lana, Sabri, and Zainah.

CONTENTS

PREFACE

In spite of the importance of soft matter as a major class of materials, and the increasing power of Raman spectroscopy as a characterization technique capable of interrogating matter on the molecular level, there was no devoted text focusing on applications of Raman spectroscopy in soft matter. I selected a number of experts from all over the world to put together this book *Raman Spectroscopy for Soft Matter Applications*, discussing, each in her/his field of expertise, how Raman spectroscopy has contributed to the generation of knowledge and the development of better and deeper understanding in each branch of soft matter. In addition, and due to the dynamic nature of the fields discussed, each expert has shared her/his ideas regarding the limitations that need to be overcome and the future developments in the field. The book is divided into three parts. In Part I, an overview of the field of soft matter is discussed as well as a concise explanation of the Raman theory and its basic instrumentation. In Part II, we discuss the Raman applications in polymers, including films, synthetic and natural fibers, polymer-based composites and nanocomposites, emulsions, liquid crystals, foams, and food analysis. We devoted Part III to the discussion of Raman spectroscopy utilization in medical applications focusing on Raman capabilities of bone imaging and its ability of cancer investigation in soft tissues.

This book is intended for graduate students and researchers working or willing to work in the field of Raman spectroscopy and its applications in any of the branches of soft matter. In each chapter, the specifics of Raman theory and instrumentation as related to the subject of the chapter were discussed. The book represents the thoughts of experts from different academic and industrial institutions from Denmark, England, France, Germany, Japan, Spain, and United States of America, hence representing different scientific ideologies and methodologies, giving the book a global scientific flavor.

MAHER S. AMER, EDITOR

Dayton, OH, USA
August 2008

CONTRIBUTORS

Maher S. Amer, Department of Mechanical and Materials Engineering, Wright State University, Dayton, OH 45435

Philippe Colomban, Universite Pierre et Marie Curie (UPMC—Paris 6) / CNRS, 2 rue H. Dunant, 94320 Thiais, France

Steve J. Eichhorn, Materials Science Centre, School of Materials, University of Manchester, Grosvenor Street, Manchester M1 7HS, UK

Oihana Elizalde, Polymer Research, BASF SE, GKD/F-B001, D-67056, Ludwigshafen, Germany

Søren Balling Engelsen, University of Copenhagen, Department of Food Science, Rolighedsvej 30, DK-1958 Frederiksberg C, Denmark

Peter Fratzl, Max Planck Institute of Colloids and Interfaces, Department of Biomaterials, Research Campus Golm, 14424 Potsdam, Germany

Sonja Gamsjäger, Ludwig Boltzmann Institute of Osteology at the Hanusch Hospital of WGKK and AUVA Trauma Centre Meidling, 4th Medical Department, Hanusch Hospital, Heinrich Collin Str. 30, 1140 Vienna, Austria

Gwénaël Gouadec, Universite Pierre et Marie Curie (UPMC—Paris 6) / CNRS, 2 rue H. Dunant, 94320 Thiais, France

Giriprasath Gururajan, Department of Chemical and Biomolecular Engineering, and Center for Advanced Engineering Fibers and Films, Clemson University, Clemson, SC 29634

Hiro-o Hamaguchi, Department of Chemistry, School of Science, The University of Tokyo, 7-1, Hongo, Bunkyo-ku, Tokyo 113-0033, Japan

Naoki Hayashi, Analysis Technology Center, FUJIFIL Corporation, 210 Nakanuma, Minami-Ashigara, Kanagawa 250-0193, Japan

Murat Kazanci, Max Planck Institute of Colloids and Interfaces, Department of Biomaterials, Research Campus Golm, 14424 Potsdam, Germany / Chemistry Department, St Francis Xavier University, Antigonish, NS, B2G 2W5, Canada

Ehiichi Kohda, Department of Radiology, Ohashi Hospital, Toho University School of Medicine, 2-17-6 Ohashi, Meguro-ku, Tokyo 153-8515, Japan

Jose Ramon Leiza, Institute for Polymer Materials (Polymat)/Department of Applied Chemistry, University of the Basque Country, Joxe Mari Korta Zentroa, Tolosa Etorbidea 72, 20018 Donostia-San Sebastian, Spain

Young-Kun Min, Department of Chemistry, School of Science, The University of Tokyo, 7-3-1 Hongo, Bunkyo-ku, Tokyo 113-0033, Japan

Satoru Naito, Tochigi Research Laboratories, Kao Corporation, 2606 Akabane Ichikai-machi, Haga-gun, Tochigi 321-3497, Japan

Amod A. Ogale, Department of Chemical and Biomolecular Engineering, and Center for Advanced Engineering Fibers and Films, Clemson University, Clemson, SC 29634

Eleftherios P. Paschalis, Ludwig Boltzmann Institute of Osteology at the Hanusch Hospital of WGKK and AUVA Trauma Centre Meidling, 4th Medical Department, Hanusch Hospital, Heinrich Collin Str. 30, 1140 Vienna, Austria

Tina Salomonsen, University of Copenhagen, Department of Food Science, Rolighedsvej 30, DK-1958 Frederiksberg C, Denmark

Frans van den Berg, University of Copenhagen, Department of Food Science, Rolighedsvej 30, DK-1958 Frederiksberg C, Denmark

Nanna Viereck, University of Copenhagen, Department of Food Science, Rolighedsvej 30, DK-1958 Frederiksberg C, Denmark

Hiroya Yamazaki, Department of Radiology, Tokyo Metropolitan Hiroo Hospital, 2-34-10 Ebisu, Shibuya-ku, Tokyo 150-0013, Japan

Robert J. Young, Materials Science Centre, School of Materials, University of Manchester, Grosvenor Street, Manchester M1 7HS, UK

I

INTRODUCTION, THEORY, AND INSTRUMENTATION

1

INTRODUCTION AND OVERVIEW

Maher S. Amer, Ph.D.
Professor of Materials Science and Engineering, Wright State University

1.1 INTRODUCTION

For many millennia, humankind has exploited matter in daily life. It has been only a few centuries since we have been able to engineer matter to control its properties. Over billions of years of evolution, nature has, however, engineered a material system to build its astonishing living constituents: "soft matter." Soft matter, as made by nature, is known to be versatile, adaptable, efficient, and—within the limits of our current knowledge—mysteriously capable of self-organizing and healing.

In spite of the fact that this book is not concerned with the theory of soft matter but rather with Raman spectroscopy and its capabilities in characterization of soft matter, we will briefly describe soft matter systems and their most important features and potentials in the following sections. Experimental characterization techniques utilized in investigating soft matter and the type of information they provide regarding the system will also be presented in Section 1.3.

Raman Spectroscopy for Soft Matter Applications, Edited by Maher S. Amer

1.2 WHAT SOFT MATTER IS

To start from the beginning, let us try to describe our universe from a materials scientist's viewpoint. We can say that the universe is made of energy and matter. These two entities are transformable into each other according to Einstein's famous equation, $E = mc^2$, and are bridged by another entity known to materials scientists as plasma. We usually classify matter into three different forms: solids, liquids, and gases. It can also be classified into two forms: hard matter (which includes metals and ceramics) and soft matter. It is scientifically safe to say that based on our deep understanding of molecular theory, many properties of hard matter are currently well understood. However, when it comes to soft matter, we have to admit that we are still on a steep learning curve.

If one asks the simple question, "what is soft matter?" the answer would not be so simple. Soft matter is a complex and flexible sort of matter referred to sometimes as "complex fluids." The term "fluids" should not be taken literally since soft matter is not necessarily fluids. Other terms such as "colloidal suspensions" and "colloidal dispersions" have also been used to describe soft matter [1]. The term describes a very wide range of matter including polymers, colloids, liquid crystals, emulsions, foams, biological tissues, as well as a wide variety of other materials such as milk, mayonnaise, and ice cream [2]. Soft matter systems can exhibit a drastic change in their mechanical, optical, and electrical properties as the result of very mild chemical or conformational changes in their structures [3]. To give a more quantitative description, it can be said that at ambient temperatures, molecular kinetic energy within a soft matter system is close to k_BT, a quantity that is much smaller than k_bT in a hard matter system[1] [4]. Considering structural features, soft matter systems exhibit a structural ordering intermediate between that of crystalline solids and liquids with a periodicity typically in the range of 1 nm to 1 μm. This makes soft matter systems ordered on the nano and/or the meso[2] length scales, depending on the system.

What really add to the importance of soft matter as a class of materials are the interesting theoretical analogies and behavior resemblances sometimes observed between them and other scientific fields. For example, S.F. Edwards [5] observed the correspondence between the

[1]Here, k_B is Boltzmann's constant and T is absolute temperature.

[2]Meso (Latin for "in between") scale covers lengths less than 1 μm and larger than 100 nm. Sometimes, it is extended down to a 10-nm limit in the literature.

conformations of a flexible chain and the trajectories of a nonrelativistic particle in the presence of external potential. He also showed that both systems are ruled by the same Schrodinger equation. This observation has been the key to all later developments in polymer statistics [3]. An interesting overlap in thought between the highbrow string theories and the descriptions of soaps has also been discussed [6].

Last, but definitely not least, it is possible to state that the most striking feature of soft matter is the ability to utilize it to create new forms of matter with properties we could only dream of a few decades ago. Microemulsions and liquid crystals are two good examples for such new forms of matter [7]. It is also expected to see, within our lifetime, more development for new forms of matter with novel engineered properties and adaptively controlled structure that we can as yet only imagine. The recently reported [8] colloidal crystals exhibiting negative Poisson's ratio that can be controlled in the range of 1 to 2 is a good example. Such ability is crucial for utilization in other advanced applications such as narrowband rejection filters, nanosecond optical switches, and sensors [9].

1.3 CHARACTERIZATION TECHNIQUES OF SOFT MATTER

A necessity for the understanding and further development of a new class of materials is a powerful characterization technique capable of interrogating such system on an appropriate length scale. A number of experimental methods were employed in investigating the structure and performance of soft matter systems. Such experimental techniques include microscopy, spectroscopy, scattering methods, rheology, calorimetry, and surface structure probes such as atomic force probe (AFP or AFM) and scanning tunneling probe (STP or STM). In the following sections, we will briefly describe these experimental techniques. The rest of the book is dedicated to Raman spectroscopy and its applications in investigating different classes of soft matter.

1.3.1 Microscopy and Surface Probes

Since the length scale of interest in soft matter is the nano-/mesoscale, optical microscopy (with a Rayleigh limit $\lambda/2$ on the spatial resolution) suffers from serious resolution limitations as a structural analysis method. However, the technique can be useful in viewing aggregated structures larger than $1\,\mu m$ formed in the system. Polarized optical microscopy was also found useful to identify birefringent structures rather than examining the structures themselves in cases such as spherulites in polymers. Certain optical imaging methods such as differential interference contrast (DIC) that relies on the interference between light waves reflected from different regions (with different birefringence or thickness) in the sample, were developed and used as identification techniques in plants and biological samples [10]. Recent development in near-field scanning optical microscopy (NSOM) techniques with a spatial resolution in the range of $\lambda/20$ has allowed better characterization of the structure in soft matter.

Electron microscopy (both in transmission and scanning modes TEM and SEM) provides a good means for examining the structure down to sub-nanometer resolution. Elaborate sample preparation, and the no fluid restriction in transmission electron microscopy (TEM) (due to high vacuum required in the microscope chamber) impose restrictions on the type of soft matter samples that can be examined using such techniques. The development of environmental scanning electron microscopy (ESEM) and field emission (FESEM) has greatly aided in investigating the structure of soft matter.

Surface probe microscopy (SPM) techniques such as atomic force microscopy (AFM) and scanning tunneling microscopy (STM), with their atomic resolution, have proven to be invaluable in characterizing surface structures in soft matter on the nanoscale. The force probing function in AFM was also used to measure local mechanical constants (such as local stiffness) in soft matter as well.

1.3.2 Scattering Methods

Scattering of light, as well as X-rays and neutrons, was used to investigate the structure of soft matter. Static light scattering, especially small angle light scattering (SALS) can be used to determine the particle size. Data analysis in this case will depend on the particle size relative to the wavelength of the light (λ). Dynamic light scattering can also be employed to measure relaxation times and translational diffusion coefficients in soft matter.

X-ray and neutron scattering measurements can also be used to probe the structure of soft matter on a much smaller level (Typically 1 Å). Wide-angle X-ray scattering (WAXS) with scattering angles (2θ) larger than $10°$ has been successfully utilized in investigating structures and crystallinity in soft mater such as polymers. Small-angle X-ray spectroscopy (SAXS) with scattering angles (2θ) smaller than $10°$ has been used to characterize structures in the range of 5 to 100 nm within soft matter as well. Small-angle neutron scattering (SANS) is very similar to SAXS in principle and has also been extensively utilized in structural investigations of soft matter. A fundamental difference between the two techniques, however, arises from the fact that X-ray scattering intensity depends on electron density distribution in the scattering medium while neutron scattering intensity depends on nuclear scattering length density. Hence, SANS can clearly distinguish between hydrogen atoms ^{1}H and deuterium atoms ^{2}H, and has been utilized extensively in characterizing soft matter structure by deuterium substitution in parts that need to be investigated.

1.3.3 Rheology and Calorimetry

Rheology was utilized to investigate mechanical deformation and flow characteristics of soft matter under stress. Flow behavior, i.e. of Newtonian, non-Newtonian, or Bingham nature, of soft matter is usually determined by rheology studies and can be correlated to its structure. Calorimetric studies, however, are used to investigate phase transitions in soft matter. Differential scanning calorimetry (DSC) is the most widely used technique to investigate soft matter phase transitions, determine transition temperatures, and to measure the associated transition enthalpy.

1.3.4 Spectroscopy

Different spectroscopic techniques have been utilized in characterizing soft matter. Techniques such as nuclear magnetic resonance (NMR), dielectric spectroscopy, infrared spectroscopy, and Raman spectroscopy have all been proven powerful techniques in characterizing and understanding soft matter. Here, the ability of such spectroscopic techniques will be briefly introduced.

NMR has been utilized to probe chain local microstructure and to determine the number average molecular weight and the dynamics of segmental motion in polymers, as well as composition of copolymers. The technique was also used to provide information on orientational ordering in liquid crystals, and to determine translational self-diffusion coefficient in amphiphiles and colloids' dilute solutions.

Dielectric spectroscopy with its very wide range of frequency coverage (10^{-5} to 10^{10} Hz) provides information complementary to that provided by rheology. Dielectric spectroscopy, however, can cover much higher frequencies than that covered in rheology and thus becomes the technique of choice when dynamic mechanical properties of soft matter need to be probed at high frequency range.

Infrared spectroscopy is known to be a complementary technique to Raman spectroscopy. The technique is used extensively for identification purposes in soft matter. In addition, the technique is used to provide information regarding chain branching, tacticity, length of unsaturated groups in polymer as well as information regarding hydrogen bond network formation in surfactant and colloidal solutions. Polarized infrared spectroscopy, as in Raman spectroscopy, can also be used to determine the orientation of a particular chemical bond within a soft mater system.

1.4 OVERVIEW OF THE BOOK

In Chapter 2, Philippe Colomban, and Gwénaël Gouadec of CNRS, Paris, France, give a simplified overview of the Raman scattering theory and elements of Raman instrumentation. In Chapter 3, Giriprasath Gururajan and Amod A. Ogale of Clemson University discuss Raman applications in polymer films and its ability to conduct *in situ*, real-time characterization of the films. Specialized instrumentation for this application and the impact of Raman characterization on better processing of polymer films are discussed. In Chapter 4, Robert Young and Steve Eichhorn of University of Manchester, England, review Raman applications in investigating synthetic and natural polymer fibers and their composites. Information provided by Raman spectroscopy regarding the structure and mechanical properties of such fibers including their composites, as well as the potential of Raman spectroscopy techniques in investigating nanocomposites, is discussed. In Chapter 5, Oihana Elizalde and Jose Ramon Leiza of BASF, Germany, and University of the Basque Country, Spain, respectively, review emulsions and emulsion polymerization, and Raman spectroscopy applications in this branch of soft matter. A study of liquid crystals and their characterizations using Raman spectroscopy is given in Chapter 6 by Naoki Hayashi of Fuji Photo Film Co., Japan. In Chapter 7, Maher S. Amer of Wright State University discusses foams, their structure, importance, and the information that can be provided by Raman spectroscopy regarding this important class of matter. In Chapter 8, Søren Balling Engelsen and his team from the University of Copenhagen, Denmark, review Raman applications in the food industry and the characterization capabilities of the Raman technique in this field.

Medical applications of Raman spectroscopy are discussed in the third section of this book. In Chapter 9, Peter Fratzl of Max Planck Institute of Colloids and Interfaces, and his coauthors S. Gamsjäger and

E.P. Paschalis of Ludwig Boltzmann Institute of Osteology, Austria, and M. Kazanci of St. Francis Xavier University, Canada, discuss the capabilities, potentials and impact of Raman spectroscopy imaging on bone characterization and investigation. Raman applications in cancer studies and the invaluable information provided by the technique are discussed in Chapter 10 by Hiro-o Hamaguchi of the University of Tokyo and his coauthors of Kao corporation, Tokyo Metropolitan Hiroo Hospital, and Toho University School of Medicine, Japan.

This book is intended as a reference for experienced researchers working in any of the aforementioned branches of soft matter, and as a guide for researchers and graduate students preparing to investigate and explore the amazing world of soft matter.

REFERENCES

1. Likos, C.N. (2001). Effective interactions in soft condensed matter physics. *Phys. Reports* 348, 267–439.
2. Hunter, R.J. (1986). *Foundations of Colloid Science*, Vol. I. Oxford: Clarendon Press.
3. de Gennes, P.G. (1992). Soft matter. *Rev. Modern Phys.* 64(3), 645–647.
4. Hamley, I.W. (1999). *Introduction to Soft Matter*. New York: Wiley.
5. Edwards, S.F. (1965). The statistical mechanics of polymers with excluded volume. *Proc. Phys. Soc. London* 85, 613.
6. Porte, G., Marignan, J., Bassereau, P., May, R. (1988). Shape transformations of the aggregates in dilute surfactant solutions: A small-angle neutron scattering study. *J. Phys. (Paris)* 49, 511.
7. Muthukumar, M., Ober, C.K., Thomas, E.L. (1997). Competing interactions and levels of ordering in self-organizing polymeric materials. *Science* 277, 1225.
8. Baughaman, R.H., Dantas, S.O., Stafsström, S., Zakhidov, A.A., Mitchell, T.B., Dubin, D.H.E. (2000). Negative Poisson's ratios for extreme states of matter. *Science* 288, 2018.
9. Weissman, J.M., Sunkara, H.B., Tse, A.S., Asher, S.A. (1996). Thermally switchable periodicities and diffraction from mesoscopically ordered materials. *Science* 274, 959.
10. Ruzin, S.E. (1999). *Plant Microtechnique and Microscopy*. New York: Oxford University Press.

2

RAMAN SCATTERING THEORY AND ELEMENTS OF RAMAN INSTRUMENTATION

Philippe Colomban, Ph.D.
Research Fellow at CNRS, Head of LADIR Laboratory
Gwénaël Gouadec, Ph.D.
Assistant Professor at LADIR Laboratory, Université Pierre et Marie Curie

2.1 INTRODUCTION

Among the many interactions of light with matter, Raman scattering is particularly suited to the multiscale analysis of ill-organized heterogeneous samples. First, the submicronic spatial resolution of optical instruments is accessible because a laser in the near infrared–near ultraviolet range is used as the excitation. Second, the probe being for interatomic bonds, the technique offers a "bottom-up" approach to nanomaterials, which best works in the case of imperfect crystals with strong covalent bonds—typically polymers.

In recent years, the development of instruments incorporating improved detection and processing devices has paved the way to a new generation of low-price, sometimes transportable, Raman spectrometers. The computer-assisted processing of the spectra now makes it

Raman Spectroscopy for Soft Matter Applications, Edited by Maher S. Amer

possible to analyze geometry, phase distribution, stress, or thermal state quantitatively, with submicronic resolution.

The purpose of this chapter is to provide nonspecialists with introductory knowledge on the theory of micro-Raman Spectroscopy (μRS) for the analysis of nonmetallic materials.

2.2 THE RAMAN EFFECT

The polarization of a sample illuminated with light (electric field $\vec{E}_0$, frequency ν_l) has the following form:

$$\vec{P} = \bar{\bar{\alpha}} \times \vec{E}_0 \cos(2\pi\nu_l t). \tag{2.1}$$

In Equation 2.1, $\bar{\bar{\alpha}}$ represents the polarizability tensor, whose dependence to matter vibrations (the atom oscillations around their equilibrium position) can be expressed as a function of normal coordinates Q,[1] using a Taylor approximation:

$$\alpha_{ij} = \alpha_{ij}^0 + \left(\frac{\partial \alpha_{ij}}{\partial Q}\right)_{Q=Q_0} \times Q, \quad \text{with} \quad (\mathrm{i},\mathrm{j}) = x, y \text{ or } z \text{ and } Q = Q_0 \cos(2\pi\nu_{vib} t). \tag{2.2}$$

The combination of Equations 2.1 and 2.2 yields the following expression for the polarization components:

$$P_i = \sum_j \alpha_{ij} \times E_j = \sum_j \left[\alpha_{ij}^0 E_{0_j} \cos(2\pi\nu_l t) + \frac{E_{0_j} Q_0}{2}\left(\frac{\partial \alpha_{ij}}{\partial Q}\right)_{Q=Q_0} \times [\cos(2\pi(\nu_l - \nu_{vib})t) + \cos(2\pi(\nu_l + \nu_{vib})t)] + \cdots\right] \tag{2.3}$$

With the scattered electric field being proportional to $\vec{P}$, Equation 2.3 thus predicts both elastic ($\nu = \nu_l$) and inelastic ($\nu = \nu_l \pm \nu_{vib}$) scattering by atomic vibrations. The former is called Rayleigh scattering, and the latter, which only occurs if vibrations change polarizability ($\partial\alpha_{ij}/\partial Q \neq 0$), is the Raman scattering. This frequency shift was predicted by Brillouin [1] and Smekal [2] but was named after Sir C.V. Raman, the Indian scientist who led one of the teams that collected first experimental evidences in 1928 [3]. Raman scattering is complementary to infrared

[1]A general result of wave mechanics is that atomic displacements associated with matter vibrations can be expressed as a function of such harmonic coordinates.

absorption (which also investigates vibrations), with the advantage of much narrower peaks,[2] the width of which can be used to characterize amorphous and low-crystallinity solids. However, as Raman scattering only affects one out of a million incident photons, its study could not generalize before the advent of lasers—powerful monochromatic excitations—in the 1970s. The availability of high-sensitivity charge-coupled devices (CCDs) designed for multichannel detection later made it reasonable, timewise, to work with very low laser powers (a necessary condition to analyze colored materials without burning them). It also reduced the acquisition time for a given power and allowed tight surface mappings or real-time monitoring of the effect of an external perturbation (such as temperature, stress, voltage)[4]. The latest trend in Raman spectroscopy is "hyperspectral" imaging, where a Raman mapping of a surface is used to image a physical or chemical property, after due processing of the spectral features by dedicated programs.

Raman spectroscopists normally refer to vibrations by their wave number, $\bar{\nu} = \nu_{vib}/c$ (c, the light speed; $\bar{\nu}$ in cm^{-1} unit), and the classical electromagnetic theory of oscillating dipoles predicts Raman peaks should have a Γ_0-wide Lorentzian shape:[3]

$$I(\bar{\nu}) \propto \frac{1}{[\bar{\nu} - \bar{\nu}_{vib}]^2 + \left(\frac{\Gamma_0}{2}\right)^2}. \tag{2.4}$$

The signal intensity is predicted with the following formula [5]:

$$I_{Raman} \propto I_{laser} \bar{\nu}_{laser}^4 \left| e_0 \,\bar{\bar{\alpha}}\, e_s \right|^2 d\Omega. \tag{2.5}$$

In Equation 2.5, e_o and e_s are unit vectors indicating the laser polarization and direction of observation, respectively, whereas $d\Omega$ represents the solid angle of light collection.

As polarizability changes drastically from one bond to another, Raman intensity may not be used to measure the relative amounts of different phases. This limitation can sometimes be an advantage when secondary phases like an enamel pigment [6] or carbon in SiC fibers [7] can be detected in very small quantities (even traces). From a

[2]The probes of infrared spectroscopy, the instantaneous dipole moments, are subject to long-distance interactions.

[3]The experimental bands are a convolution between this natural line shape, the instrumental transfer function, and the disorder-induced distribution of vibrators. It is often taken as a Gaussian or Voigt function (a perfectly symmetric convolution of a Lorentzian function with a Gaussian function).

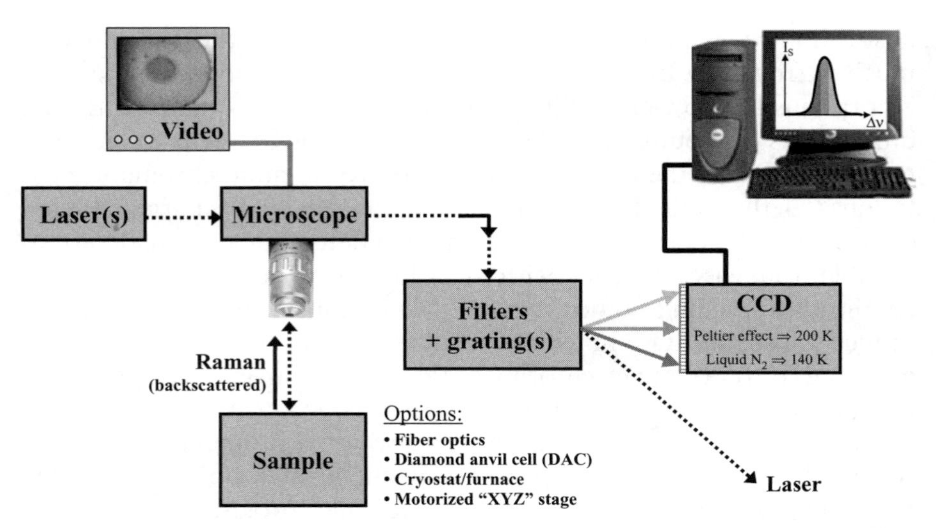

Figure 2.1 Principle of a conventional micro-Raman spectrometer, in backscattering geometry.

general point of view, elements with high atomic numbers from the right side of the periodic table (covalent materials) exhibit high Raman scattering efficiency, whereas ionic structures are difficult to analyze with Raman spectroscopy. As for metals, their surface plasmons limit light penetration and their Raman signal is therefore extremely weak.

2.3 ELEMENTS OF RAMAN INSTRUMENTATION

Figure 2.1 shows the basic architecture of a Raman microspectrometer. An up-to-date equipment now includes holographic "notch" or "edge" filters for the rejection of the elastic scattering (laser line), one or several gratings for Raman signal dispersion, and a liquid nitrogen or Peltier effect-cooled CCD mosaic for detection. The laser sources are often built-in, but light coming from external excitation sources can also be used. A higher laser wavelength improves the dispersion of the gratings, hence the spectral resolution (~0.5 cm^{-1} can easily be achieved in red), but also drastically reduces the signal intensity (according to Equation 2.5). The choice of the laser line eventually depends on whether a limited acquisition time or a high spectral resolution is more important to the user.[4] The 514.5-nm line of an Ar^+ ion laser and the

[4]In case of colored materials, the wavelength choice will depend on whether resonance (see further) must be favored (for improved selectivity of the resonant phase) or limited (so as not to mask the signal from nonresonant phases).

532-nm line of a less-expansive frequency-doubled Nd:YAG laser generally offer a good compromise.

The main options of Raman microspectrometers are motorized stages for XY(Z) mapping and optical fiber plugs for connection to remote optical heads. Some spectrometers have a "macro" setup where the laser beam (section ~0.1 mm^2) is sent directly to the sample, rather than through the microscope, and the scattered light is collected at 90° by a camera lens. The last generation of equipments is transportable and the use for quality control is developing.

2.4 THE SPATIAL RESOLUTION OF CONVENTIONAL RAMAN MICROSPECTROSCOPY

Due to the wave nature of light, it is impossible to distinguish points closer than half the wavelength λ that is used to probe a sample. This results from diffraction effects and places the theoretical limit of μRS lateral resolution in the 0.2-μm range (~400-nm excitation). The actual value depends on the numerical aperture, and the lateral resolution of μRS is usually said to be around 1 micron [8]. Attempts to perform Raman near-field scanning optical microscopy (RNSOM) [9,10] have improved the lateral resolution (Figure 2.2) but at the cost of a sensitivity reduction and experimental setting difficulties [11] that preclude, so far, the analysis of most materials [12].

An estimate of the axial resolution Δz of μRS (Figure 2.2) is given by half the width of the axial intensity profile. This parameter, commonly referred to as the depth of field, is generally around twice the lateral resolution [13].

2.5 THE INTERPRETATION OF RAMAN SPECTRA

According to Equation 2.3, each vibration contributes twice to Raman scattering depending on whether it is excited (Stokes side, $\nu_{scattered} = \nu_l - \nu_{vib}$) or annihilated (anti-Stokes side) by the laser. The intensity ratio of the Stokes to anti-Stokes contributions obviously depends on the vibrational population and is used, on occasion, to measure the temperature of a sample (through Boltzmann statistics) [14]. However, Stokes and anti-Stokes contributions carry the same vibrational information, and only the most intense Stokes contribution is usually shown on Raman spectra. A schematic is given in Figure 2.3, with the different levels of information that can be accessed. Note that

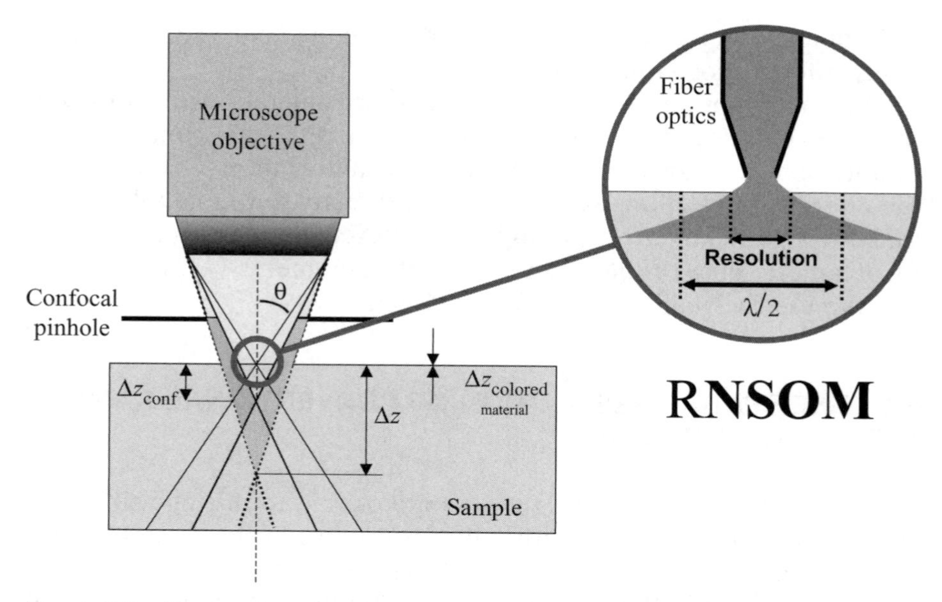

Figure 2.2 Observation of a sample through a Raman microscope. A confocal hole rejects the shadowed light and facilitates a more accurate in-depth analysis ($\Delta z_{conf} < \Delta z$). The insert shows one way of performing Raman near-field scanning optical microscopy (RNSOM) by using the tapered end of a metal-coated fiber optics for light delivery. (Adapted from Gouadec, G., Colomban, Ph. [2007]. *Prog. Cryst. Growth Charact. Mater.*, 53, 1–56.)

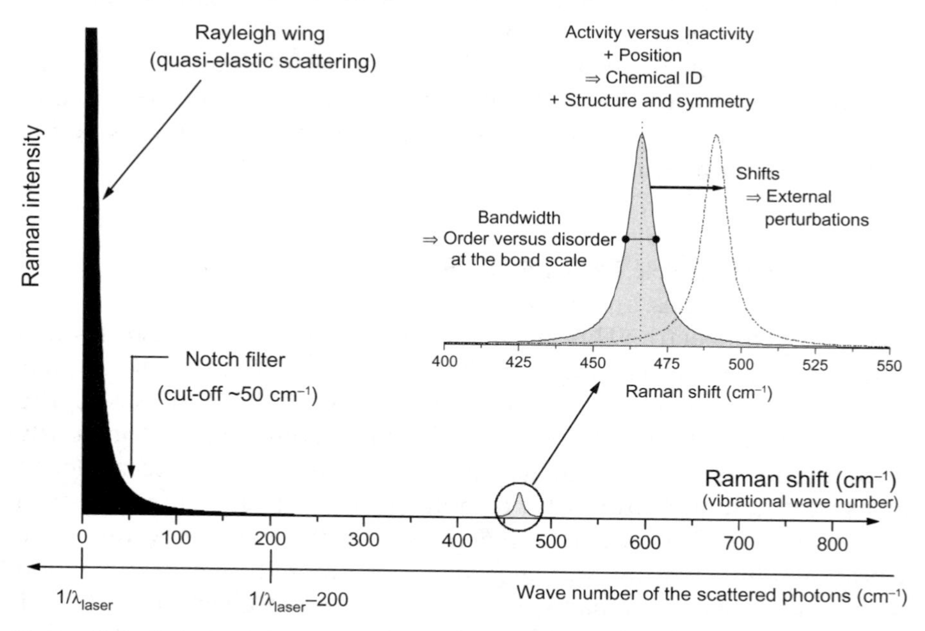

Figure 2.3 Schematic representation of Stokes Raman scattering. The anti-Stokes side (negative Raman shifts) is not shown because Raman intensity is weaker.

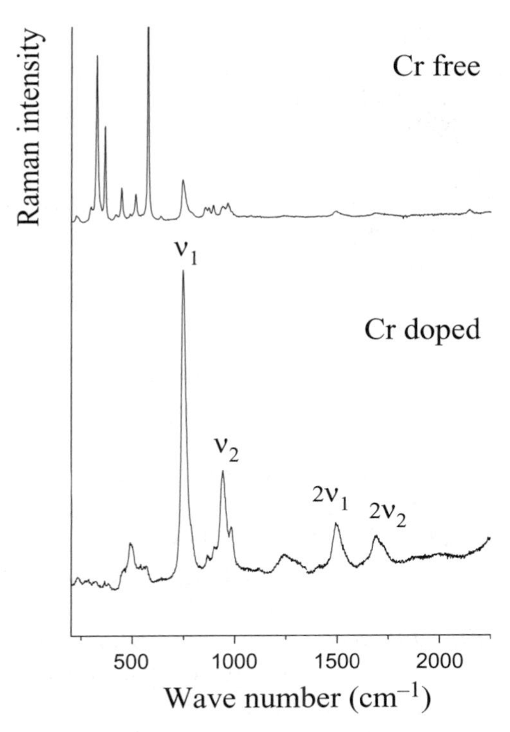

Figure 2.4 Raman spectra recorded at 1-mW power illumination on Cr-free and Cr-doped tin-based sphene (malayite) pink pigment with λ = 531.8-nm excitation. ν_1 and ν_2 are fundamental Sn–O stretching bands. (Adapted from Faurel, X., Vanderperre, A., Colomban, Ph. [2003]. *J. Raman Spectrosc.*, 34, 290–294.)

the Raman shifts (the wave number difference between the scattered and incident photons) are directly equivalent to the vibration wave numbers.

In colored samples, the energy of visible lasers approaches that of electronic transitions, and Raman spectra may be mingled with photoluminescence spectra. This is called *resonance Raman spectroscopy* (RRS) because the consequence is a strong enhancement of some vibrational modes [15,16]. For instance, it is obvious from Figure 2.4 that the few percent Cr doping of sphene (a mixed calcium–tin silicate) used to color enamel pigments in pink has a very strong effect on the intensity of all Raman peaks involving the chromophore.[5] The light

[5]The probed chemical bonds are in a markedly disturbed state in the case of RRS, and all vibrational modes involving the chromophore are enhanced, including harmonic ($2\nu_{vib}$, $3\nu_{vib}$, ...) and combination ($\nu_1 + \nu_2$, $\nu_1 - \nu_2$, ...) modes which would otherwise be invisible.

penetration is reduced to a few tens of nanometers only in case of RRS ($\Delta z_{\text{colored}}$ in Figure 2.2), and using excitation wavelengths close to the electronic absorption threshold can help separate surface from bulk Raman contributions [17] or discriminate layers in samples polished with the appropriate angle [18].

Equation 2.3 points out the sensitivity of Raman scattering to both the electrical (α_{ij}) and mechanical (ν_{vib}) properties of the investigated materials. Two kinds of parameters will therefore influence the spectrum:

(1) Parameters governing vibration-induced charge transfers (iono-covalency, band structure, electronic insertion) will set Raman scattering intensity. Although very few authors have commented on that specific point, Raman intensity can thus be very informative. One way of enhancing Raman intensity is to disperse the sample on metallic surfaces (either roughened wafers or colloidal solutions). The photon–plasmon interaction results in a huge signal enhancement, and the technique, referred to as *surface-enhanced Raman spectroscopy* (SERS), is often used to observe diluted samples (for instance, to study a single molecule-to-single molecule interaction) [19,20].

(2) Parameters governing the bond "dynamics," in first approximation the reduced mass (μ) and bond strength (k) of a harmonic modeling, will set the peak positions (the eigenfrequencies of matter vibrations)[6]:

$$\bar{\nu} = \frac{1}{2\pi c}\sqrt{\frac{k}{\mu}}. \tag{2.6}$$

Realistic bonds are actually anharmonic since they weaken in tension (they even end up dissociating) and strengthen in compression (electronic clouds repel each other). Vibrational wave numbers turn out to be sensitive to the actual bond lengths, and the stretching (compression) of a solid results in a decrease (increase) of Raman wave numbers [21], which is usually reported to be linear.

$$\Delta\bar{\nu} = S^{\varepsilon} \times \Delta\varepsilon = S^{\varepsilon} \times \frac{\Delta\sigma}{E} = S^{\sigma} \times \Delta\sigma. \tag{2.7}$$

[6]The system geometry (actual interatomic distances, atomic substitutions) can obviously influence the effective strength of a bond formed with two given atoms.

In Equation 2.7, S^{ε} and S^{σ} are negative coefficients,[7] and E, Young's modulus, is the proportionality factor between the macroscopic strain (ε) and the macroscopic stress (σ). Micro-Raman extensometry (μRE) uses Equation 2.7 to measure stress fields using Raman spectra, and it proves particularly useful when the stress is subject to variations at the micrometric scale, for instance, in composites [22,23][8] or in microelectronics devices [24]. A reliable measurement necessitates an absolute reference (unstressed sample), a careful calibration of the $S^{\varepsilon/\sigma}$ coefficients,[9] and a consideration of possible thermal effects under laser illumination [25].[10]

2.5.1 Vibrations in Crystalline Solids

All atomic positions in a crystal derive from those in a reference unit cell by translational symmetry. Because of this long-distance periodicity, only collective harmonic vibrations, the so-called phonons, can propagate. In a crystal containing N unit cells, each formed with p atoms, the total number of such phonons is given by (3pN-6),[11] and their wave vectors $(\vec{k})$, which indicate the direction of wave propagation, all belong to a volume of the reciprocal space (energy versus wave vector representation) called the Brillouin zone (BZ) [26]. In this zone, phonons are distributed along p "dispersion branches," depending on whether neighboring atoms vibrate in phase (acoustic branch) or out of phase (optical branch), and either parallel (longitudinal branch) or perpendicular (transversal branch) to $\vec{k}$ (see Figure 2.5 for an example of longitudinal and transverse optical phonons).

The scattering of one photon (wave vector $\vec{k}_{h\nu}$) by one phonon $(\vec{k}_{vib})$ is governed by the momentum conservation rule and because $\vec{k}_{h\nu}$ is almost nil, only Brillouin zone center (BZc) phonons (long-

[7]These coefficients have different symbols in the literature. In crystals, they are often measured for uniaxial stresses applied along the main crystallographic directions.
[8]A truly *in situ* analysis of the fiber is possible only in translucent matrix composites. The alternative is to observe a fiber entering the matrix (a matrix that appears opaque to the eye can be translucent through the first few tens of microns) or to work on polished cross sections.
[9]$S^{\varepsilon} \sim -10$ cm^{-1} %$^{-1}$ for the first-order modes in high-crystallinity carbon fibers.
[10]If the absorption is very high (presence of resonant species), temperature may rise at the point of laser impact, even for only a few μW/μm^2 incident power. This effect often results in a wave number shift or a chemical degradation of the sample (oxidation, decomposition, etc.) It can be reduced if the sample is dispersed in a heat-conducting matrix and put in a rotating cell or in a low-temperature cryostat.
[11]The six modes that are removed correspond to rotations and translations of a rigid molecule, which are not considered proper vibrations.

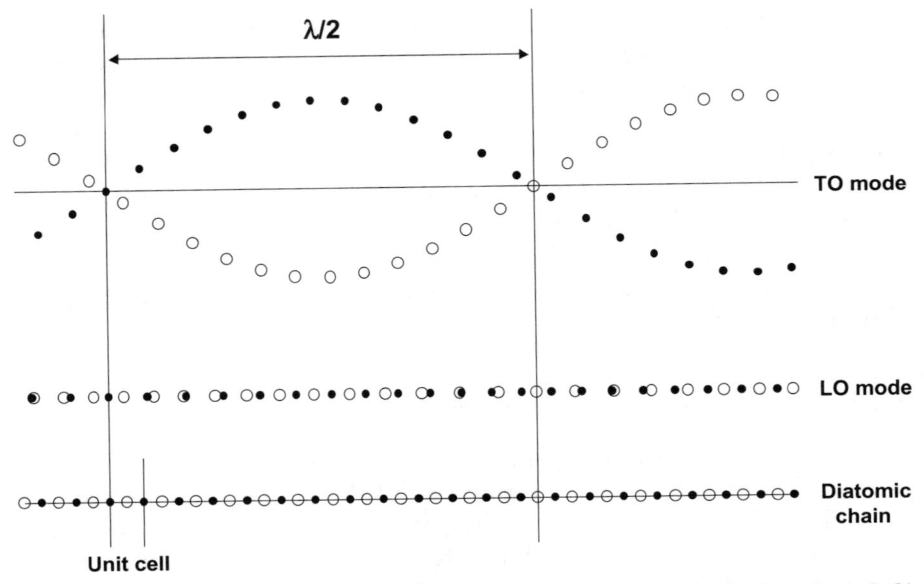

Figure 2.5 Examples of transverse optical (TO) and longitudinal optical (LO) phonons in a 1-D diatomic lattice.

wavelength phonons) can be active in first-order Raman spectra. Besides, acoustic vibrations vanish at BZc and only zone center optical modes can therefore be observed. The condition for one such mode to be active ($\partial\alpha_{ij}/\partial Q \neq 0$ in Equation 2.3) is governed by the symmetry of the crystal, and Raman activity can eventually be predicted through Group Theory [27].

However, nonagreement between the crystallography-found symmetry and the observations on the Raman spectra happens for internal modes (and, on occasion, external modes) whenever the structure is not perfect at the scale of the nanoregions probed with Raman spectroscopy. For instance, no Raman scattering would be expected at all if the high-temperature phase of the perovskite piezoceramic studied in Figure 2.6 had the exact cubic symmetry found with X-ray scattering [28]. The low residual intensity that is detected above the transition temperature indicates some degree of imperfection in the structure.

2.5.2 Vibrations in Nanocrystals

The translational symmetry in a crystal is lost at any grain boundary, which results in the appearance of specific surface and interface vibrational modes [29]. In the case of nanocrystals, where the concentration

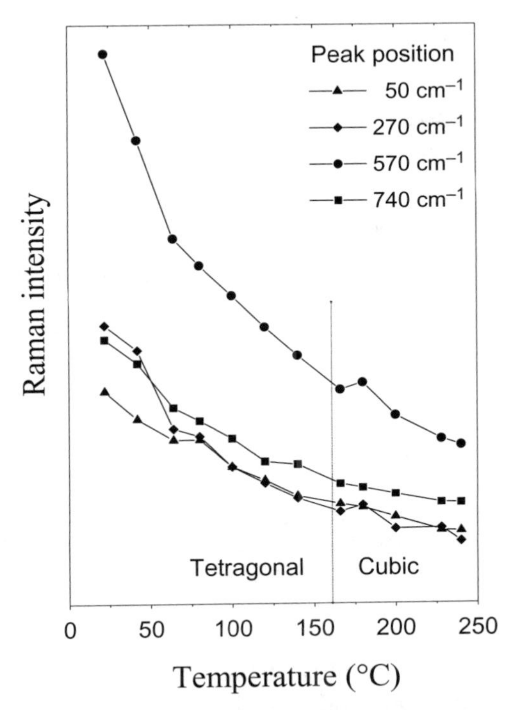

Figure 2.6 Room temperature-measured intensity of four Raman peaks in a PMN-34.5%PT textured perovskite ceramic previously heated up to 240 °C. The dotted line indicates the ferroelectric (tetragonal) to paraelectric (cubic) transition temperature. (Adapted from Slodczyk, A., Barré, M., Romain, F., Colomban, Ph., Pham-Thi, M. [2007]. *J. Phys.: Conf. Ser.*, 92, 012159.)

of grain boundaries is very high, these "nonbulk" modes can contribute very significantly to the spectra. Besides, the outer atomic layers of nanocrystals often react with neighboring species (lattice reconstruction, passivation/corrosion layers, contamination) to form interphases, with spectral contributions that become more visible as crystal size decreases (Figure 2.7).

Yet, as long as grains remain larger than ~10–15 nm, the Raman spectrum of nanocrystals often remains sufficiently similar to that of the corresponding single crystal to facilitate direct identification of the phases and, sometimes, to monitor their transitions.[12] The restriction of the phonon propagation only introduces some uncertainty on the wave number, which activates theoretically forbidden modes in the vicinity of BZc. In the most frequent situation where optical branch maxima

[12]The interpretation of Raman spectra from particles smaller than 10 nm is discussed in Reference [11] and in references therein.

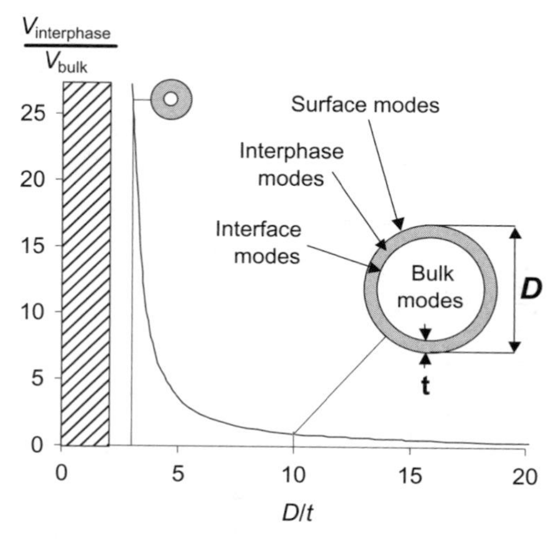

Figure 2.7 The interphase-to-bulk volume ratio in spherical nanocrystals with outer interphase thickness *t* and diameter *D*. (Adapted from Gouadec, G., Colomban, Ph. [2007]. *Prog. Cryst. Growth Charact. Mater.*, 53, 1–56.)

stand at BZc, this results in a low wavenumber side asymmetry and the broadening of the crystal modes, as is obvious for ceria (cubic cerium oxide) in Figure 2.8. The band broadening can be amplified by the high defect density encountered in many nanograined materials. For instance, zirconia exhibits a much wider signature in Figure 2.8 than the same structure ceria because it hosts vacancies which, despite the high crystallinity, break the vibrational periodicity [30].

2.5.3 Vibrations in Amorphous Matter

In amorphous matter, the total lack of periodicity impinges the propagation of collective waves, and phonons are replaced by more or less localized bond vibrations. When structures made of strong covalent bonds can be isolated in the atomic network, one can expect to recognize the spectral features of the equivalent "molecule-like" vibrational unit on the Raman spectrum. If these "molecular" unit cells are sufficiently connected, low-frequency modes should also appear, showing the very slight rotational and translational oscillations of a single atom (R′/T′) or a group of vibrational units (librational/lattice modes) that the structure will allow [31].

In a "molecular" description of solids, vibrations are classified as either stretching (ν), bending (δ), or torsional (τ) modes, the width of which is highly sensitive to the slightest short-distance disorder in the

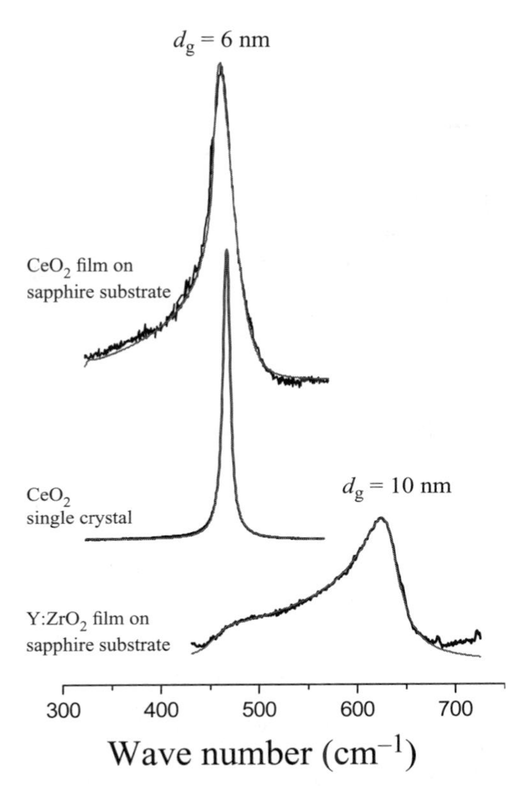

Figure 2.8 Effect of the grain size (d_g) on ceria and yttrium-doped zirconia Raman spectra. (Adapted from Kosacki, I., Petrovsky, V., Anderson, H.U., Colomban, Ph. [2002]. *J. Am. Ceram. Soc.*, 85[11], 2646–2650.)

first (0.1–0.5 nm) and second (0.5–5 nm) atomic shells.[13] This is the reason why μRS is sometimes more powerful than X-ray analysis for detecting and monitoring crystallization/amorphization processes in covalent materials [32,33].

2.6 SELECTED EXAMPLES OF POLYMER RAMAN SPECTRA

Figure 2.9 shows how polyaniline chain length has little effect on its Raman signature because the vibrational units are very small and, as in most polymers, vibrate independently [34]. On the other hand, the

[13]Raman bending modes are specifically sensitive to local geometric misorientation, and Raman stretching modes to the neighboring disorder (particularly atoms from other sublattices or electric defects resulting from substitutions/vacancies).

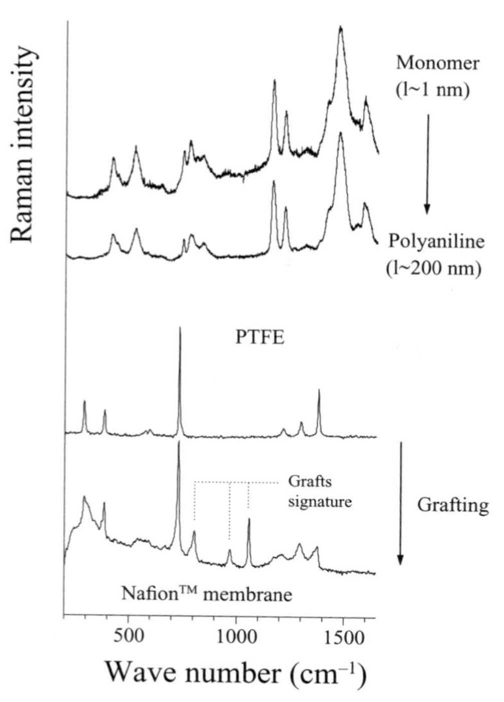

Figure 2.9 Raman spectra from polyaniline precursor monomer, one emeraldine base form of polyaniline, polytetrafluoroethylene (PTFE) and Nafion™ membrane (PTFE grafted with sulfonic groups). (Adapted from Colomban, Ph., Folch, S., Gruger, A. [1999]. *Macromolecules* 32, 3080–3092 and Gruger, A., Regis, A., Schmatko, T., Colomban, Ph. [2001]. *Vib. Spectrosc.*, 26, 215–225).

grafting of sulfonic groups on the polytetrafluoroethylene (PTFE) framework of Nafion™ protonic conduction membranes produces obvious effects below 400 cm^{-1}, due to an orientational medium range disorder [35]. Besides, the grafts act like new independent vibrational units, and their specific signature makes it easy to follow the grafting process.

Diffraction techniques can discriminate "periodical" and "disordered" domains but fail to characterize the progressive orientational disorder (paracrystal; pp. 615–620 in Ref. [36]) encountered in organic and inorganic polymers presenting interlocking submicron "crystalline" and "amorphous" conformational domains [37,38]. In that case, μRS proves useful, and the simplest way to picture the problem is to fit the low-frequency components, which show the collective chain movements, using a narrow Lorentzian lineshape for the "crystalline" state and a broader Gaussian lineshape, representing distributed configurations, for the "amorphous" state.

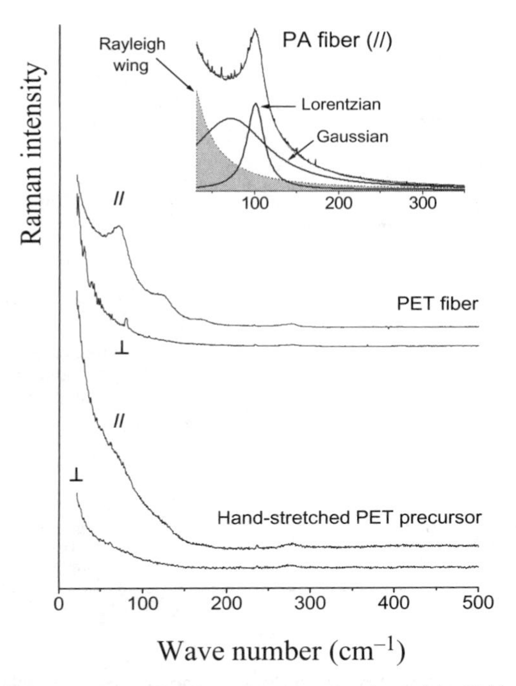

Figure 2.10 Low wavenumber Raman spectra of polyamide (PA) and polyethylene-terephthalate (PET) fibers or precursor with the laser electric field polarized either parallel (//) or perpendicular (⊥) to the fiber axis. (Adapted from Colomban, Ph., Herrera-Ramirez, J.-M., Paquin, R., Marcellan, A., Bunsell, A. [2006]. *Eng. Fract. Mech.*, 73[16], 2463–2475.)

Figure 2.10 illustrates this point in the case of a polyamide (PA) fiber. Fits like the one that is shown were used to demonstrate that the mechanical fatigue of the PA66 fiber results from the progressive transformation of its amorphous phase [39]. On the same figure, the strong variation of the low wavenumber region in a strained polyethylene-terephthalate (PET) polymer (fiber extrusion) is obvious. The axial orientation of the macromolecular chains involved by the combined temperature/stress treatments of the fiber leads to a polarization of the spectra (as evidenced by the difference between "//" and "⊥" spectra), and the increase of crystallinity is visible with the narrowing of the lattice modes.

Figure 2.11 shows that the variations of Young's modulus (the slope on the strain–stress curve) in a PET fiber can be correlated with the Raman shifts of the low-frequency contributions corresponding to the amorphous and crystalline nanosubstructures. Depending on the nature of the polymer, orientation and crystallization are more or less coupled. In PET fibers, most of the stress is obviously transferred to the

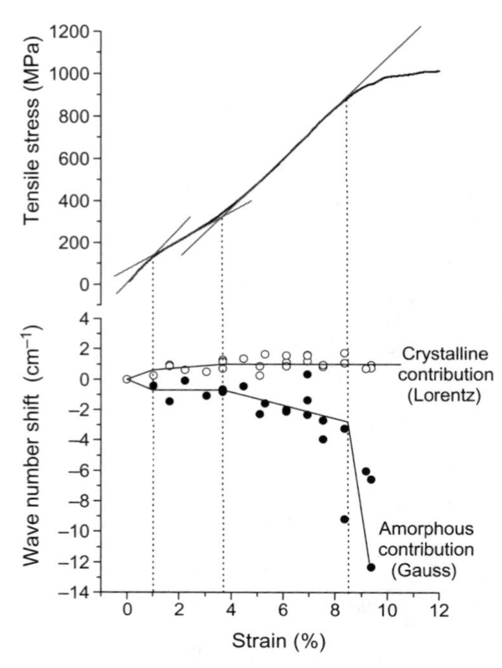

Figure 2.11 Comparison of the strain dependency of the crystalline and amorphous low-frequency Raman contributions with the stress–strain tensile curve for polyethylene-terephthalate (PET) fibers. (Adapted from Colomban, Ph., Herrera-Ramirez, J.-M., Paquin, R., Marcellan, A., Bunsell, A. [2006]. *Eng. Fract. Mech.*, 73[16], 2463–2475.)

amorphous matrix, whereas nanocrystallites are only slightly compressed by Poisson's effect.

In PA fibers, both the amorphous and crystalline moieties accommodate the stress because the nanostructure puts them "in series". Wave number–strain curves thus start with a plateau that corresponds to the disentanglement of the polymer chains (viscoelastic regime) and wave numbers start downshifting as soon as "knots" block this process (elastic regime) [40].

REFERENCES

1. Brillouin, L. (1922). Diffusion of light and X-rays by a transparent homogeneous body. The influence of thermal agitation. *Ann. Phys. (Paris)*, 17, 88.
2. Smekal A. (1923). The quantum theory of dispersion. *Naturwissenschaften*, 11, 873.

3. Raman, C.V., Krishnan, K.S. (1928). A new type of secondary radiation. *Nature*, 121, 501.
4. Lavrencic-Stangar, U., Orel, B., Groselj, N., Colomban, Ph., Statthatos, E., Lianos, P. (2002). In-situ resonance Raman studies of a dye-sensitized photoelectrochemical cell with a sol-gel electrolyte. *J. Electrochem. Soc.*, 149(11), E413.
5. Nafie, L.A. (2001). Theory of Raman scattering. In: Lewis IR, Edwards HGM, eds. *Handbook of Raman Spectroscopy—from the Research Laboratory to the Process Line*. New York: Marcel Dekker Inc., 1.
6. Faurel, X., Vanderperre, A., Colomban, Ph. (2003). Pink pigment optimization by resonance Raman spectroscopy. *J. Raman Spectrosc.*, 34, 290.
7. Gouadec, G., Colomban, Ph. (2001). Non-destructive mechanical characterization of SiC fibers by Raman spectroscopy. *J. Eur. Ceram. Soc.*, 2, 1249.
8. Colomban, P. (2003). Raman analyses and "smart imaging" of nanophases and nanosized materials. *Spectroscopy Europe*, 15(6), 8.
9. Sun, W.X., Shen, Z.X. (2003). Near-field scanning Raman microscopy using apertureless probes. *J. Raman Spectrosc.*, 34, 668.
10. Hallen, H.D., Jahncke, C.L. (2003). The electric field at the apex of a near-field probe: Implications for nano-Raman spectroscopy. *J. Raman Spectrosc.*, 34, 655.
11. Lee, N., Hartschuh, R.D., Mehtani, D., Kisliuk, A., Foster, M.D., Sokolov, A.P., Maguire, J.F., Green, M. (2007). High contrast scanning nano-Raman spectroscopy of silicon. *J. Raman Spectrosc.*, 38.
12. Gouadec, G., Colomban, Ph. (2007). Raman spectroscopy of nanomaterials: How spectra relate to disorder, particle size and mechanical properties. *Prog. Cryst. Growth Charact. Mater.*, 53, 1.
13. Bruneel, J.-L., Lassègues, J.-C., Sourisseau, C. (2002). In-depth analysis by confocal Raman microspectroscopy. *J. Raman Spectrosc.*, 33(10), 815.
14. Laplant, F., Laurence, G., Ben Amolz, D. (1996). Theoretical and experimental uncertainty in temperature measurement of materials by Raman spectroscopy. *Appl. Spectrosc.*, 50, 1034.
15. Schrader, B. (1995). Resonance Raman spectrocopy (§6.1.2). In: Schrader B, ed. *Infrared and Raman Spectroscopy—Methods and Applications*. Weinheim, Germany: VCH, 466.
16. Kürti, J., Kuzmany, H. (1991). Resonance Raman scattering from finite and infinite polymer chains. *Phys. Rev.*, B44, 597.
17. El-Khalki, A., Gruger, A., Colomban, Ph. (2003). Bulk / surface nanostructure and defects in polyaniline films and fibres. *Synth. Met.*, 139, 215.
18. Gouadec, G., Colomban Ph., Bansal, N.P. (2001). Raman Study of Hi-Nicalon fiber reinforced Celsian composites, Part 1: Distribution and nanostructure of different phases. *J. Am. Ceram. Soc.*, 84, 1129.

19. Cotton, T.M., Kim, J.-H., Chumanov, G.D. (1991). Application of surface-enhanced Raman spectroscopy to biological systems. *J. Raman Spectrosc.*, 22, 729.

20. Leona, M., Lombardi, J.R. (2007). Identification of berberine in ancient and historical textiles by surface-enhanced Raman scattering. *J. Raman Spectrosc.*, 38(7), 853.

21. Anastassakis, E., Burstein, E. (1971). Morphic effects II-Effects of external forces on the frequencies of the q ~ 0 optical phonons. *J. Phys. Chem. Solids*, 32, 563.

22. Beyerlein, I.J., Amer, M.S., Schadler, L.S., Phoenix, S.L. (1998). New methodology for determining in-situ fibre, matrix and interfaces stresses in damaged multifiber composites. *Sci. Eng. Comp. Mater.*, 7(1–2), 204.

23. Gouadec, G., Colomban, Ph., Bansal, N.P. (2001). Raman study of Hi-Nicalon fiber reinforced Celsian composites, Part 2: Residual stress in the fibers. *J. Am. Ceram. Soc.*, 84, 1136.

24. De Wolf, I. (1999). Stress measurements in Si Microelectronics Devices using Raman spectroscopy. *J. Raman Spectrosc.*, 30(10), 877.

25. Bollas, D., Parthenios, J., Galiotis, C. (2006). Effect of stress and temperature on the optical phonons of aramid fibers. *Phys. Rev.*, B73, 094103.

26. Kittel, C. (1996). *Introduction to Solid State Physics*, 7th Edition. New York: John Wiley and Sons.

27. Poulet, H., Mathieu, J.P. (1976). *Vibration Spectra and Symmetry of Crystals*. New York: Gordon and Breach.

28. Pasto, A.E., Condrate, R.A. (1973). The laser Raman spectra of several perovskite zirconates. In: Mathieu, J.P., ed., *Advanced Spectroscopy*. London: Heyden & Sons Ltd.

29. Sernelius, B.E. (2001). *Surface Modes in Physics*. Berlin, Germany: Wiley-VCH.

30. Kosacki, I., Petrovsky, V., Anderson, H.U., Colomban, Ph. (2002). Raman spectroscopy of nanocrystalline ceria and zirconia thin films. *J. Am. Ceram. Soc.*, 85(11), 2646.

31. Malyarevich, A.M., Posledovich, M.R. (1996). The assignment of lattice vibrations in triglycine sulfate-type crystals. *J. Mol. Struct.*, 375, 43.

32. Wang, Y., Tanaka, K., Nakaoka, T., Murase K. (2002). Effect of nanophase separation on crystallisation process in Ge-Se glasses studied by Raman scattering. *Physica B*, 568, 316.

33. Mortier, M., Monteville, A., Patriarche, G., Mazé, G., Auzel, F. (2001). New progresses in transparent rare-earth doped glass-ceramics. *Opt. Materials*, 16, 255.

34. Colomban, Ph., Folch, S., Gruger, A. (1999). Vibrational study of short-range order and structure of polyaniline bases and salts. *Macromolecules*, 32, 3080.

35. Gruger, A., Regis, A., Schmatko, T., Colomban, Ph. (2001). Nanostructure of Nafion membranes at different states of hydration. An IR and Raman study. *Vib. Spectrosc.*, 26, 215.
36. Guinier, A. (1956). *Théorie et Technique de la Radiocristallographie*. Paris, France: Dunod.
37. Guinet, Y., Denicourt, T., Hedoux, A., Descamps, M. (2003). The contribution of the Raman spectroscopy to the understanding of the polyamorphism situation in triphenil phosphate. *J. Mol. Struct.*, 507, 651–653.
38. Stuart, B.H. (1996). Polymer crystallinity studied using Raman spectroscopy. *Vib. Spectrosc.*, 10, 79.
39. Colomban, Ph., Herrera-Ramirez, J.-M., Paquin, R., Marcellan, A., Bunsell, A. (2006). Micro-Raman Study of the fatigue and fracture behaviour of single PA66 fibres. Comparison with single PET and PP fibres. *Eng. Fract. Mech.*, 73(16), 2463.
40. Marcellan, A., Bunsell, A.R., Piques, R., Colomban, Ph. (2003). Micromechanisms, mechanical behavior and probabilistic fracture analysis of PA 66 fibers. *J. Mater. Sci.*, 38(10), 2117.

II

POLYMER, COLLOIDAL, AND FOOD APPLICATIONS

3

RAMAN APPLICATIONS IN POLYMER FILMS FOR REAL-TIME CHARACTERIZATION

Giriprasath Gururajan, Ph.D. and Amod A. Ogale, Ph.D.
Department of Chemical and Biomolecular Engineering and Center for Advanced Engineering Fibers and Films, Clemson University

3.1 INTRODUCTION

Since the discovery of the Raman scattering effect in the late 1920s, several research articles [1–14], review papers [15–17], books [18–21], and industrial patents [22–24] have been published on the application of Raman spectroscopy for polymers. Typically, the technique is used to determine the polymer type and content [1–3], polymerization reaction kinetics [4,5], amount of pigment additives [6,7], amount of degradation [8,9], change in the morphology of the polymer such as crystallinity, and molecular orientation [10–14]. A vast majority of research on Raman spectroscopy of polymers [1,6,7,10–14] has involved analysis of the polymer product.

The introduction of laser excitation sources in the 1960s and charge-coupled detection devices in the 1990s enabled development of sophisticated Raman instrumentation. This includes fiber-optics coupling that not only increased the signal-to-noise ratio but also reduced

Raman Spectroscopy for Soft Matter Applications, Edited by Maher S. Amer

fluorescence effect. Further, data collection times have been reduced to durations as little as few seconds. Consequently, the technique is proving to be a powerful real-time polymer process monitoring tool [25–27].

Conventional manufacturing uses off-line measurements on the processed polymer to empirically modify the process to obtain the desired properties. This trial-and-error procedure is time-consuming and expensive. Therefore, there is a growing demand to characterize materials during their production by using real-time analytical tools.

Techniques such as small-angle light scattering (SALS) [28], birefringence [29,30], wide-angle X-ray diffraction (WAXD) [31–33], and infrared dichroism (IR) [34] have been applied to polymers as process monitoring tools in a laboratory environment. Bullwinkel et al. [28] used simultaneous online SALS and infrared temperature measurements to study the microstructure evolution during linear low-density polyethylene (LLDPE) blown film extrusion. They related the change in average scattered intensity of the light to the crystallization process. Nagasawa et al. [29] reported the first online measurements of orientation development during blown film extrusion using birefringence. Additional studies were conducted by Ghaneh-Fard et al. [30] during the film blowing of an LLDPE. Real-time microstructural measurement during film casting of isotactic polypropylene (PP) is reported by Lamberti and Brucato [34] using Fourier transform infrared (FTIR) spectroscopy.

These techniques, however, are not conducive to rapid real-time measurements in industrial environments that involve elevated temperatures, high pressure, and humidity. For instance, SALS and birefringence techniques are very sensitive to the thickness of the polymeric sample, whereas WAXD requires long exposure times, safety precautions, and expensive instrumentation.

It is known that the performance of a polymer product depends on the molecular architecture of the polymer and the history of its formation. The molecular architecture is determined by factors such as molecular weight distribution (MWD) and copolymer composition, and is controlled during extrusion and fabrication by factors such as thermal and strain-rate history. Raman spectroscopy offers a powerful means for *in situ* monitoring of polymer molecular architecture. Raman spectroscopy offers distinct advantages over SALS, WAXD, birefringence, and other vibrational spectroscopy (FTIR) because it is not affected by moisture in the environment and is very amenable to fiber-optics coupling using low-cost silica fibers. This feature allows one to remotely analyze samples and to conduct real-time, *in situ* measurements even in the harsh environment of a polymer processing line. The technique requires virtually no sample preparation and is independent of sample size and shape. The implementation of real-time Raman spectroscopy to obtain information about the polymer has been adopted in various petrochemical plants [24] and polymer processing industries [25,26] to enhance the control of production parameters.

3.2 REAL-TIME CHARACTERIZATION DURING POLYMER PROCESSING

Real-time measurements of microstructure during polymer processing have gained significant attention from both modeling [35–37] and experimental perspectives [28,30–34]. The orientation and crystallinity of the finished polymeric product typically determine its mechanical and physical characteristics. During its formation, crystallinity, orientation, and morphology develop depending not only on the polymer characteristics, but also on the processing conditions used. The morphology, in turn, determines the tensile, tear, optical, and barrier properties of films [38–40].

Synthetic polymers are converted into films and fibers by melt extrusion of polymer melt through a die. After the melt exits the die, it experiences extensional force while it is drawn down. Processing conditions such as melt temperature, drawing force, and cooling rate control the final polymer morphology.

The microstructure characterization of fibers and films has been conducted using X-ray diffraction (XRD) [41,42], infrared spectroscopy [43,44], differential scanning calorimetry (DSC) [45,46], and electron microscopy [47,48]. Since the 1960s, Raman spectroscopy has also been used for characterization of fibers and films of different polymers such as PE [10,49], PP [11,50], polyethylene terephthlate (PET) [12,51], and nylon [52,53]. However, these studies were conducted as post-processing analyses.

Although the potential of Raman spectroscopy for characterizing polymers has been recognized for a number of years, systematic studies on technique for real-time microstructure development are relatively few. Recently, Paradkar et al. [54] utilized real-time Raman spectroscopy for crystallinity measurements at different points along the moving fiber spinning line. Their study demonstrated the feasibility of using Raman spectroscopic technique to monitor the development of crystallinity in melt-spun high-density polyethylene (HDPE) fibers. The real-time Raman spectra from a single HDPE fiber, obtained as a function of distance from the spinneret, were used to study the effect of process parameters including throughput, quench rate, and take-up speed during fiber spinning.

In a companion study, real-time Raman spectroscopy measurements during blown film extrusion were conducted [22,27,36,55–57]. Blown film extrusion is one of the major processes used to produce films, ranging from simple monolayer films for bags to complex multilayer structures used in specialized applications such as food and medical packaging. A typical blown film line is shown schematically in Figure 3.1. It consists of three units: an extruder, a die, and a take-up unit. The extruder does the job of melting the polymer and pumping it through a die. The extruder screw is driven by a motor whose speed can be adjusted to obtain the desired flow rate through the die. The molten extrudate exiting the annular die is blown as a bubble by expanding in the transverse direction (TD) by air fed through the center of the die. The film is also pulled longitudinally in the machine direction (MD) by rollers. The bubble so formed is simultaneously cooled using quench air that is blown along the periphery of the bubble. The solidified film is then collapsed, flattened by nip rolls, and wound using rollers in the take-up unit.

Besides melt temperature, there are three important parameters that can be controlled during the blown film extrusion process. The rate at

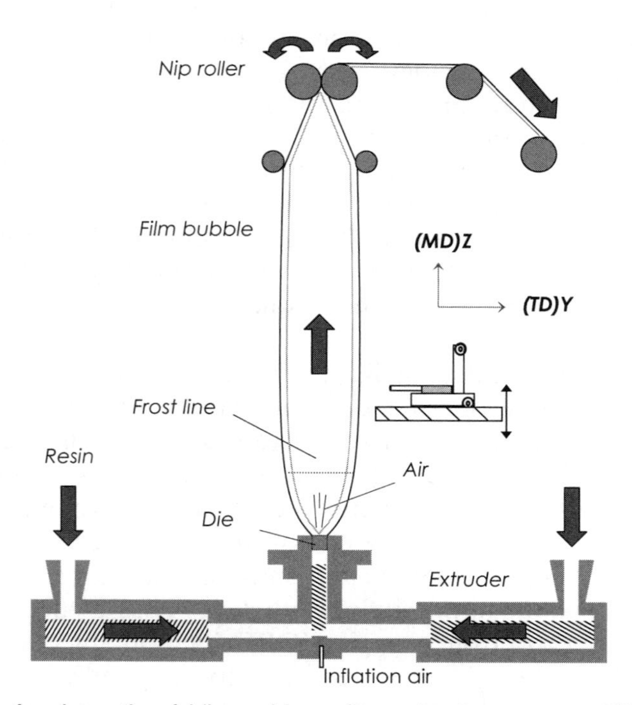

Figure 3.1 A schematic of bilayer blown film extrusion process. TD, transverse direction; MD, machine direction.

which the bubble is stretched in the longitudinal (machine) direction is defined by the take-up ratio (TUR), i.e. the ratio of the velocity of the take-up roller (V_f) to the velocity of the extrudate at the exit of the die (V_0). The blow-up ratio (BUR) is the ratio of the final bubble diameter to the die diameter, and ranges from 1.5 to 4. This transverse expansion of the bubble is primarily controlled by the inflation air pressure or the volume of the bubble. The quench air around the bubble cools the molten bubble and locks further expansion of the bubble after a certain distance above the die, which is defined as the frost-line height (FLH).

The dynamics of film formation are more complex than those for fiber spinning because film blowing involves biaxial stretching of the film. Bubble kinematics and temperature are often measured for correlating them to mechanical properties of the final film. The velocity and bubble diameter profiles are typically obtained using the video tracer and image analysis techniques [58], and are used to solve the momentum equations to obtain strain rates in the bubble. The stress experienced by the bubble during film blowing is proportional to the strain rate.

Figure 3.2 presents the variation of strain rates with changes in BUR from 0.4 to 2.0, while the TUR is kept constant at 3.8. The strain rates along the TD (e_2) progressively switch from negative to positive values (biaxial distribution), while the strain rates in the machine direction (e_1) do not show significant variation. As compared with a small BUR

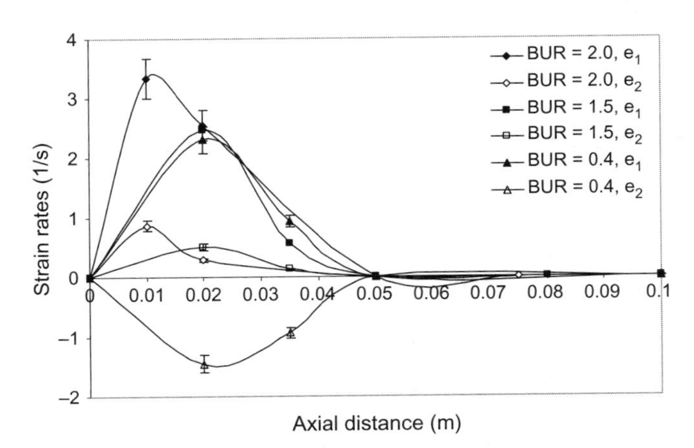

Figure 3.2 Machine direction (e_1) and transverse direction strain rates (e_2) in the linear low-density polyethylene (LLDPE) bubble during blown film extrusion for blow-up ratios (BURs) of 2.0, 1.5, and 0.4, at a constant take-up ratio (TUR) of 3.8 (curves represent trends, not model predictions).

of 0.4, the larger BURs of 1.5 and 2.0 lead to larger bubble diameters. This results in thinner films, which causes an increase in stresses along the transverse direction.

The wide variation possible in processing conditions can lead to variations in microstructure of the film, and thence to its properties. Therefore, experimental and theoretical developments addressing real-time microstructural measurements during polymeric film processing are of topical importance. Experimental measurements can help in real-time control of film microstructure so that product properties can be controlled. Experimental results can also help to develop realistic process models so that future film processing can be guided by simulation results.

3.3 REAL-TIME MEASUREMENTS DURING BLOWN FILM EXTRUSION

Figure 3.3a shows a photograph of a typical online experimental setup for film blowing process. Nonpolarized Raman probes are usually used for crystallinity measurements, whereas polarized probes (ZZ, YY, ZY) are needed for orientation measurements. The experiments should be conducted in a dark environment to avoid saturation of the charge-coupled device (CCD) detector. Starting from a location near the die, spectra can be recorded along the longitudinal direction by a fine adjustment of a micrometer stage. The collection of the scattered Raman signal can be either in the backscattered mode (180°) or right-angle mode (90°), as shown by the schematic in Figure 3.3b.

3.4 ESTIMATION OF CRYSTALLINITY IN POLYMER FILMS

Figure 3.4 presents a typical Raman spectrum of a solid PE film in the 1000–1500 cm^{-1} spectral range. Literature studies [10,59] have reported the spectrum to be a superposition of three components: an orthorhombic crystalline phase (α_c), a melt-like amorphous phase (α_a), and an intermediate disordered phase of anisotropic nature (α_b), where chains are oriented without any lateral order. The content of the three phases was derived from the integrated intensities of the characteristic bands, the weight of the orthorhombic crystalline component being proportional to the $-CH_2$ wagging band at 1418 cm^{-1}.

To determine the crystalline content of polyethylene, the integrated intensity of the 1418-cm^{-1} band should be normalized using a standard

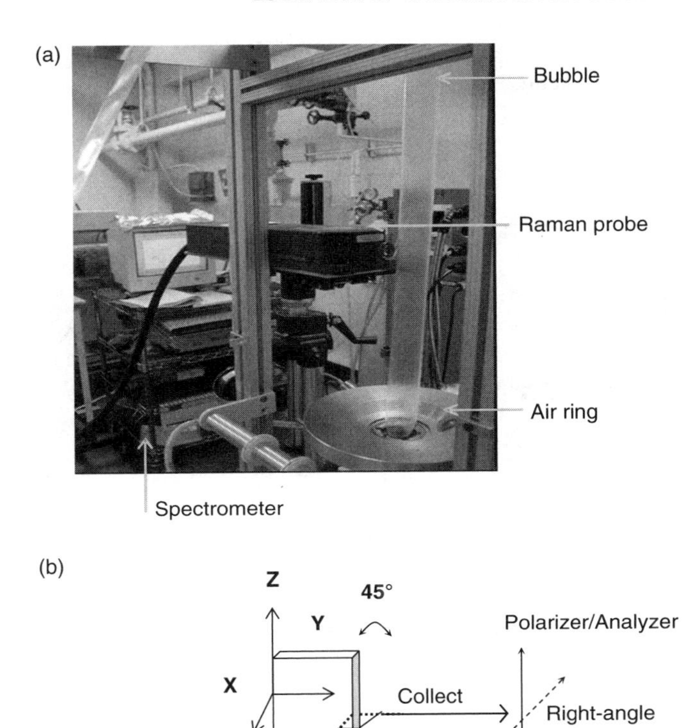

Figure 3.3 (a) Experimental setup used for real-time Raman spectroscopy during blown film extrusion. (b) A schematic of backscattering and right-angle scattering geometries used for Raman spectroscopy.

reference band [10], namely, the CH_2 twisting region near $1300\,cm^{-1}$, which is invariant with the state of the polymer. The calculated integrated intensity ratio of the final film samples $I_{1418/1300}$ can be calibrated against crystallinity values of the final film obtained by DSC and WAXD techniques according to the procedure reported by Strobl and Hagedorn [10].

A typical Raman spectrum of polypropylene film is depicted in Figure 3.5. The peaks of interest (highlighted with broken lines) are due to mixed $-CH_2$ rocking and C–C stretching vibration at $809\,cm^{-1}$ and pure $-CH_2$ rocking vibration at $841\,cm^{-1}$ representing the helical

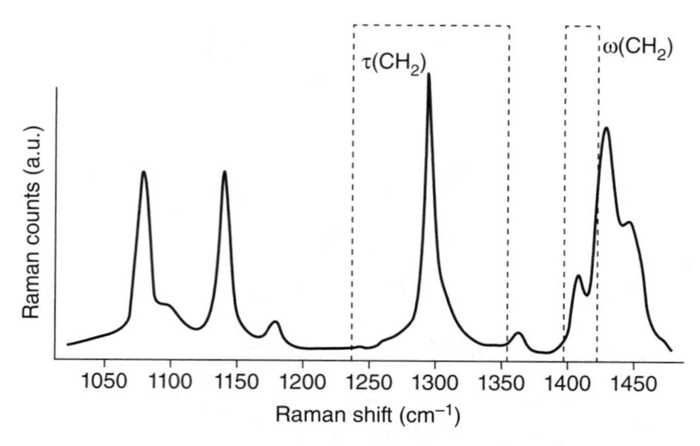

Figure 3.4 Raman spectrum of polyethylene film in 1000–1500 cm^{-1} range.

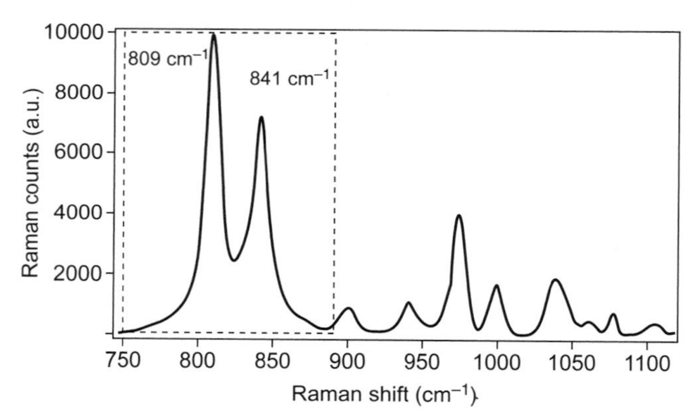

Figure 3.5 Raman spectrum of an isotactic polypropylene film in 750–1150 cm^{-1} range.

conformation of the chains in the crystalline region and the isomeric defect phase, respectively. These two bands arise from the nonhelical conformation at 830 cm^{-1} present in melt or amorphous phase. Nielsen et al. [11] reported that the sum of band intensities at 809, 830, and 841 cm^{-1} ($I_{809} + I_{830} + I_{841}$) is independent of chain conformation or crystallinity. Therefore, the total integrated intensity under the 809- to 841-cm^{-1} band is invariant of the state of the polymer, and is used as a standard.

The crystallinity values of PP were calculated by dividing the integrated intensity of 809-cm^{-1} peak (helical chain conformation) by the total integrated intensity under the 809- to 841-cm^{-1} band:

$$X_c = I_{809}/(I_{809} + I_{830} + I_{841}). \quad (3.1)$$

The 830-cm^{-1} peak was found to be very weak and does not contribute significantly to the crystallinity values. The final crystallinity values calculated from Nielsen's method [11] correlated with WAXD results as $X_{c,Raman} = 0.9\,X_{cWAXD}$.

3.4.1 Crystallinity Measurements During Single-Layer Film Extrusion

Real-time Raman spectra along the axial distance are displayed in Figure 3.6 for LLDPE films for one set of processing conditions. The spectrum at the die exit is that for a melt, and consists of three broad-bands at 1080, 1305, and 1440 cm^{-1}, all for the amorphous melt. As the polymer melt cools down, the crystallization causes the intensity of C–C stretching vibrations at 1060 and 1130 cm^{-1} to increase at the expense of an amorphous peak at 1080 cm^{-1}. Also, the broad amorphous $-CH_2$ twisting vibration at 1300 cm^{-1} splits into a narrow crystalline band at 1296 cm^{-1} and a broad amorphous band at 1305 cm^{-1}. Finally, the $-CH_2$ bending vibration (1440 cm^{-1}) in the amorphous region is transformed into three bands (1418, 1440, and 1460 cm^{-1}), of which the $-CH_2$ wagging vibration at 1418 cm^{-1} is used to estimate the orthorhombic crystalline content of PE.

Figure 3.6 also shows that as the polymer travels from a location below the frost line to that above, the intensity of the 1418-cm^{-1} peak steadily increases as a consequence of crystallization. As the melt is subjected to an undercooling, nucleation occurs followed by crystallization. Ultimately, these processes slow down as the bubble temperature decreases to ambient. The crystallinity profile, presented in Figure 3.7, shows a steep initial increase immediately after the frost line, but then plateaus at higher axial distance. A real-time temperature profile, obtained using an IRCON (Santa Cruz, CA) infrared pyrometer, shows a plateau at ≈104 °C as a consequence of the exothermic heat of crystallization.

Figure 3.8a–f display Raman spectra obtained at various TURs, BURs, and FLHs. The region in the vicinity of 1418 cm^{-1} is highlighted to note the effect of process variable on crystalline development. Spectra obtained for three different TURs of 3.8, 11.5, and 19.2, with a BUR of 1.5 and an FLH of 0.03 m, are shown in panels (a–c) of Figure 3.8, respectively. Panels (a), (d), and (e) present the spectra obtained at BURs of 1.5, 0.2, and 2.0, respectively, for a TUR of 3.8 and an FLH of 0.03 m. It was observed that at a particular axial distance,

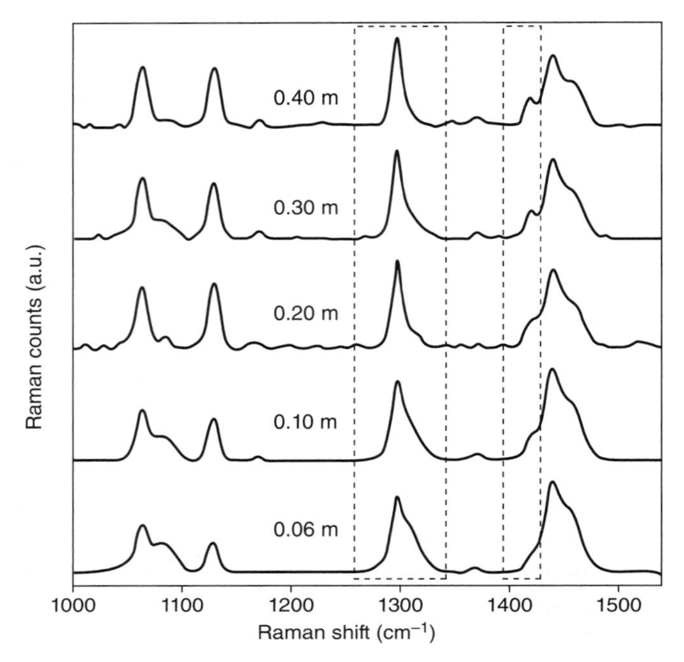

Figure 3.6 Raman spectra obtained at various axial positions along the blown film line for linear low-density polyethylene.

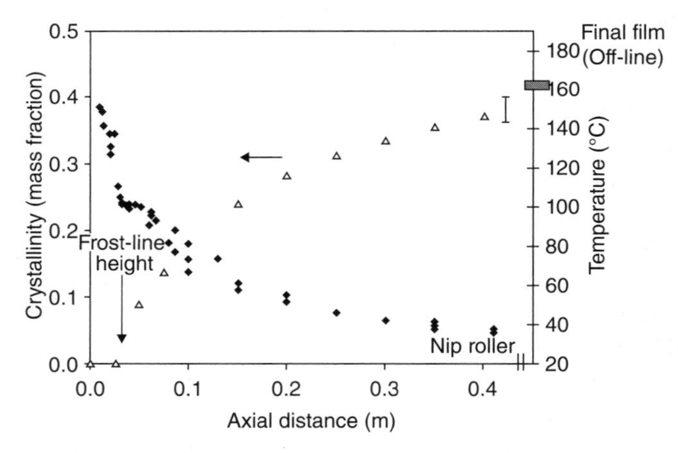

Figure 3.7 Crystallinity (left axis) and temperature (right axis) profiles for linear low-density polyethylene (LLDPE) as a function of axial distance along the blown film line.

the intensity of the 1418-cm^{-1} peak is lower for a BUR of 0.2 as compared with that at 1.5 or 2.0.

The development of crystallinity appears significantly different when plotted as a function of *process time*, in contrast to that observed as a

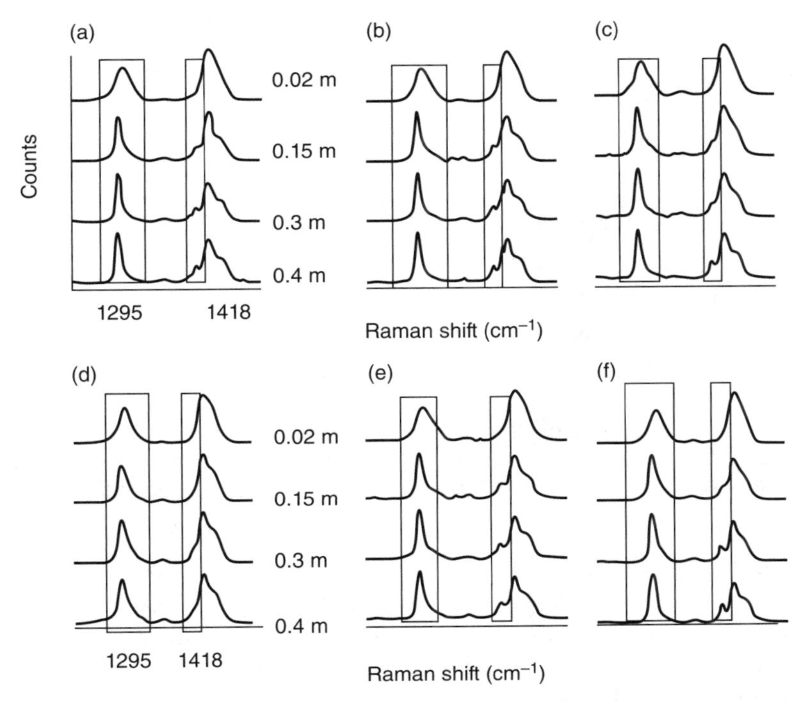

Figure 3.8 Raman spectra at axial positions between 0.02 and 0.4 m above the die exit at various processing conditions: (a) TUR = 3.8, BUR = 1.5, FLH = 0.03 m; (b) TUR = 11.5, BUR = 1.5, FLH = 0.03 m; (c) TUR = 19.2, BUR = 1.5, FLH = 0.03 m; (d) TUR = 3.8, BUR = 0.2, FLH = 0.03 m; (e) TUR = 3.8, BUR = 2.0, FLH = 0.03 m; (f) TUR = 3.8, BUR = 2.0, FLH = 0.06 m [27]. TUR, take-up ratio; BUR, blow-up ratio; FLH, frost-line height.

function of axial distance. The process time is defined as the time taken for a particle exiting the die to reach a particular location along the film line. Figure 3.9 displays the crystallinity profiles as a function of process time for TURs of 3.8, 11.5, and 19.2, and at a BUR of 1.5 and FLH of 0.03 m. The curves indicate that the crystallization process starts as the extrudate reaches the FLH, increases along the film line, and finally plateaus. The increase in the crystallization growth rate with an increase in TUR is an indication of flow-enhanced crystallization [60,61].

Real-time Raman spectroscopy was also used for other polyolefins such as HDPE, low-density polyethylene (LDPE), and isotactic polypropylene (i-PP), since these polyolefins are also converted to blown films. A comprehensive real-time study of the film blowing process using different polymers also helped in the generalization of

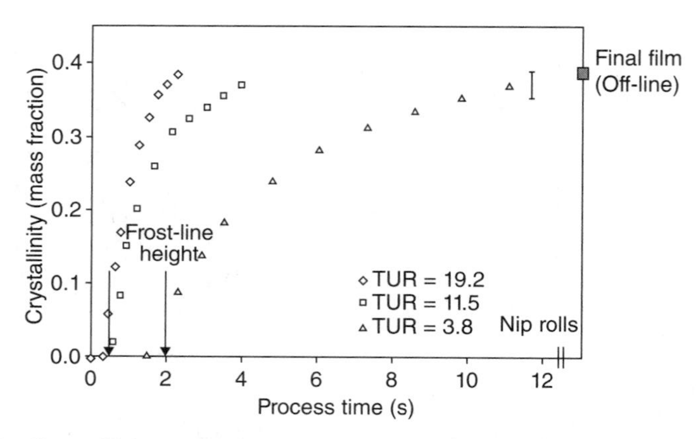

Figure 3.9 Crystallinity profiles of linear low-density polyethylene (LLDPE) at take-up ratios (TURs) of 3.8, 11.5, and 19.2 at a constant blow-up ratio (BUR) of 1.5 and a frost-line height (FLH) of 0.03 m. Solid symbol represents data collected off-line on the films.

the experimental protocol and stress-induced crystallization concepts [60,62]. Raman spectra for i-PP along the machine direction of the blown film line are displayed in Figure 3.10. The development of structure with axial distance is evident from the increase in intensity of the characteristic bands for crystallinity, i.e. increase of 809 cm^{-1} relative to 841 cm^{-1}. Crystallinity profiles were obtained using the method explained in the previous section for polypropylene.

Figure 3.11 displays the crystallinity profiles plotted as a function of process time for TURs of 3.8, 11.4, and 18.9, and at a constant BUR of 1.6 and FLH of 0.025 m. The curves indicate that the crystallization process starts as the extrudate reaches the FLH, increases along the film line, and finally plateaus.

These real-time crystallinity measurements enabled us to estimate the non-isothermal crystallization half-times [55]. The different crystallization kinetics had a significant effect on the resulting microstructure and mechanical properties of polyethylenes and i-PP blown films. Details of these results, which are outside the scope of this chapter, are described elsewhere [55,56].

Like any other vibrational spectroscopic technique, Raman spectroscopy is not a primary measurement technique to obtain crystallinity and orientation in fibers or films. Therefore, integrated intensities from Raman spectra are calibrated using primary measurement techniques such as DSC, density measurements, or XRD [10]. Recently, a comparison between real-time crystallinity measurements from WAXD and Raman spectroscopy was reported [63]. The orthorhombic crystalline content from WAXD spectrum was based on prominent (110) and

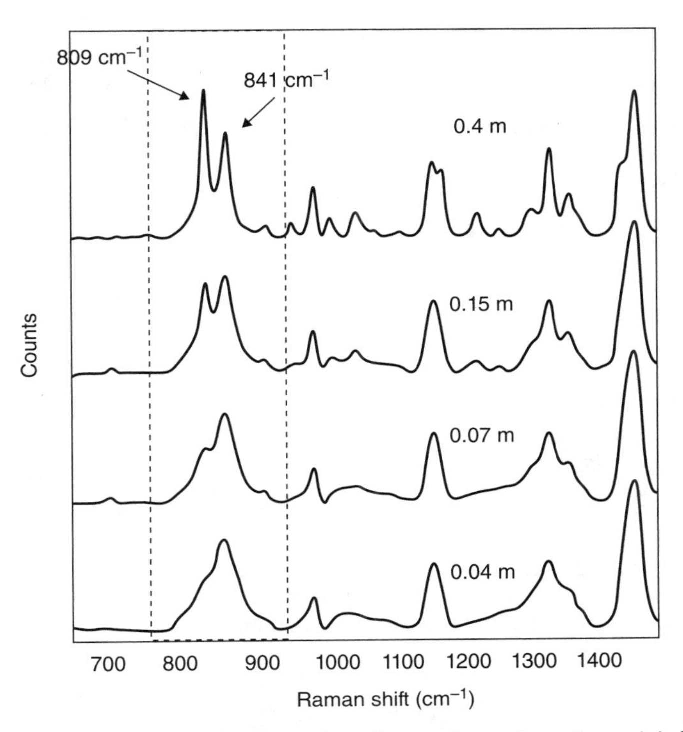

Figure 3.10 Raman spectra for isotactic polypropylene along the axial distance of the blown film line.

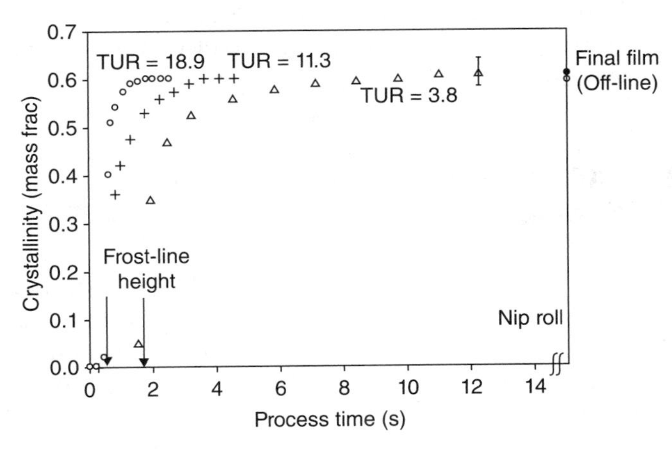

Figure 3.11 Crystallinity of polypropylene (PP) as a function of process time for different take-up ratios (TURs) at a constant blow-up ratio (BUR) of 1.6 and frost-line height (FLH) of 0.025 m.

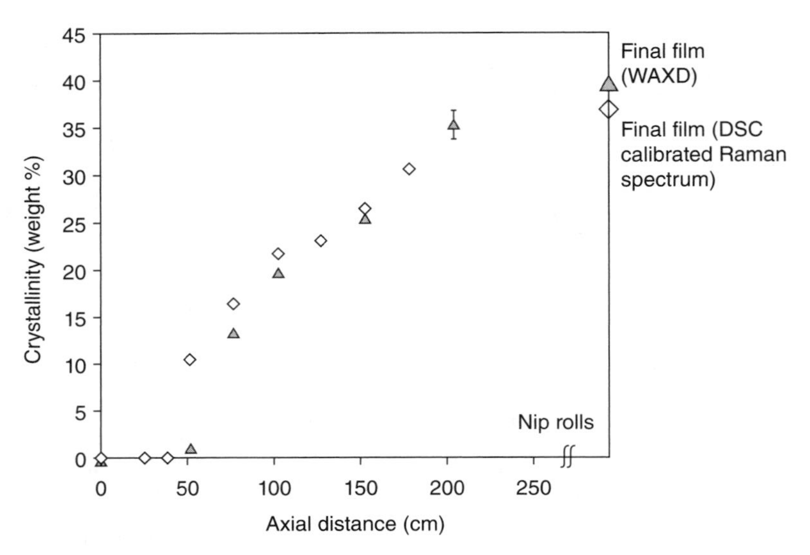

Figure 3.12 Comparison of crystallinity values measured by real-time Raman spectroscopy (open diamonds) and online WAXD (solid triangles). WAXD, wide-angle X-ray diffraction; DSC, differential scanning calorimetry.

(200) crystalline peaks [64], while the crystallinity from Raman spectrum was based on the method described earlier in Section 3.4. Figure 3.12 displays the crystallinity values as function of axial distance. The crystallinity profiles from WAXD and Raman spectroscopy were found to be consistent (within ±3 wt %).

Similar consistency between WAXD and Raman spectroscopy during polypropylene fiber spinning has been reported in a comparison study [65]. Thus, Raman spectroscopy offers an effective and inexpensive way to measure crystallinity during fiber and film formation.

3.5 ORIENTATION MEASUREMENTS DURING FILM PROCESSING

Techniques such as WAXD, birefringence, fluorescence polarization, infrared spectroscopy (IR), and nuclear magnetic resonance spectroscopy (NMR) are typically used to characterize molecular orientation in polymer samples. Unlike polarized IR, Raman spectroscopy allows determination of both the second ($\langle P_2(\cos\theta)\rangle$) and the fourth ($\langle P_4(\cos\theta)\rangle$) moment of the orientation distribution function. It also provides orientation information about individual amorphous and crystalline regions in a semicrystalline polymer. In contrast, a technique such as birefringence provides overall molecular orientation. Limited studies have reported on the use of off-line polarized Raman spectroscopy to estimate orientation in films [12,14,66] and fibers [13,67,68]. The detailed procedure for determining the orientation of the polymer specimens by polarized Raman spectroscopy was presented by Bower [69]. Later studies [12,13,67] proposed approaches to simplify the experimental complexities.

Paradkar et al. [23] demonstrated the use of Raman spectroscopy for measuring molecular orientation during fiber spinning of i-PP. They used the ratio of Raman peaks at 841 and 809 cm^{-1}, which is sensitive to the molecular anisotropy in PP, to estimate a semiquantitative molecular orientation in the fiber line for different throughput and draw ratios. Figure 3.13 shows the real-time ZZ polarized Raman spectra for PP obtained at three different positions along the fiber line. The ratio of the 841- to 809-cm^{-1} peak was reported to decrease as the

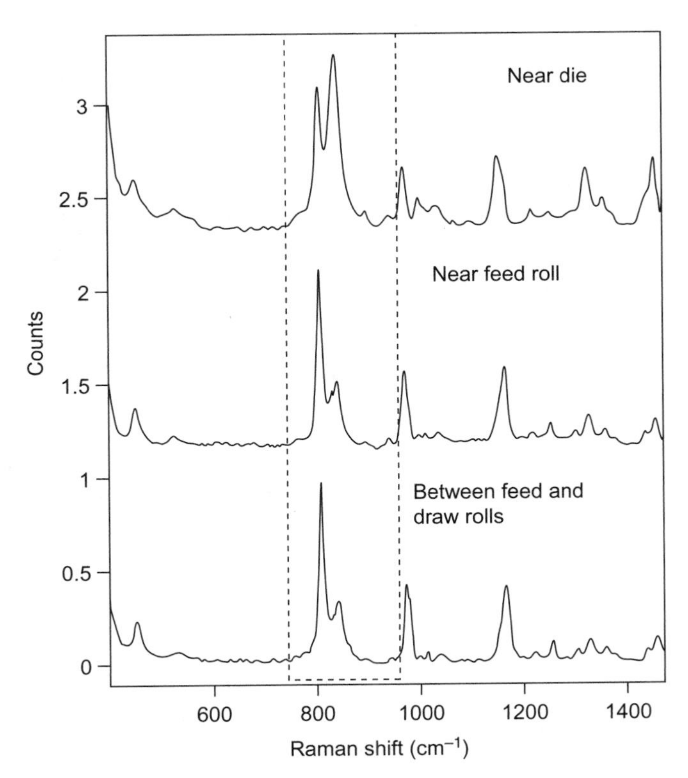

Figure 3.13 Polarized Raman spectra (ZZ) representing development of molecular orientation in polypropylene at three different locations in the fiber-spinning line. (Reproduced from Paradkar et al. [23].)

polymer travels from a position near the die to the position near the draw roll. The intensity ratios of the Raman bands for different take-up speeds were reported to linearly correlate with its birefringence values [23].

Real-time polarized Raman spectroscopy to measure molecular orientation development during blown film extrusion of an LDPE was reported [70]. Figure 3.14 displays the spectra obtained in ZZ and YY polarization modes at different locations along the blown film line, starting from the molten state near the die and extending up to the solidified state near nip rolls. The *trans* C–C symmetrical stretching vibration of PE at 1130 cm^{-1} was analyzed for films possessing uniaxial symmetry. The 1130-cm^{-1} peak in ZZ and YY spectra matched in the melt, indicating its isotropic nature. The 1130-cm^{-1} peak became more intense in ZZ direction than in YY direction at other locations in the film line, indicating the orientation in the machine direction.

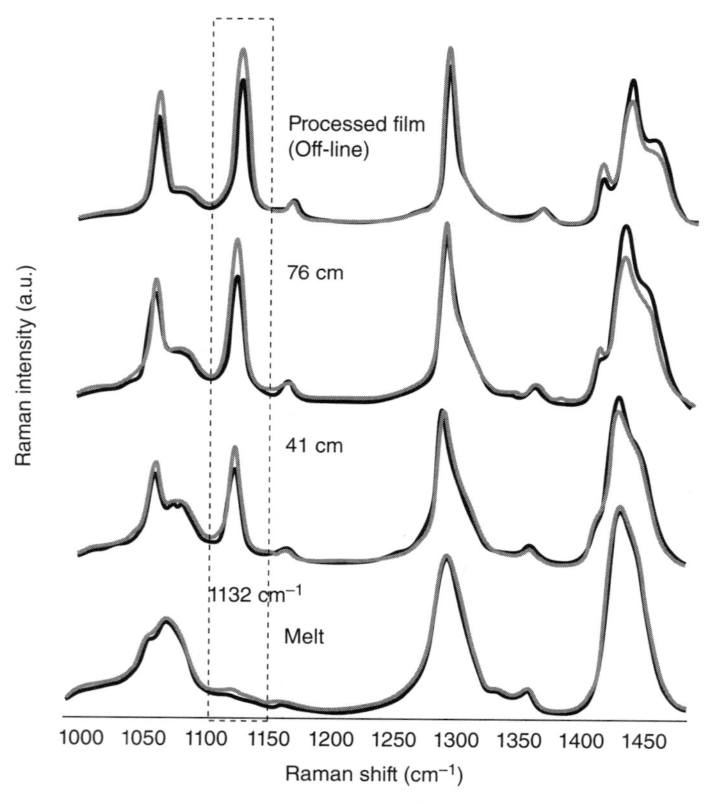

Figure 3.14 Real-time polarized Raman spectra of polyethylene (PE) obtained at different axial positions in the blown film line for two states of polarization: ZZ (gray line) and YY (dark line).

Figure 3.15 presents the integrated intensity ratio I_{1132}/I_{1064} from backscattered ZZ polarized spectra, which can be used to obtain a semiquantitative estimate of the molecular orientation in LDPE. The symmetrical C—C stretching band at 1132 cm^{-1} has $A_g + B_{1g}$ symmetry mode and intensifies parallel to the stretching direction (ZZ) [14,68,71], while the asymmetrical C—C stretching band at 1064 cm^{-1} has $B_{2g} + B_{3g}$ symmetry mode that intensifies only for cross-polarized (ZY) spectra. The ratio I_{1132}/I_{1064} was found to increase with the axial distance, indicating an increase in chain axis orientation in the machine direction.

The principal axis of the Raman tensor at 1130 cm^{-1} is coincident with the *c*-axis of the orthorhombic crystal and can be used to solve a set of intensity ratio equations to obtain second (and fourth) moments of the orientation distribution function [13,14]. As shown in Figure 3.16, P_2 ($\langle P_2(\cos\theta)\rangle$) and P_4 ($\langle P_4(\cos\theta)\rangle$) orientation parameters were found to increase along the axial distance in the film line even past the FLH. The P_2 values also showed an increasing trend with crystalline evolution during extrusion consistent with past observations [30] that molecular orientation takes place even after the blown film diameter is locked into place. The values for P_4 fall within the range of the most probable ones estimated by Bower [72]: $35P_2^2 - 10P_2 - 7)/18 \le P_4 \le (5P_2 + 7)/12$.

It was also noted that the Herman's orientation factor f_c of processed films from WAXD were smaller than those from Raman measurements. Such quantitative differences in spectroscopic (FTIR) and XRD have also been reported in a recent literature study [43].

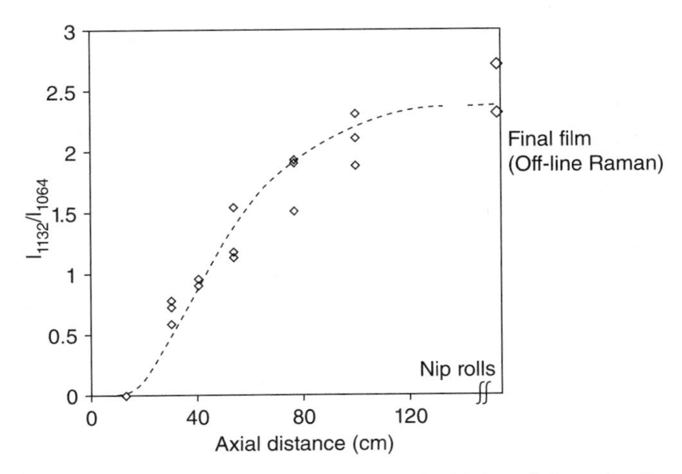

Figure 3.15 Integral intensity ratio of 1132 and 1064 cm^{-1} from backscattered ZZ polarized spectra as a function of axial distance (dotted curve represents the trend).

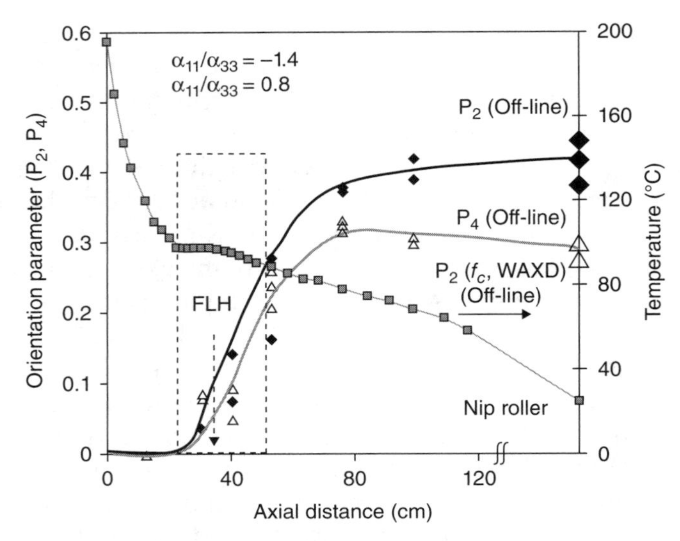

Figure 3.16 Molecular orientation parameters, P_2 (diamond) and P_4 (triangle), obtained from real-time Raman spectroscopy as function of axial distance (lines represents the trends). Large symbols are values from off-line Raman measurements on the processed films. The real-time temperature profile is also presented (right axis). FLH, frost-line height.

3.6 REAL-TIME CHARACTERIZATION OF MULTILAYERED FILMS

Finally, the use of real-time Raman spectroscopy for monitoring microstructural evolution within individual components of a bilayer PP/LDPE blown film is discussed. Figure 3.17 displays typical off-line Raman spectra for a single-layer polyethylene (LDPE) film, a single-layer polypropylene (PP) film, and a bilayer LDPE/PP blown film [22].

As observed in Figure 3.17a, the –CH2 rocking and C–C stretching peaks (809–841 cm^{-1}) that are used for crystallinity measurement in PP are characteristic of the helical structure present in PP, but are not present in polyethylene. Also, as observed in the highlighted blocks in Figure 3.17b, the methylene-bending peak at 1418 cm^{-1}, used for calculation of orthorhombic content in PE, is characteristic only of PE. The 1296– to 1305-cm^{-1} peak, it turns out, is present in both PE and PP. However, its intensity is reported to be rather weak in PP as compared with that in PE. Studies by Quintana and coworkers on PP/PE blends [73] and Markwort et al. [74] on heterogeneous polymer systems used Raman spectroscopy, and reported the 1296–to 1305-cm^{-1} band as char-

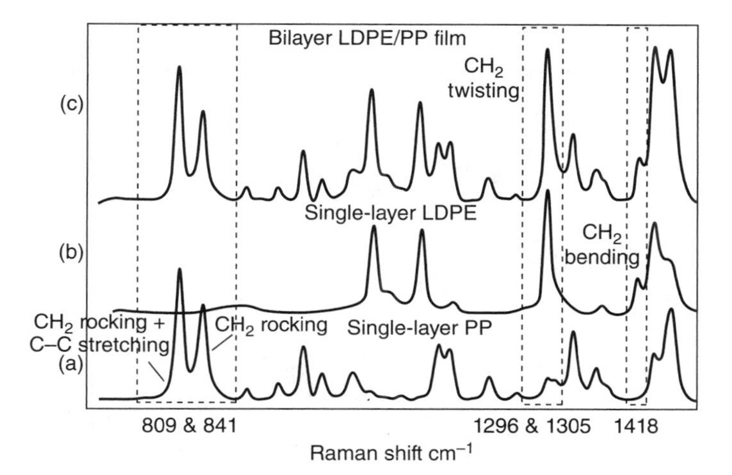

Figure 3.17 Raman spectra for (a) single-layer polypropylene, (b) single-layer polyethylene, and (c) bilayer low-density polyethylene (LDPE)/polypropylene (PP) films. (Reproduced from Gururajan and Ogale [26]).

acteristic of PE and only slightly influenced by the PP component in PP/PE blends. Figure 3.17c displays the spectrum obtained off-line from a bilayer LDPE/PP film with contributions from both components. The spectrum is generally consistent with that reported for LDPE/PP blends. The crystallinity of PP and PE was independently estimated up to a limited thickness without significant interference of PP-over-PE or vice versa.

Figure 3.18 displays representative real-time Raman spectra obtained along the axial distance during bicomponent blown film extrusion. The Raman spectrum of the melt is devoid of any characteristic crystalline peaks. As the melt is subjected to cooling, crystalline growth starts. Near the FLH, a characteristic peak appears at 809 cm^{-1}, indicating the onset of crystallization of PP.

Further up the line, PE starts to crystallize as identified by the peak at 1418 cm^{-1}. The onset of crystallization of PE can also be confirmed from the emergence of the crystalline peak at 1130 cm^{-1}. The crystalline peaks of PP and PE gradually intensify along the axial direction and saturate at some distance in the film line as the bubble temperature approaches ambient. The crystallinity of the components was estimated based on the methods described earlier in Section 3.2.

Figure 3.19 displays the crystallinity of PP and PE components plotted as a function of axial distance in the film line for a TUR of 10 and a BUR of 1.5. The crystallization temperature for PP and LDPE were observed to be 121 °C and 99 °C, respectively. As expected,

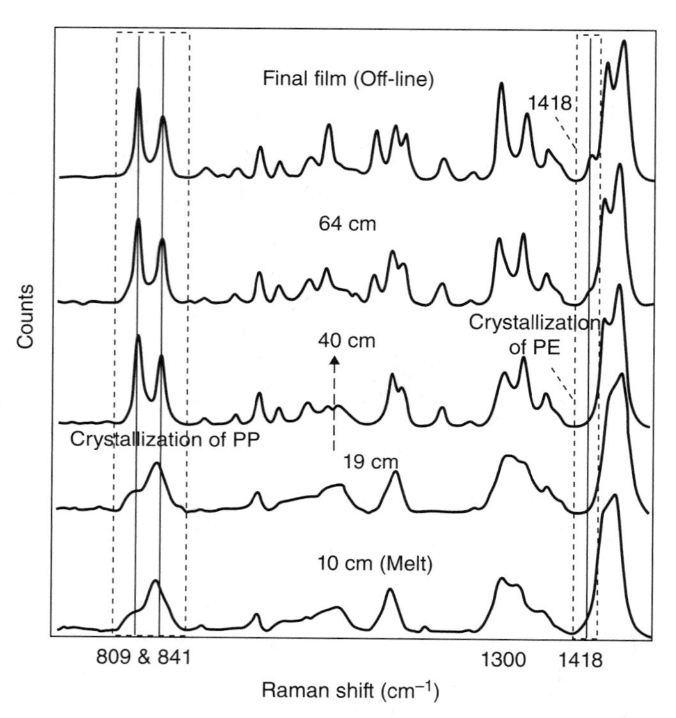

Figure 3.18 Real-time Raman spectra at different axial positions between die exit and nip rolls during polypropylene (PP)/low-density polyethylene (LDPE) bilayer blown film extrusion (Reproduced from Gururajan and Ogale [22]).

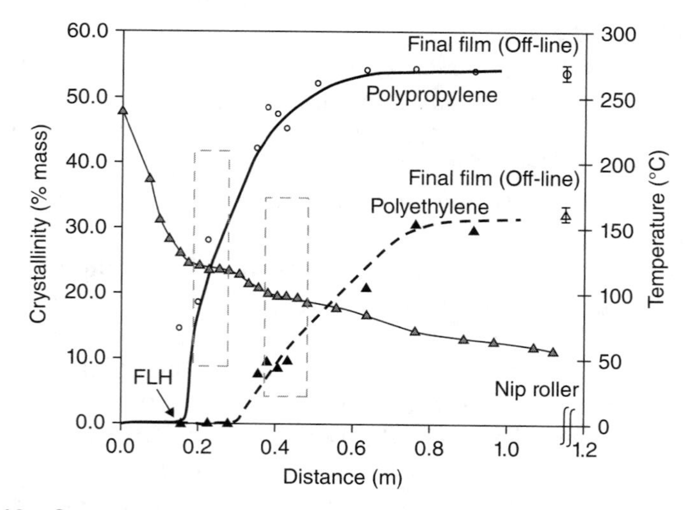

Figure 3.19 Crystallinity values obtained from real-time Raman spectra plotted as a function of axial distance for TUR = 10, BUR = 1.5 and moderate cooling condition during PP/LDPE blown film extrusion. Real-time temperature data are presented on the right axis. Solid and dashed curves represent trends. FLH, frost-line height.

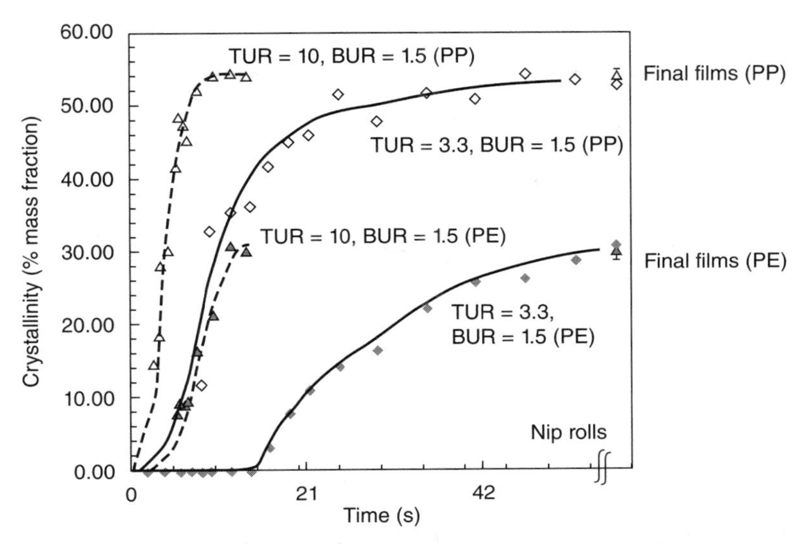

Figure 3.20 Crystallinity values of polypropylene (PP) and polyethylene (PE) plotted as a function of process time for two different take-up ratios (TURs): 3.3 and 10 at constant blow-up ratio (BUR) and cooling conditions (curves represent trends).

crystallization for PP starts closer to the die as compared with that for LDPE, i.e. at a higher temperature. Once the crystallization process starts, the crystallinity profiles for PP and LDPE show a steep growth, which is consistent with the double plateaus in temperature profile observed as a result of the opposing phenomena of exothermic crystallization (heat generation) and external cooling (heat loss).

The crystallinity of PP and LDPE components are plotted against process time in Figure 3.20. The effect of the take-up speed was studied by keeping the BUR constant and varying the TUR from 3.3 to 10. The profile for the film subjected to a high TUR is slightly steeper and plateaus faster compared with that for the bubble subjected to lower TUR, indicating higher crystalline growth rates at higher speeds due to higher stresses.

3.7 SIGNIFICANCE OF REAL-TIME RAMAN SPECTROSCOPY

During the film blowing process, polymer molecular characteristics couple with processing conditions to produce the film morphology that ultimately determines the end-use properties. Figure 3.21 displays the real-time crystallinity profiles as function of the process time during PP/LDPE bilayer blown film extrusion. PP crystallizes first ($T_c \approx 122\,°C$)

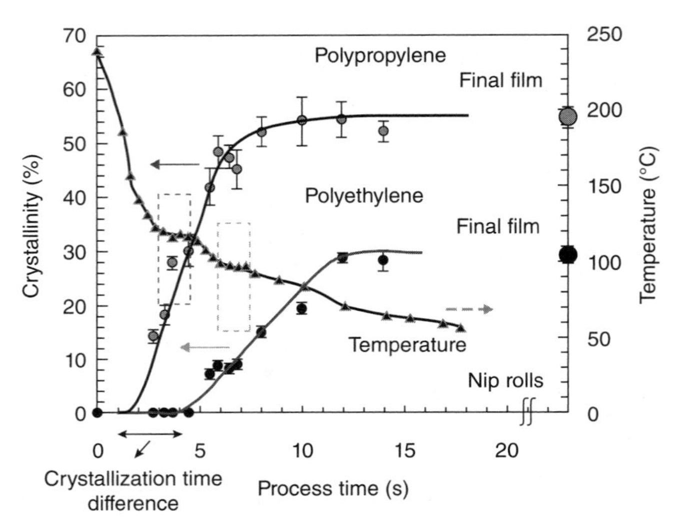

Figure 3.21 Real-time crystallinity profiles of polypropylene (PP) and low-density polyethylene (LDPE) during PP/LDPE blown film extrusion as function of process time superposed with temperature profile for blow-up ratio (BUR) = 1.5, take-up ratio (TUR) = 10, and moderate cooling (curves represent trends).

with a steep increase in crystallinity followed by a plateau. PE displays a similar trend but at a later time. The process-time difference between the onset of crystallization of PP and LDPE components was measured. For bicomponent (PP/LDPE) blown films, process dynamics are complex compared with single-layer blown films as also noted for HDPE/EVOH blown film [75]. During the film formation, after the first layer freezes, the stresses are borne by the solidified layer (PP), while the second layer (LDPE) is still in the molten state. The molecular chains in molten LDPE tend to relax, and the extent depends on the crystallization time difference between the two components. For PE, the extent of row nucleation depends on the stress applied during processing. At very low stress levels, lamellae are disordered and row-nucleated structure is less distinct, whereas at high stresses, significant row-nucleated structure is formed.

This difference can be attributed to the partial molecular relaxation that occurs in long-chain fibrils, which reduces the extent of row nucleation in LDPE. It is known that the maximum stress occurs near the frost line and the extent of stress-induced crystallization is determined by the number of extended long chains during the onset of crystallization. Thus, a difference in orientation and extent of row nucleation was observed in single-layer LDPE film versus co-extruded LDPE film as

Table 3.1 Comparison of Herman's crystalline orientation factors (f_a, f_b) for single-layer and co-extruded PP and LDPE

	Single PP		Co-extruded PP		Single PE		Co-extruded PE	
TUR	f_c (c-axis)	f_b (b-axis)	f_c (c-axis)	f_b (b-axis)	f_a (a-axis)	f_b (b-axis)	f_a (a-axis)	f_b (b-axis)
3.5	0.11	−0.22	0.14 ± 0.02	−0.21 ± 0.01	0.12	−0.07	0.06 ± 0.02	−0.03 ± 0.02
10	0.21	−0.31	0.22 ± 0.02	−0.32 ± 0.01	0.27	−0.18	0.19 ± 0.02	−0.11 ± 0.01

PP, polypropylene; PE, polyethylene; LDPE, low-density polyethylene; TUR, take-up ratio.

shown in Table 3.1 and Figure 3.22. As noted earlier, this difference arises from relaxation of high-molecular-weight LDPE chains, which could have acted as fibrillar nuclei for row-nucleated structure. Thus, crystallization time difference for the onset of crystallization of PP and LDPE plays an important role in determining the orientation and morphology of the components during PP/LDPE bilayer blown film extrusion [56,57]. The present results show that crystallinity development in individual components has a profound effect on the microstructure and properties of the multilayer films relative to that observed in single-layer films. Thus, Raman spectroscopy can serve as a powerful tool for estimation of microstructure development during the production of multilayer films that are of paramount importance in industrial applications.

3.8 CONCLUDING REMARKS

In this chapter, the application of Raman spectroscopy for real-time process monitoring during single- and multilayer blown film extrusion was demonstrated. The crystallinity and orientation evolution during single-layer blown film extrusion of polyolefins helped in the formulation of processing–structure–property (PSP) relationships and also validating process model predictions. For bilayer (PP/LDPE) blown film extrusion, online Raman spectroscopy was used to investigate the effect of co-extrusion on the microstructure of the components during the process. The crystallinity values obtained from Raman spectroscopy were found to be comparable to those obtained from XRD. Thus, it can be said that Raman spectroscopy offers an inexpensive, powerful, and viable option for real-time characterization of polymer films during processing.

(a)

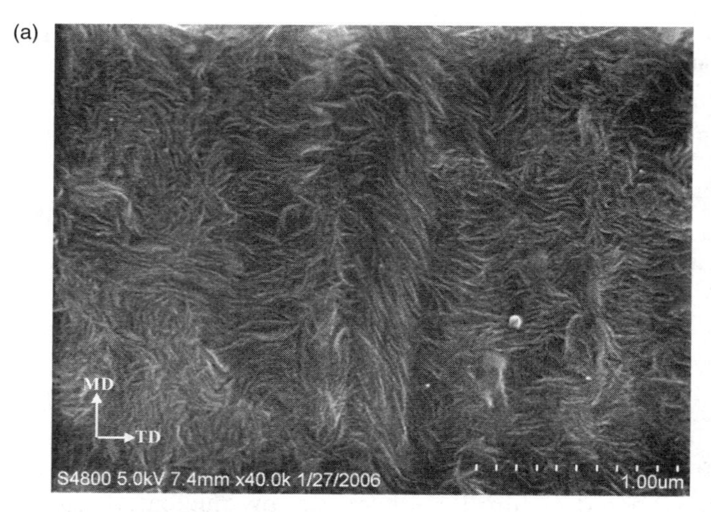

(b)

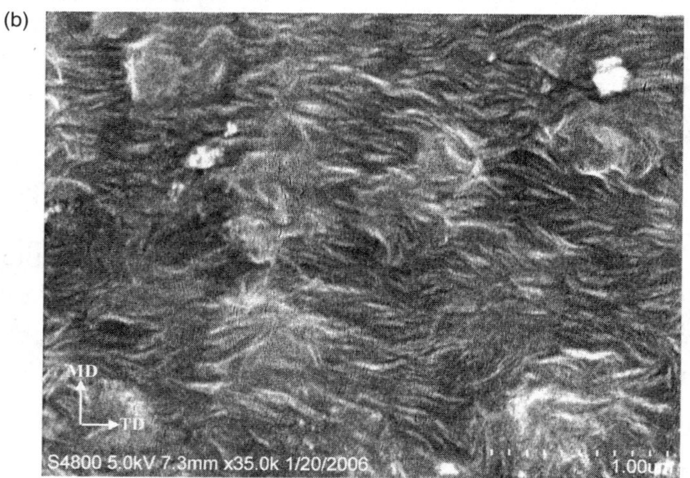

Figure 3.22 Morphology of low-density polyethylene (LDPE) film extruded as (a) single-layer film and (b) co-extruded with polypropylene (PP) layer. TD, transverse direction; MD, machine direction.

REFERENCES

1. van Overbeke, E., Devaux, J., Legras, R., Carter, J.T., McGrail, P.T., Carlier, V. (2001). Raman spectroscopy determination of the thermoplastic content within epoxy resin-copolyethersulfone blends. *Applied Spectroscopy*, 55(11), 1514–1522.
2. Edwards, H.G.M., Johnson, A.F., Lewis, I.R., Ward, N.J., Wheelwright, S. (1993). Preparation and vibrational spectroscopic characterization of

copolymers of 2,3-dimethylbutadiene and methyl-methacrylate. *Journal of Raman Spectroscopy* 24(8), 495–500.

3. Tanto, B., Guha, S., Martin, C.M., Scherf, U., Winokur, M.J. (2004). Structural and spectroscopic investigations of bulk poly [bis(2-ethyl) hexylfluorene]. *Macromolecules*, 37(25), 9438–9448.
4. De Santis, A. (2005). Photo-polymerisation effects on the carbonyl C=O bands of composite resins measured by micro-Raman spectroscopy. *Polymer*, 46(14), 5001–5004.
5. Parnell, S., Min, K., Cakmak, M. (2003). Kinetic studies of polyurethane polymerization with Raman spectroscopy. *Polymer*, 44(18), 5137–5144.
6. Cappitelli, F., Vicini, S., Piaggio, P., Abbruscato, P., Princi, E., Casadevall, A., Nosanchuk, J.D., Zanardini, E. (2005). Investigation of fungal deterioration of synthetic paint binders using vibrational spectroscopic techniques. *Macromolecular Bioscience*, 5(1), 49–57.
7. Zieba-Palus, J., Borusiewicz, R. (2006). Examination of multilayer paint coats by the use of infrared, Raman and XRF spectroscopy for forensic purposes. *Journal of Molecular Structure*, 792, 286–292.
8. Williams, K.P.J., Klenerman, D. (1992). UV resonance Raman-spectroscopy using a quasi-continuous-wave laser developed as an industrial analytical technique. *Journal of Raman Spectroscopy*, 23(4), 191–196.
9. Oldak, D., Kaczmarek, H., Buffeteau, T., Sourisseau, C. (2005). Photo- and bio-degradation processes in polyethylene, cellulose and their blends studied by ATR-FTIR and Raman spectroscopies. *Journal of Materials Science*, 40(16), 4189–4198.
10. Strobl, G.R., Hagedorn, W. (1978). Raman-spectroscopic method for determining crystallinity of polyethylene. *Journal of Polymer Science Part B—Polymer Physics*, 16(7), 1181–1193.
11. Nielsen, A.S., Batchelder, D.N., Pyrz, R. (2002). Estimation of crystallinity of isotactic polypropylene using Raman spectroscopy. *Polymer*, 43(9), 2671–2676.
12. Jarvis, D.A., Hutchinson, I.J., Bower, D.I., Ward, I.M. (1980). Characterization of biaxial orientation in poly(ethylene-terephthalate) by means of refractive-index measurements and Raman and infrared spectroscopies. *Polymer*, 21(1), 41–54.
13. Citra, M.J., Chase, D.B., Ikeda, R.M., Gardner, K.H. (1995). Molecular-orientation of high-density polyethylene fibers characterized by polarized Raman-spectroscopy. *Macromolecules*, 28(11), 4007–4012.
14. Pigeon, M., Prudhomme, R.E., Pezolet, M. (1991). Characterization of molecular-orientation in polyethylene by Raman-spectroscopy. *Macromolecules*, 24(20), 5687–5694.
15. Everall, N., King, B. (1999). Raman spectroscopy for polymer characterization in an industrial environment. *Macromolecular Symposia*, 141, 103–116.

16. Tanaka, M., Young, R.J. (2006). Polarised Raman spectroscopy for the study of molecular orientation distributions in polymers. *Journal of Materials Science* 41(3), 963–991.
17. Craver, C.D., Carraher, C.E. (2000). *Applied Polymer Science: 21st Century*. Burlington, MA: Elsevier Science & Technology Books.
18. Pelletier, M.J. (1999). *Analytical Applications of Raman Spectroscopy*. Oxford and Malden, MA: Blackwell Science, vii.
19. Lewis, I.R., Edwards, H.G.M. (2001). *Handbook of Raman Spectroscopy: From the Research Laboratory to the Process Line*. New York and Basel: Marcel Dekker, xiii.
20. Bower, D.I., Maddams, W.F. (1989). *The Vibrational Spectroscopy of Polymers (Cambridge Solid State Science Series)*. New York: Cambridge University Press.
21. Siesler, H.W., Holland-Moritz, K. (1980). *Infrared and Raman Spectroscopy of Polymers (Practical Spectroscopy Series)*. New York: Marcel Dekker Inc.
22. Marrow, D.G., Cochran, A.M., Roger, S.T. (2004). On-line measurement and control of polymer product properties by Raman spectroscopy. *PCT Int. Appl.*, 45 pp. WO 2004063234 A1 2004-07-29.
23. Marrow, D.G., Yahn, D.A., Veariel, Thomas R. (2007). On-line properties analysis of a molten polymer by Raman spectroscopy for control of a mixing device. *U.S. Pat. Appl. Publ.*, 20 pp. US 2007019190 A1 2007-01-25.
24. Leugers, M.A., Shepherd, A.K, Weston, J.W., Sun, Z. (1999). Online measurement of crystallinity in semicrystalline polymers by Raman spectroscopy. *PCT Int. Appl.*, 18 pp. WO 9927350 A1 1999-06-03.
25. Paradkar, R.P., Sakhalkar, S.S., He, X.J., Ellison, M.S. (2001). On-line estimation of molecular orientation in polypropylene fibers using polarized Raman spectroscopy. *Applied Spectroscopy*, 55(5), 534–539.
26. Gururajan, G., Ogale, A.A. (2005). Real-time Raman spectroscopy during LDPE/PP bilayer blown-film extrusion. *Plastics Rubber and Composites*, 34(5–6), 271–275.
27. Cherukupalli, S.S., Ogale, A.A. Online measurements of crystallinity using Raman spectroscopy during blown film extrusion of a linear low-density polyethylene. *Polymer Engineering and Science*, 44(8), 1484–1490.
28. Bullwinkel, M.D., Campbell, G.A., Rasmussen, D.H., Krexa, J. and Brancewitz, C.J. (2001). Crystallization studies of LLDPE during tubular blown film processing. *International Polymer Processing*, 16(1), 39–47.
29. Nagasawa Toshio, M.T., Hoshino, S. (1973). Film forming process of crystalline polymer—1. Factors inducing a molecular orientation in tubular blown film. *Applied Polymer Symposium*, 20, 275–293.
30. Ghaneh-Fard, A., Carreau, P.J., Lafleur, P.G. (1997). On-line birefringence measurement in film blowing of a linear low density polyethylene. *International Polymer Processing*, 12(2), 136–146.

31. Dees, J.R., Spruiell, J.E. (1974). Structure development during melt spinning of linear polyethylene fibers. *Journal of Applied Polymer Science*, 18(4), 1053–1078.
32. Spruiell, J.E., White, J.L. (1975). Structure development during polymer processing—studies of melt spinning of polyethylene and polypropylene fibers. *Polymer Engineering and Science*, 15(9), 660–667.
33. Hsiao, B.S., Barton, R., Quintana, J. (1996). Simple on-line X-ray setup to monitor structural changes during fiber processing. *Journal of Applied Polymer Science*, 62(12), 2061–2068.
34. Lamberti, G., Brucato, V. (2003). Real-time orientation and crystallinity measurements during the isotactic polypropylene film-casting process. *Journal of Polymer Science Part B-Polymer Physics*, 41(9), 998–1008.
35. Muke, S., Connell, H., Sbarski, I., Bhattacharya, S.N. (2003). Numerical modelling and experimental verification of blown film processing. *Journal of Non-Newtonian Fluid Mechanics*, 116(1), 113–138.
36. Henrichsen, L.K., McHugh, A.J., Cherukupalli, S.S., Ogale, A.A. (2004). Microstructure and kinematic aspects of blown film extrusion process: II. Numerical modelling and prediction of LLDPE and LDPE. *Plastics Rubber and Composites*, 33(9–10), 383–389.
37. Han, C.D., Park, J.Y. (1975). Studies on blown film extrusion. 1. Experimental determination of elongational viscosity. *Journal of Applied Polymer Science*, 19(12), 3257–3276.
38. Krishnaswamy, R.K., Lamborn, M.J. (2000). Tensile properties of linear low density polyethylene (LLDPE) blown films. *Polymer Engineering and Science*, 40(11), 2385–2396.
39. Patel, R.M., Butler, T.I., Walton, K.L., Knight, G.W. (1994). Investigation of processing-structure-properties relationships in polyethylene blown films. *Polymer Engineering and Science*, 34(19), 1506–1514.
40. Gupta, A., Simpson, D.M., Harrison, I.R. (1993). A morphological-study of Hdpe-blown films using small-angle X-ray-scattering. *Journal of Applied Polymer Science*, 50(12), 2085–2093.
41. Alexander, L.E. (1979). *X-ray Diffraction Methods in Polymer Science*. Huntington, NY: Krieger Publishing Company, xv.
42. Krishnaswamy, R.K., Sukhadia, A.M. (2000). Orientation characteristics of LLDPE blown films and their implications on Elmendorf tear performance. *Polymer*, 41(26), 9205–9217.
43. Ajji, A., Zhang, X., Elkoun, S. (2005). Biaxial orientation in HDPE films: Comparison of infrared spectroscopy, X-ray pole figures and birefringence techniques. *Polymer*, 46(11), 3838–3846.
44. Kissin, Y.V. (1992). Infrared method for measuring orientation in polyethylene films. *Journal of Polymer Science Part B-Polymer Physics*, 30(10), 1165–1172.

45. Lu, J.J., Sue, H.J., Rieker, T.P. (2000). Morphology and mechanical property relationship in linear low-density polyethylene blown films. *Journal of Materials Science*, 35(20), 5169–5178.
46. Yuksekkalayci, C., Yilmazer, U., Orbey, N. (1999). Effects of nucleating agent and processing conditions on the mechanical, thermal, and optical properties of biaxially oriented polypropylene films. *Polymer Engineering and Science*, 39(7), 1216–1222.
47. Tagawa, T., Mori, J. (1978). Observation of internal lamellar structure in polyethylene films by ion-etching and SEM technique. *Journal of Electron Microscopy*, 27(4), 267–274.
48. Zhang, X.M., Elkoun, S., Ajji, A., Huneault, M.A. (2004). Oriented structure and anisotropy properties of polymer blown films: HDPE, LLDPE and LDPE. *Polymer*, 45(1), 217–229.
49. Bentley, P.A., Hendra, P.J. (1995). Polarised FT Raman studies of an ultra-high modulus polyethylene rod. *Spectrochimica Acta Part A—Molecular and Biomolecular Spectroscopy*, 51(12), 2125–2131.
50. deBaez, M.A., Hendra, P.J., Judkins, M. (1995). The Raman spectra of oriented isotactic polypropylene. *Spectrochimica Acta Part A—Molecular and Biomolecular Spectroscopy*, 51(12), 2117–2124.
51. Yang, S.Y., Michielsen, S. (2004). Analysis of the 1030-cm(–1) band of poly(ethylene terephthalate) fibers using polarized Raman microscopy. *Journal of Polymer Science Part B-Polymer Physics*, 42(1), 47–52.
52. Michielsen, S. (1994). Effect of moisture and orientation on the fracture of nylon 6,6 fibers. *Journal of Applied Polymer Science*, 52(8), 1081–1089.
53. Sato, H., Sasao, S., Matsukawa, K., Kita, Y., Ikeda, T., Tashiro, H., Ozaki, Y. (2002). Raman mapping study of compatibilized and uncompatibilized polymer blends of Nylon 12 and polyethylene. *Applied Spectroscopy*, 56(8), 1038–1043.
54. Paradkar, R.P., Sakhalkar, S.S., He, X., Ellison, M.S. (2003). Estimating crystallinity in high density polyethylene fibers using online Raman spectroscopy. *Journal of Applied Polymer Science*, 88(2), 545–549.
55. Cherukupalli, S.S., Gottlieb, S.E., Ogale, A.A. (2005). Real-time Raman spectroscopic measurement of crystallization kinetics and its effect on the morphology and properties of polyolefin blown films. *Journal of Applied Polymer Science*, 98(4), 1740–1747.
56. Cherukupalli, S.S., Ogale, A.A. (2004). Integrated experimental-modelling study of microstructural development and kinematics in a blown film extrusion process: I. Real-time Raman spectroscopy measurements of crystallinity. *Plastics Rubber and Composites*, 33(9-10), 367–371.
57. Gururajan, G., Ogale, A.A. (2007). Effect of coextrusion on PP/LDPE bicomponent blown films: Raman spectroscopy for real-time microstructural measurements. *Journal of Plastic Film & Sheeting*, 23(1), 37–49.

58. Campbell, G.A., Babel, A.K. (1996). Physical properties and processing conditions correlations of the LDPE/LLDPE tubular brown films. *Macromolecular Symposia*, 101, 199–206.
59. Clas, S.D., Heyding, R.D., Mcfaddin, D.C., Russell, K.E., Scammellbullock, M.V., Kelusky, E.C., Stcyr, D. (1988). Crystallinities of copolymers of ethylene and 1-alkenes. *Journal of Polymer Science Part B-Polymer Physics*, 26(6), 1271–1286.
60. Kolnaar, J.W.H., Keller, A., Seifert, S., Zschunke, C., Zachmann, H.G. (1995). In-situ X-ray studies during extrusion of polyethylene. *Polymer*, 36(20), 3969–3974.
61. Mchugh, A.J. (1982). Mechanisms of flow induced crystallization. *Polymer Engineering and Science*, 22(1), 15–26.
62. Southern, J.H. (1976). Special issue on stress-induced crystallization—Southern, JH, Guest Editor—Based on National American-Chemical-Society Symposium on Stress-Induced Crystallization, sponsored by Division-of-Polymer-Chemistry-Inc, and presented at Chicago, Illinois, August 24–29, 1975. *Polymer Engineering and Science*, 16(3), 125–125.
63. Giri Gururajan, A.A.O. (2008). *Real-time X-ray diffraction during polyethylene blown film extrusion.*
64. Aggarwal, S.L., Tilley, G.P. (1955). Determination of crystallinity in polyethylene by X-ray diffractometer. *Journal of Polymer Science Part B-Polymer Physics*, 18, 17–26.
65. Pandana, D.V. (2007). Study of structure development during melt spinning of isotactic polypropylene fibers using Raman spectroscopy and wide angle x-ray diffraction simultaneously. *Materials Science and Engineering*, p. 153. Clemson University, Clemson, SC, USA.
66. Kip, B.J., Vangurp, M., Vanheel, S.P.C., Meier, R.J. (1993). Orientational order in polyethylene foils—a polarized Raman-spectroscopic study. *Journal of Raman Spectroscopy*, 24(8), 501–510.
67. Lesko, C.C.C., Rabolt, J.F., Ikeda, R.M., Chase, B., Kennedy, A. (2000). Experimental determination of the fiber orientation parameters and the Raman tensor of the 1614 cm(-1) band of poly(ethylene terephthalate). *Journal of Molecular Structure*, 521, 127–136.
68. Lefevre, T., Rousseau, M.E., Pezolet, M. (2006). Orientation-insensitive spectra for Raman microspectroscopy. *Applied Spectroscopy*, 60(8), 841–846.
69. Bower, D.I. (1972). Investigation of molecular orientation distribution by polarized Raman scattering and polarized fluorescence. *Journal of Polymer Science Part B-Polymer Physics*, 10, 2135–2153.
70. Gururajan, G., Ogale, A.A. (2008). Molecular orientation evolution during low-density polyethylene blown film extrusion using real-time Raman spectroscopy. *Journal of Raman Spectroscopy*, available online, 10 Oct.

71. Lu, S., Russell, A.E., Hendra, P.J. (1998). The Raman spectra of high modulus polyethylene fibres by Raman microscopy. *Journal of Materials Science*, 33(19), 4721–4725.

72. Bower, D.I. (1981). Orientation distribution-functions for uniaxially oriented polymers. *Journal of Polymer Science Part B—Polymer Physics*, 19(1), 93–107.

73. Quintana, S.L., Schmidt, P., Dybal, J., Kratochvil, J., Pastor, J.M., Merino, J.C. (2002). Microdomain structure and chain orientation in polypropylene/polyethylene blends investigated by micro-Raman confocal imaging spectroscopy. *Polymer*, 43(19), 5187–5195.

74. Markwort, L., Kip, B., Dasilva, E., Roussel, B. (1995). Raman imaging of heterogeneous polymers—a comparison of global versus point illumination. *Applied Spectroscopy*, 49(10), 1411–1430.

75. Morris, B.A. (1999). The effect of coextrusion on bubble kinematics, temperature distribution and property development in the blown film process. *Journal of Plastic Film & Sheeting*, 15(1), 25–36.

4

RAMAN APPLICATIONS IN SYNTHETIC AND NATURAL POLYMER FIBERS AND THEIR COMPOSITES

Robert J. Young, Ph.D. and Steve J. Eichhorn, Ph.D.
Materials Science Centre, School of Materials, University of Manchester

4.1 INTRODUCTION

Aerospace and other advanced industries are driving the need to develop lightweight materials with high stiffnesses, strengths, and toughness. Progress over the last 30 years into the production of high-performance polymeric fibers has been rapid with these properties ultimately in mind. Coupled with this progress has been the development of the necessary tools to follow the deformation processes in polymeric fibers to understand better how their performance can be improved. Particular highlights have been the development of Raman spectroscopy and X-ray diffraction (more recently from synchrotron sources), with which the molecular and crystal deformation of polymeric fibers can be followed in detail. This approach has been particularly useful for studying the micromechanics of composites, which are becoming increasingly important as high-performance materials. A clear distinction can be made between fibers derived by man-made

Raman Spectroscopy for Soft Matter Applications, Edited by Maher S. Amer

routes, and those of natural origin, in terms of the responses of their microstructural stress and strain to external deformation.

In this chapter, the study of the deformation mechanisms of polymer fibers using Raman spectroscopy will be covered firstly from a theoretical viewpoint. The original applications upon synthetic fibers will then be discussed and more recent studies of natural polymer-based systems will be described. The application of the technique to study deformation mechanisms in composites will also be described.

4.2 DEFORMATION MECHANISMS IN POLYMERIC FIBERS

It is well known that there is a good correlation between optical properties and molecular orientation in polymeric materials. Treloar first discussed the effect of the reorientation of a polymeric material (rubber), and the subsequent changes observed in birefringence measurements [1]. Later, others showed experimentally [2–4] that the optical birefringence in a wide range of polymeric materials could be used to follow the deformation of oriented molecules, and could also be related to their crystalline/amorphous structure. Hermans [5] used a similar approach to these investigations for cellulose fibers, and derived orientation functions that could be obtained from simple birefringence measurements. Ward subsequently developed models [6] that interrelated structure, orientation, and elastic modulus in idealized semicrystalline polymeric materials. Ward [6] also showed that the relationship between mechanical anisotropy and orientation was much more complex than that between the static birefringence and orientation, and derived an expression relating the birefringence and the orientation function in a similar manner to Hermans [5].

4.2.1 Fiber Mechanics

There have been numerous investigations into the nature of the semicrystalline structure of fibers and filamentous materials. Much of this research is based upon the work of Halpin and Kardos [7], who for the first time gave upper and lower bands for the predictions of mechanical stiffness of semicrystalline polymeric materials based on composite theory. The lower bound due to Reuss [8] assumes that the elements in the structure (i.e. the amorphous and crystalline units) are lined up in series and experience the same stress. The modulus of the polymer in this case, E_p will be given by an equation of the form:

$$1/E_p = V_c/E_c + (1 - V_c)/E_a. \tag{4.1}$$

where V_c is the volume fraction of crystals, and E_c and E_a are the moduli of the crystals and amorphous material, respectively. The second model due to Voigt [8] assumes that all the elements are lined up in parallel, and so experience the same strain. The modulus for the polymer for the Voigt model is given by the rule of mixtures as

$$E_p = E_c V_c + E_a(1 - V_c). \tag{4.2}$$

These upper and lower bands are represented schematically in Figure 4.1, where the upper band reflects a *uniform strain* structure and the lower band a *uniform stress* structure. This study by Halpin and Kardos [7] allowed the mechanical stiffness of a polymer to be derived on the basis of high and low aspect ratio crystals in an amorphous matrix, with different mechanical properties (i.e. high modulus crystals and a much lower modulus matrix) without the need for the presence of tie molecules.

Aggregate models, where crystalline domains are considered in series with amorphous domains, were first derived by Ward [6], and developed later by Northolt and van der Hout [9] for highly crystalline fibers. In this series arrangement, the stress in the crystals and amorphous regions is assumed to be *uniform*. Takayanagi et al. [10] reported the use of a combined parallel-series model in which the stress in the crystals and amorphous regions is *nonuniform*, with the former overstressed and the latter understressed. In this sense, one could consider, in a fiber with such a morphology, that a stress-transfer process is taking place, and given the dimensions of the crystals, that the structure is that of a nanocomposite.

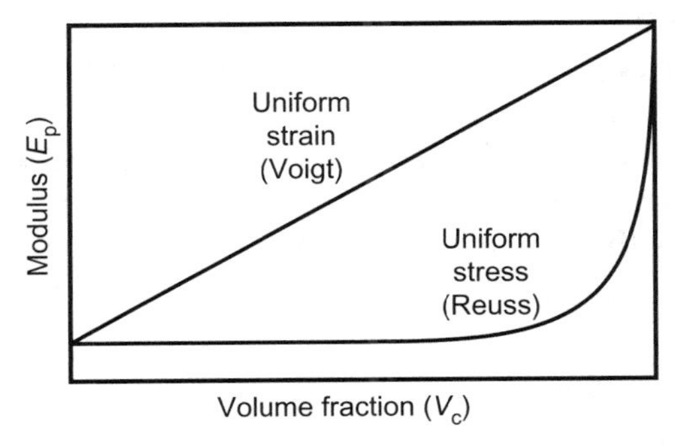

Figure 4.1 The form of the upper and lower bands of fiber modulus for uniform strain (Voigt) and stress (Reuss) models.

4.2.2 Raman Spectroscopy and Deformation

A number of noncontact and techniques with high spatial resolution have been utilized to better understand the nature of local stress and its development during deformation in polymeric fibers. This chapter will focus on Raman spectroscopy, which is useful for determining molecular deformation and orientation. X-ray diffraction, on the other hand, is sensitive to crystal and amorphous interactions. Orientation analysis, of the crystalline fraction only, is also possible using X-ray diffraction, and this is covered in more detail elsewhere [11]. It is desirable to use a combination of both techniques to obtain a complete understanding of the micromechanics of deformation of both semicrystalline polymeric fibers and composites.

The Raman spectrum of a single aramid fiber obtained using a HeNe laser is shown in Figure 4.2a and the effect of tensile stress upon the 1610-cm^{-1} Raman band is shown in Figure 4.2b. It has been established [12] that this change in Raman wave number, $\Delta\nu$, during the deformation of high-performance fibers, is a result of chain stretching and related directly to the stress on the crystalline reinforcing units, σ_r, such that for an increment of stress,

$$d\Delta\nu \propto d\sigma_r. \tag{4.3}$$

If it is assumed that the stress on the reinforcing units is the same as that on the fiber as a whole (*uniform stress*), then σ_r equals σ_f, the fiber stress, and so Equation 4.3 becomes

$$\frac{d\Delta\nu}{d\varepsilon_f} \propto \frac{d\sigma_f}{d\varepsilon_f} = E_f, \tag{4.4}$$

by dividing by an increment of fiber strain, ε_f , where E_f is the Young's modulus of the fiber. In the case of *uniform stress* since σ_r equals σ_f, then it follows from Equation 4.3 that $d\Delta\nu/d\sigma_f$ should be independent of fiber modulus.

An alternative situation that can be employed to model the behavior is the uniform strain model in which the strain on the reinforcing units, ε_r, is the same as the overall fiber strain, ε_f, which leads to

$$\frac{\sigma_r}{E_r} = \frac{\sigma_f}{E_f} \tag{4.5}$$

where E_r is the Young's modulus of the reinforcing units. Rearranging and using Equation 4.3 gives

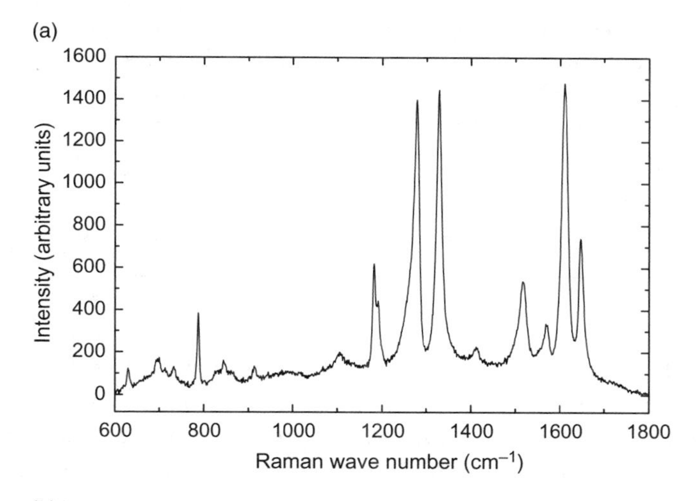

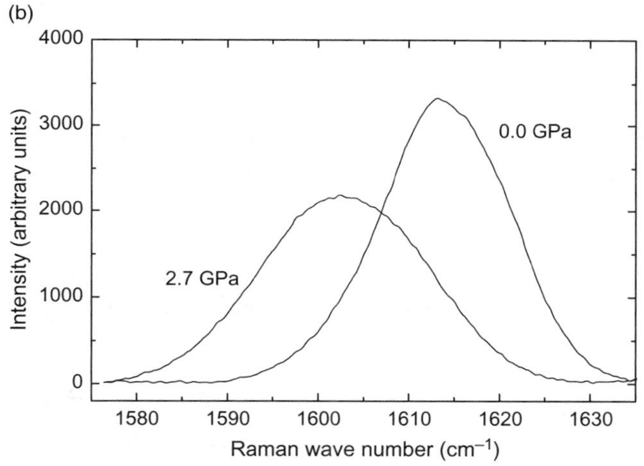

Figure 4.2 (a) Raman spectrum obtained from a single aramid polymer fiber. (b) Shift of the 1610-cm^{-1} Raman band with tensile stress. (Data replotted from Young [12].)

$$\frac{d\Delta\nu}{d\sigma_f} \propto \frac{E_r}{E_f}. \tag{4.6}$$

As it can be assumed that the modulus of the reinforcement, E_r, is constant, then

$$\frac{d\Delta\nu}{d\sigma_f} \propto \frac{1}{E_f} \tag{4.7}$$

Table 4.1 Predicted dependence of the Raman band shift rates with both strain and stress for the different models of fiber structure

Band shift	Uniform stress	Uniform strain
$\frac{d\Delta\nu}{d\varepsilon_f}$	$\propto E_f$	Independent of E_f
$\frac{d\Delta\nu}{d\sigma_f}$	Independent of E_f	$\propto \frac{1}{E_f}$

It is therefore predicted that in the situation where there is a uniform strain in the fiber, $d\Delta\nu/d\sigma_f$ should be proportional to the reciprocal of the fiber modulus, $1/E_f$. If the band shift is measured as function of strain, then using Equation 4.3 and since the strain is uniform,

$$\frac{d\Delta\nu}{d\varepsilon_f} = \frac{d\sigma_f}{d\varepsilon_r} = E_r. \tag{4.8}$$

Hence since E_r is constant, then $d\Delta\nu/d\varepsilon_f$ is constant. Table 4.1 summarizes these predictions for Raman band shifts in the cases of uniform stress and uniform strain.

4.3 RAMAN SPECTROSCOPY AND MOLECULAR DEFORMATION

Before laser sources became available, because the Raman effect is very weak [13,14], infrared spectroscopy of polymers was far more popular as a technique for the characterization of polymers [15–19]. Even in these early stages of development, however, Raman spectra were reported from polymeric materials such as polystyrene [20]. Until the development of laser technology, and particularly the emergence of new optical rejection filters [21] for enhancing Raman Stokes and anti-Stokes spectral peaks and the microscope or microprobe systems for high spatial resolution [22], detailed studies of polymers were not possible. Interest then began into using vibrational spectroscopy to follow local changes in the structure of polymer fibers. Further interest in the use of Raman spectroscopy to characterize polymers has increased with the development of polymers with conjugated backbones which undergo resonance Raman scattering and give strong, very well-defined spectra. After the initial studies into the characterization of static samples, people began to use these techniques to understand deformation processes and the progress of this work will now be reviewed.

4.3.1 Background and Theory

The predecessor of the observations of shifts in the positions of Raman bands of high-polymer fibers was the discovery of a similar effect using infrared spectroscopy [23–26]. It became clear during the early 1970s that orientation changes in polymeric materials could be followed using infrared and *in situ* deformation [27]. In addition, and during the same period, Wool showed that the intensity maxima in a number of polymers shifted toward a lower wave number upon external deformation [28,29]. Initially Zhurkov et al. proposed that the rate of the shifts of different bands with strain within a polymeric material was constant [23]. Wool showed, however, that this was not true, and that this factor depended on the molecular stress distribution via the morphology of the sample [29].

An important discovery was made in 1977 by Mitra et al. [30], who showed that when conjugated polydiacetylene single-crystal fibers were deformed in tension, Raman bands located at 1498 cm^{-1} (corresponding to the –C=C– backbone moiety) and 2104 cm^{-1} (corresponding to the –C≡C– backbone moiety) shifted in position toward a lower wave number [30]. These shifts are shown in Figure 4.3 [30]. This effect was

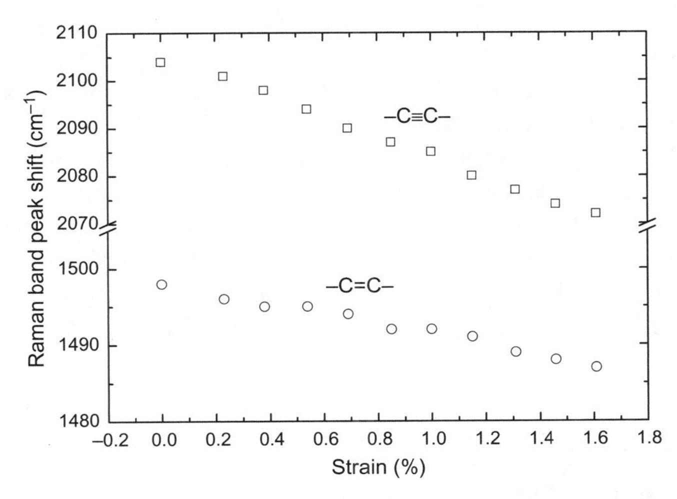

Figure 4.3 Raman band shifts as a function of applied tensile strain for two Raman bands for polydiacetylene single crystals. (Key: □ – 2104 cm^{-1} and ○ – 1498 cm^{-1} Raman bands corresponding to –C≡C– and –C=C– moieties, respectively. Data replotted from Mitra et al. [30].)

also modeled [30] by using an analysis based on anharmonic force constants of bonds within an idealized polymeric chain and Badger's rule [31], which interrelates the force constant with the equilibrium separation of neighboring atoms [31]. In this sense, Mitra et al. [30] showed that the shift was due to direct stretching of the polydiacetylene polymer backbone, and the subsequent change therefore in the force constants of the bond associated with this structural form. Batchelder and Bloor [32] also predicted the band shifts in polydiacetylene based on a model of point masses and anharmonic spring constants.

Previous studies of the plastic deformation of crystals of bis(*p*-toluene sulfonate) 2,4-hexadiyene-1,6-diol (TSHD) had shown that the predominant stress-relieving process in such materials was a twinning or kinking of the polymer chains [33]. However, since this would not occur in tension, it was assumed by Batchelder and Bloor [32] that polydiacetylene single crystals could sustain, in theory, large tensile stresses along their backbones without the occurrence of plastic deformation [32]. It is worth noting at this point that Raman spectroscopic studies of the molecular deformation of all types of polydiacetylene single crystals reveal that the band shift rate per unit strain $d\Delta\nu/d\varepsilon_f$ is invariant (~−20 $cm^{-1}/\%$), irrespective of the side groups that are present along the main chain [34]. The polydiacetylene crystal or fiber modulus is controlled by the size of these side groups and this observation that $d\Delta\nu/d\varepsilon_f$ is independent of E_f means that, with reference to Table 4.1, the reinforcing units (the polydiacetylene backbones) are subjected to the same strain as the fiber as a whole (i.e. uniform strain).

Based upon these observations upon polydiacetylene single crystals, initial attempts to follow band shifts in rigid-rod polymers, such as Kevlar, however, proved to be unsuccessful, and the lack of an effect was put down to only a small part of the structure being under sufficient stress [35]. In 1985, band shifts were observed for the first time in stressed Kevlar fibers [36], and after some initial evidence to the contrary [37], it has been shown conclusively that this method is viable for determining local deformation, provided that precautions to avoid radiation damage (e.g. the choice of laser type and power) are taken into account [38].

4.3.2 Synthetic Polymeric Fibers

After the initial studies on diacetylenes, Raman band shifts in high-performance polymer fibers such as poly(*p*-phenylene benzobisthiazole) (PBT) fibers were reported [39], although infrared band shifts had already been observed for these materials [40]. Significant breakthroughs were then made in obtaining morphological information from stressed polymers, the most notable around this time being the work of Grubb on high-modulus polyethylene fibers [41]. Not only were band shifts observed for the 1063-cm^{-1} peak, corresponding to the C–C asymmetric stretch mode, but they were also observed to broaden asymmetrically, which was something not seen for the X-ray diffraction layer lines for the same material under tensile deformation. This asymmetric broadening was attributed to "taut-tie" molecules within the amorphous regions of the fibers, and it was found to be only an initial, partially reversible effect, with relaxation occurring over time. The crystals were found to go into compression when the load was removed from the fibers, suggesting that the relaxation of the taut-tie chains was slow. Ultrahigh-molecular-weight gel-spun fibers were also investigated using the same techniques by another research group [42]. A later paper [43] found that 40% of the Raman-active crystalline phase was subjected to high strain although the mechanisms and morphology proposed were different from those of Grubb [41].

An important contribution to the understanding of the deformation of polymeric fibers using Raman spectroscopy was a series of papers on rigid-rod materials; namely PpPTA, PIPD, and others [39,43,44]. The first of these studies detailed only PBT and PBO [39,43] but a later paper outlined molecular deformation in more detail for poly-aromatic rigid-rod fibers [44]. This paper [44] showed that for a range of different fibers (Kevlar, Twaron, Technora, PBO, and PET), the band shift rates per unit strain varied with fiber properties (shown in Figure 4.4a for Kevlar fibers [44]). The stress-induced Raman band shift for the 1610-cm^{-1} peak corresponding to the aromatic ring stretch was, however, invariant and equal to a value of roughly $-4.0\,cm^{-1}\,GPa^{-1}$ (shown in Figure 4.4b for Kevlar fibers [44]). This result was in spite of the fact that some of the fibers were produced using different procedures resulting in different microstructural morphologies. Stress-induced band broadening was also reported for all fibers (Figure 4.4c shows this result for Kevlar fibers [44]), which was shown to decrease with increasing fiber modulus for the PpPTA (Kevlar and Twaron) fibers, approaching zero for a fully crystalline structure. Symmetrical broadening of the 1610-cm^{-1} peak was reported for Kevlar, Twaron, Technora and PBO fibers, whereas PET showed asymmetric broadening. This difference was reported to be due to the fact that some of bonds in PET must be overstressed, as had been detailed previously [45].

The deformation of all fibers was shown to follow the continuous chain model of Northolt et al. [46] wherein the structure deforms by a chain extension, orientation, and sequential plastic deformation due to

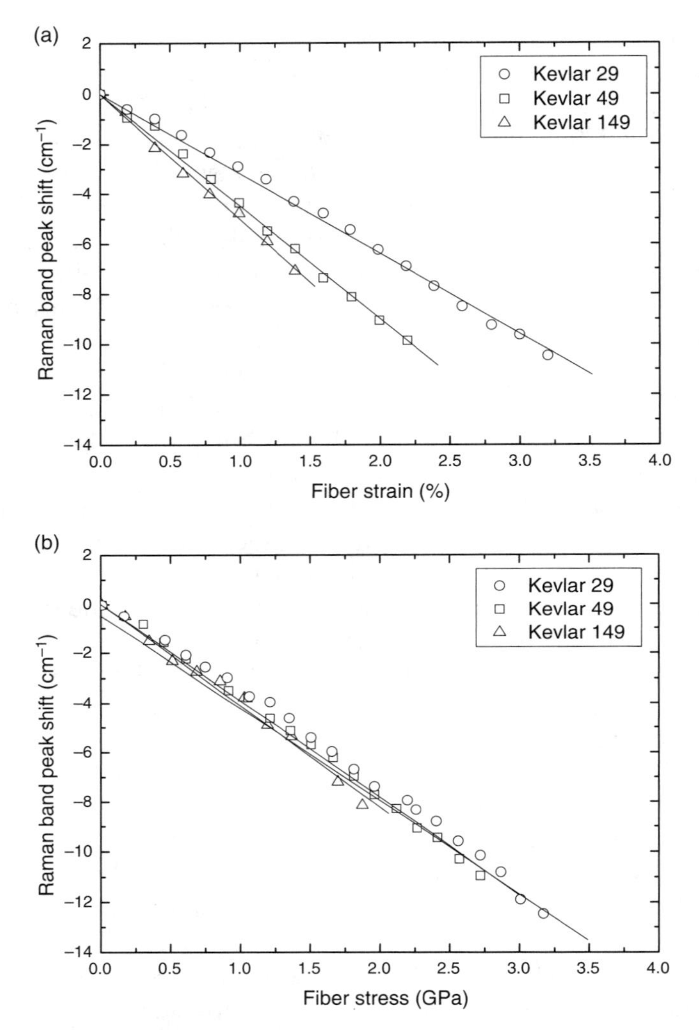

Figure 4.4 Raman band shifts as a function of (a) tensile stress, (b) tensile strain for a range of Kevlar fibers (29, 49 and 149). (c) Stress-induced Raman band broadening as a function of fiber modulus for a range of Kevlar and Twaron fibers. (Data replotted from Yeh and Young [44]. Original data were not fitted with an error weighted linear line. If this is done, then the predicted modulus is reduced to 180 GPa.)

shear between the crystallites. This model is similar to an approach by Zhang et al. [47], who showed that a number of thermotropic liquid crystalline polyesters and polyamides deformed by this mechanism. One outcome of this type of deformation is the increase in the fiber modulus as the deformation proceeds into the high-strain region [46,47],

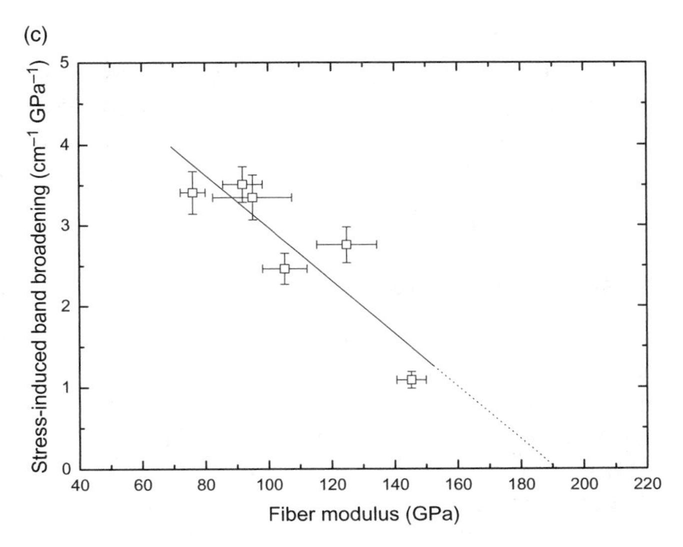

Figure 4.4 *Continued*

which was seen for all the rigid-rod fiber samples analyzed by Yeh and Young [44]. This type of strain stiffening has been seen more clearly from the molecular deformation of a number of cellulose fibers by Kong and Eichhorn [48]. The molecular deformation in PBO fibers was also reported [49,50] wherein it was shown for a range of fibers (AS, HM, and HM+) that band shifts in the 1618-cm^{-1} band, corresponding to the aromatic stretch, could be used for this purpose. It was also found [49] that reduced changes in the FWHM (full-width-half-maximum) of the 1618-cm^{-1} peak for HM+ fibers could be accounted for, due to their more homogenous structure. A second paper in this series [50] showed that less hysteresis in the HM+ fibers supported this hypothesis. Skin-core orientations were also inferred from the distributions of the radial 2θ intensity profiles by selective area electron diffraction (SAED) where the core of the fibers was found to be less oriented than the skin. This type of skin-core orientation had been previously reported for PpPTA fibers [51], where it was shown that different Raman band shift rates could be obtained from fibers with different skin orientations as the Raman signal is recorded only from this region of the fibers.

A recent paper by Galiotis et al. [52] has returned to a more detailed description of the effect of stress-dependent shifts of two Raman optical phonons ($u_1 = 1611\,\text{cm}^{-1}$ and $u_2 = 1648\,\text{cm}^{-1}$) for PpPTA fibers at different temperatures. The fibers were deformed inside a furnace, and spectra recorded under isothermal conditions. It was found that u_1 was moderately anharmonic and u_2 harmonic and that both soften with axial deformation, under isothermal conditions, irrespective of the fiber modulus. However, under increased temperatures (above 25 °C), the phonons appeared to harden. This result was also predicted, based on a theoretical approach that treats the fibers as molecular wires with negative expansion coefficients.

4.3.3 Natural Polymeric Fibers

The characterization of natural fibers using Raman and infrared spectroscopy has, in recent times, developed into an area that is gaining interest worldwide. It must be noted, however, at this point that spectroscopic characterization of natural materials is fraught with difficulty. It is found that light scattering from such samples is often not very intense and they sometimes undergo fluorescence, and hence longer exposure times are required if one is to obtain well-resolved Raman spectra. Resonance Raman also does not occur for natural polymers such as cellulose, and hence the high intensities seen in polymers such as the diacetylenes are not possible. Nevertheless, there have been developments in laser technology that have led to the availability of high-power near-infrared lasers that are found to reduce fluorescence significantly. Figure 4.5 shows Raman spectra obtained from both natural and regenerated cellulose fibers using a near-infrared laser.

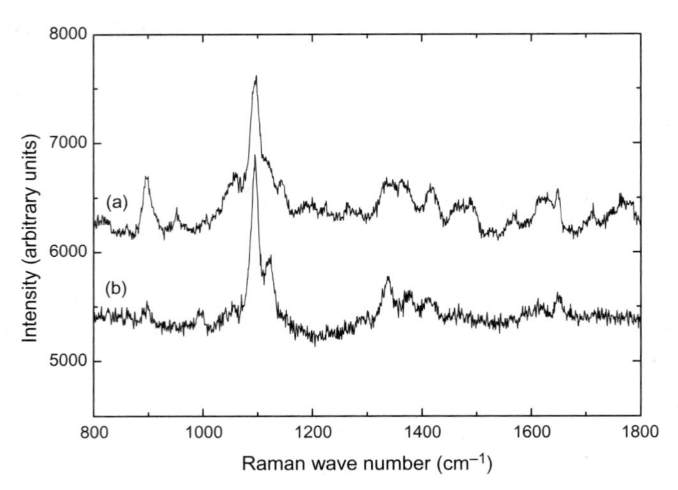

Figure 4.5 Raman spectra of (a) a flax fiber and (b) a regenerated cellulose fiber. (Data replotted from Eichhorn et al. [66].)

One of the first reported Raman (and infrared) studies on plant materials (cellulose) was the work by Blackwell et al. in the early 1970s [53]. Since the 1940s, people had been using infrared spectroscopy to conduct structural characterizations of hydrogen bonding in plant-based cellulose [54]. It is interesting to note that even at this very early stage, Ellis and Bath [54] were able to see changes in the hydrogen bond system of the cellulose structure before and after chemical treatment. After this initial period, a number of detailed papers were published on using infrared spectroscopy for structural characterization of celluloses by Marchessault and Liang [55–57].

Blackwell et al. [53] obtained Raman and infrared spectra from the cell walls of the algae *Valonia ventricosa* using an argon ion laser and a Spex-1400 double monochromator, and assigned a number of bands to particular parts of the structure. Later on in the 1970s, Lin and Koenig also reported Raman spectra for silk, wool, hair, and feather [58]. A great deal of effort was made during the 1970s by Wiley and Atalla to use Raman spectroscopy to characterize cellulosics, most of which has been extensively reviewed elsewhere [59]. Equipment in those days was reliant on photomultiplier tubes for recording the spectra, but the development of charge-coupled devices (CCDs) for recording signals from spectrometers has greatly reduced exposure times [60]. Finally, perhaps the most important body of work on the characterization of natural fibers has been the work of Edwards, including detailed band assignments of cellulosics [61], silks [62], wool [63], and hair [64]. However, it was only in the late 1990s that people began to look at the deformation of natural fibers using spectroscopic techniques.

The first study on the molecular deformation of cellulose fibers using Raman spectroscopy was on regenerated forms by Hamad and Eichhorn [65]. This showed for the first time that band shifts, as reported for synthetic fibers, were observable for cellulosic materials. However, there were problems with resolving the shifts with lower-modulus fibers, and indeed this study [65] showed that larger band shifts per percent tensile strain applied could be obtained with stiffer fibers. This was indeed achieved with natural plant fibers (flax and hemp) a few years later [66] and Figure 4.6a shows the shift of two Raman bands with strain. It was also shown that the rate of band shift for the two bands with respect to strain was directly proportional to the stiffness of the fibers (Figure 4.6b). More studies of these effects then followed [67,68] investigating the deformation of wood and other plant fibers in detail. In recent times, these types of experiments have been repeated by two independent groups based at the University of Halle and at the Max

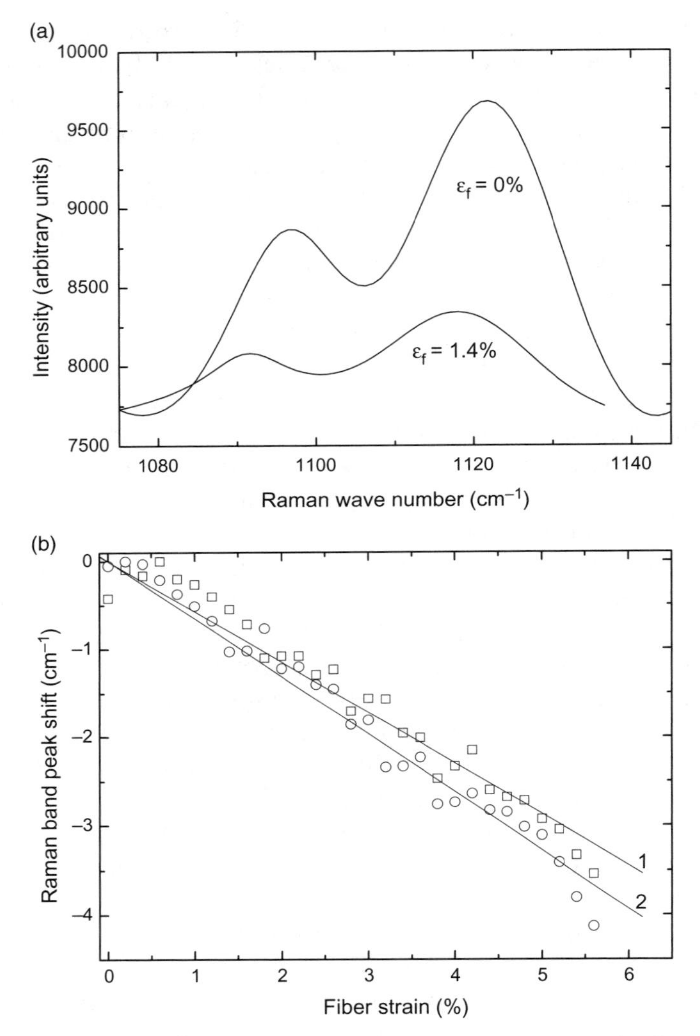

Figure 4.6 Raman band shifts from the maximum for a flax fiber. (a) Spectra in the 1070- to 1150-cm^{-1} region before and after straining, (b) 1- to 1122-cm^{-1} peak and 2- to 1095-cm^{-1} peak [66]. (Data replotted from Eichhorn et al. [66])

Planck Institute [69–71]. The first of these studies showed that it was indeed possible to detect band shifts in plant material with different chemical and enzymatic treatments [70]. Unlike the original work by Eichhorn et al. [66,67], it was not necessary to pre-bleach the fibers using hydrogen peroxide to overcome problems with fluorescence, as a Fourier transform (FT) Raman system was used.

Other studies on the deformation micromechanics of cellulose fibers have used infrared spectroscopy. Of particular note has been the work of Salmén and coworkers on the use of dynamic infrared spectroscopy to elucidate the complex hydrogen bond interactions during the loading of the cellulosic chains during deformation [72,73]. Without recourse to a full review of the area, the role of the -OH moiety in cellulose spectra is notoriously difficult to interpret. We now know, from detailed studies on the structure of cellulose polymorphs [74–77], that the disorder along the chains, and the interdigitation of disorder, may play a key role in mechanics and structural interconversion [78], and hence our understanding of the mechanics of this aspect of the structure is vital. Indirect evidence of the breakdown in hydrogen bonding has recently been obtained from Raman spectroscopic studies of deformed cellulose-II fibers [48], but it is difficult to ascertain within which part of the structure (crystalline or amorphous) this is occurring [48]. However, distinct plateaus in the band shift were observed for both the 1095- and 1414-cm^{-1} bands for fibers produced using different draw ratios (see Figure 4.7), which was attributed to a breakdown in the hydrogen bonding.

There have also been reports of the use of Raman spectroscopy to probe the deformation of collagen (in the dry state) [79] and silk [80,81]. The work on collagen fibers [79] showed that it is possible to obtain Raman band shifts, and therefore indications of molecular deformation. The Raman spectra of both silkworm and spider silk are shown in Figure 4.8. Raman deformation studies on both the silkworm and spider forms [81] have indicated that the structure responds in a near-uniform stress fashion. The 1095-cm^{-1} band of spider silk shifts toward a lower wave number with stress. This shift is shown in Figure 4.9. It can be seen that the band shift is approximately linear with fiber stress.

The exact conformational structure of silk fibers is still in debate, but most people accept that there are β-pleated sheets (hard segments) within an entangled amorphous network (soft segments) as originally suggested by Termonia [82]. This structure has been confirmed by others in light of further experimental evidence [83,84]. The exact location of the amorphous segments relative to the crystalline β-pleated sheet segments will determine whether the material undergoes uniform stress or strain deformation, but Termonia's model [82] would appear to be a uniform strain structure.

Various attempts have been made to understand the morphology of silk, starting with modeling of the chain stiffness. The chain stiffness of an alanine–glycine dipeptide structure has been found to be ~135 GPa

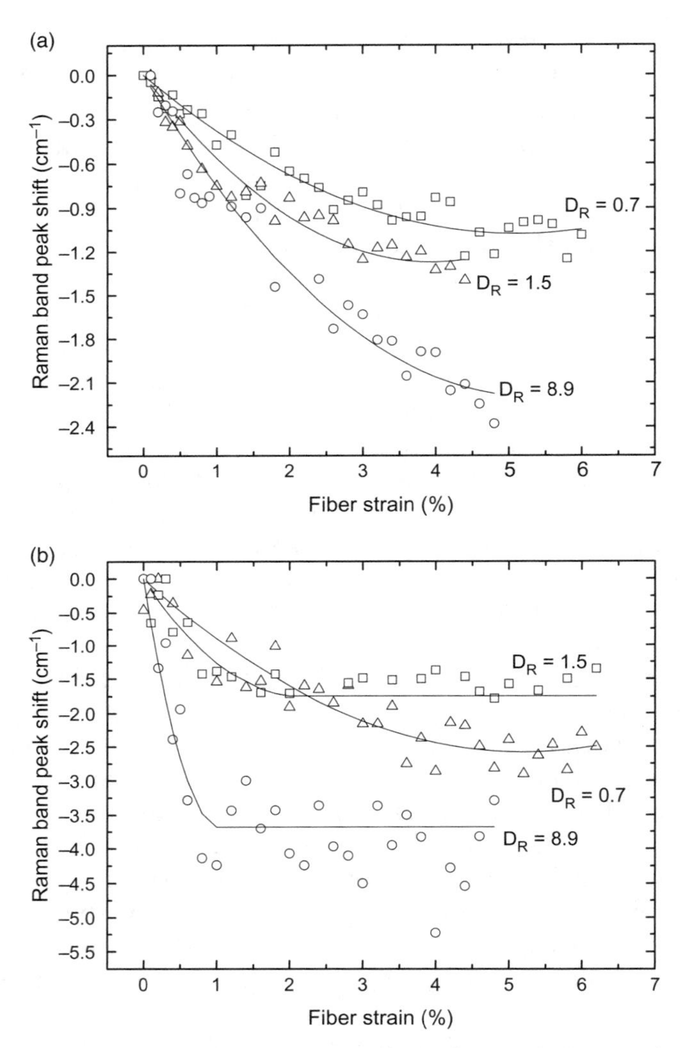

Figure 4.7 Raman band shifts of the (a) 1095-cm^{-1} and (b) 1414-cm^{-1} peaks as a function of applied strain for regenerated cellulose fibers produced using a range of draw ratios (D_R = 0.7, 1.5, and 8.9). (Data replotted from Kong and Eichhorn [48].)

[85]. In spite of this, experimental determinations of the crystal modulus have been comparatively lower in magnitude (in the range 20–28 GPa) [86,87]. The deficiency of the uniform stress assumption for silk fibers has in fact been recently highlighted [86], where it was found that the crystal modulus varied as a function of the crystallinity of a range of fibers. The values of crystal modulus reported in this study [86] are apparently comparable with theoretical values when using seemingly

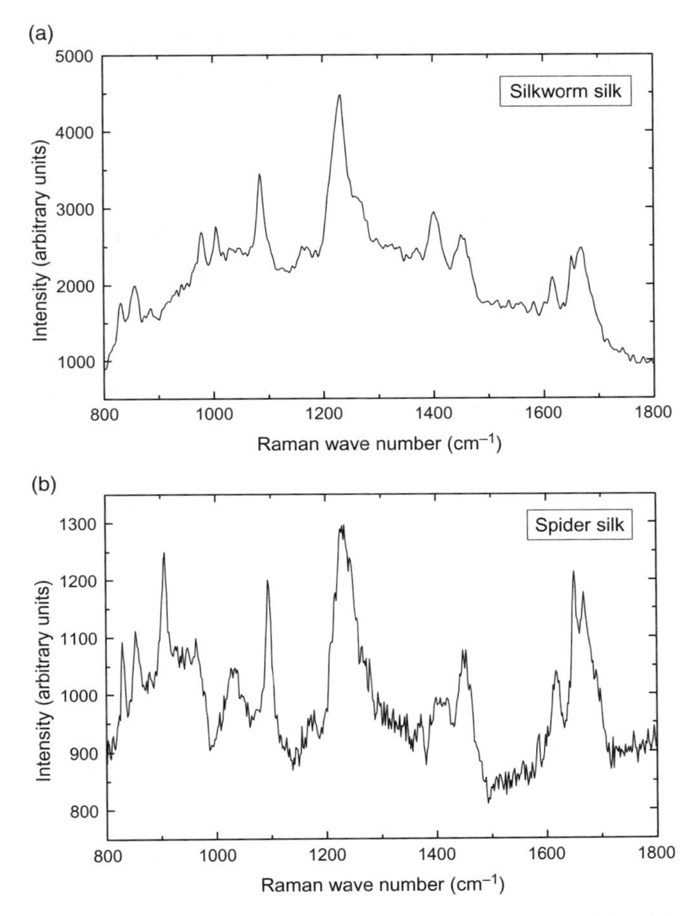

Figure 4.8 Raman spectra of silk fibers. (a) Silkworm silk and (b) spider silk. (Data replotted from Sirichaisit et al. [81].)

improved computational analysis, and the acknowledgment of relaxation of the fiber stress is taken into account [88]. However, differences could still be due to the fact that there is *an inherent assumption that the structure is subjected to a uniform stress* during the measurement of crystal modulus. If this were not true, and the structure was more akin to a nanocomposite, subjected to uniform strain, then one might expect that the crystals would be overstressed, with an understressed matrix component. This has recently been experimentally proven for silk fibers [89], and some evidence for this type of microstructure has also been found in cellulose-II fibers [90]. The work recently reported for silk fibers [89] showed that the Raman band shifts as a function of stress ($d\Delta\nu/d\sigma_f$) shown in Figure 4.9b are proportional to the inverse of the

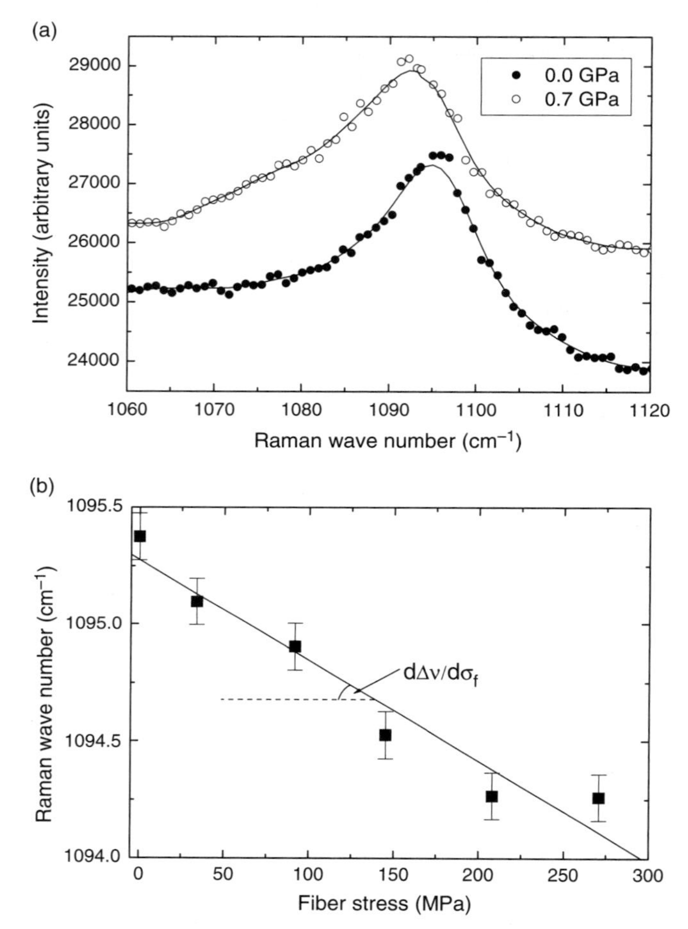

Figure 4.9 Raman band shifts of spider silk fiber. (a) Spectra in the 1060- to 1120-cm^{-1} region before and after straining. (b) Shift of the 1095-cm^{-1} peak with fiber stress [66]. (Data replotted from Brookes et al. [89])

fiber modulus ($1/E_f$), i.e. conforming with the uniform strain model (Table 4.1). This result is in direct contrast to the rigid-rod polymer fibers where there is proportionality between the modulus (E_f) and ($d\Delta\nu/d\varepsilon_f$) (e.g. [45]), i.e. uniform stress model (Table 4.1).

4.4 MODELING STRESS-INDUCED RAMAN BAND SHIFTS IN POLYMERIC FIBERS

One of the first attempts to explain a Raman band shift in theoretical terms was by Mitra et al. [30], who gave an empirical relationship for the effect based on anharmonic potentials, and predicted the shifts seen in polydiacetylene single-crystal fibers (see Figure 4.5). Considerable work has been undertaken by Tashiro [88] upon the prediction of spectroscopic (particularly Raman) band shifts, including work on polyoxymethylene (POM) and polyethylene (PE) chains. Using general expressions for the strain energy and potential functions of asymmetric and symmetric vibrations, Tashiro [88] was able to obtain good predictions of the rate of his experimentally determined Raman band shifts. This type of approach was later applied to a PBO chain by Kitagawa et al. [91] and more rigorously for the case of a semicrystalline structure [92].

This approach is generally adequate in the case where nonbonded interactions do not significantly contribute toward the deformation of the structure. Little or no work exists, however, on modeling Raman band shifts using molecular force fields. This type of approach uses a normal mode analysis to predict vibrations of minimized structures. Normal mode analyses on static theoretical cellulose structures have been previously performed [93], but not for deformed polymer chains. In recent times, Eichhorn and others [90,94] have used a commercial force field, a molecular mechanics approach, and normal mode analysis to predict band shifts in cellulose polymorphs (both I and II). Each structure was taken from recently published atomic coordinates [74–76] and minimized within a commercial force field (COMPASS™) under restraint. The normal mode analysis used predicted infrared intensities, and by a process of elimination a number of bands were chosen which best represented the experimentally observed Raman band shifts. Good agreement with the experimental data was obtained, and also with chain stiffness data obtained from X-ray diffraction, a subject to which we will now turn.

4.5 COMPOSITES

4.5.1 Fiber-Reinforced Composites

The observation of Raman band shifts indicating the molecular deformation of polydiacetylene single-crystal fibers was the first report of its kind [30] and it opened up the possibility of using Raman spectroscopy to study composite micromechanics. This was followed a few years later by Galiotis et al. [95], who showed that it was possible to follow the interfacial adhesion between two lap-jointed polydiacetylene fibers made using an epoxy resin adhesive. They found that strong well-defined Raman spectra could be obtained from the fibers and that the strains, as determined by a shift in a polydiacetylene Raman peak, were different in each fiber and within the lap-jointed region [95]. This was the first study to show that it was possible to follow the interface between a fiber and a resin using Raman spectroscopy, but a later paper showed that a shear-lag type stress transfer occurred over the ends of a fully embedded fiber [96] (see Figure 4.10).

The principle of using Raman spectroscopy to map local deformation in such systems relies on the fact that one can use the local molecular deformation of a fiber or phase, as revealed by the shift in the position of a Raman peak, as a sensor of the local stress state within a composite, provided that the resin one uses is transparent to the laser or does not interfere spectroscopically with the fiber or phase. The technique allows direct mapping of fiber stress or strain along the fiber and has revolutionized the study of composite micromechanics.

A large number of fiber-matrix systems have been studied as described in detail elsewhere [97–99]. A typical example is shown in Figure 4.11 for a single PIPD M5 high-performance polymer fiber embedded in an epoxy resin matrix. Distributions of fiber strain along the fiber are shown for different levels of matrix strain, ε_m. It can be seen in Figure 4.11a that the fiber strain builds up from the fiber ends to become equal to the matrix strain in the middle of the fiber. The form of the strain distribution is similar to that given by shear-lag analysis [100–104]. At the highest level of matrix strain (Figure 4.11b),

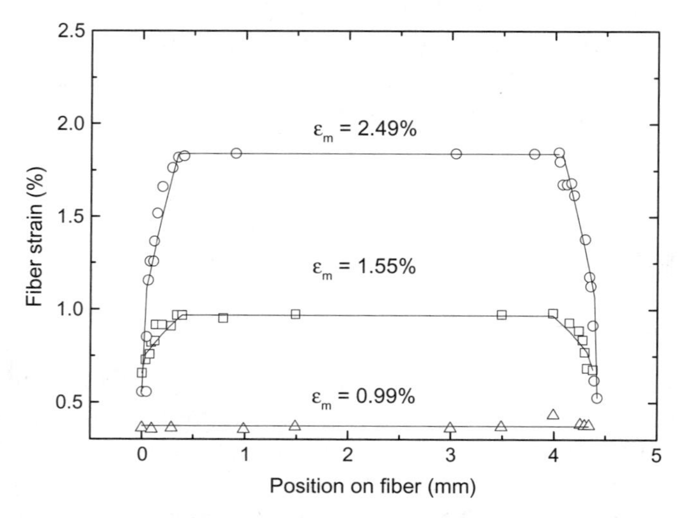

Figure 4.10 Axial strain of single a poly-1,6-di-(*N*-carbazoyl)-2,4-hexadiyene (pDCH) fiber as a function of position along the fiber in a model composite at different applied matrix strains (ε_m) (Data replotted from [96].)

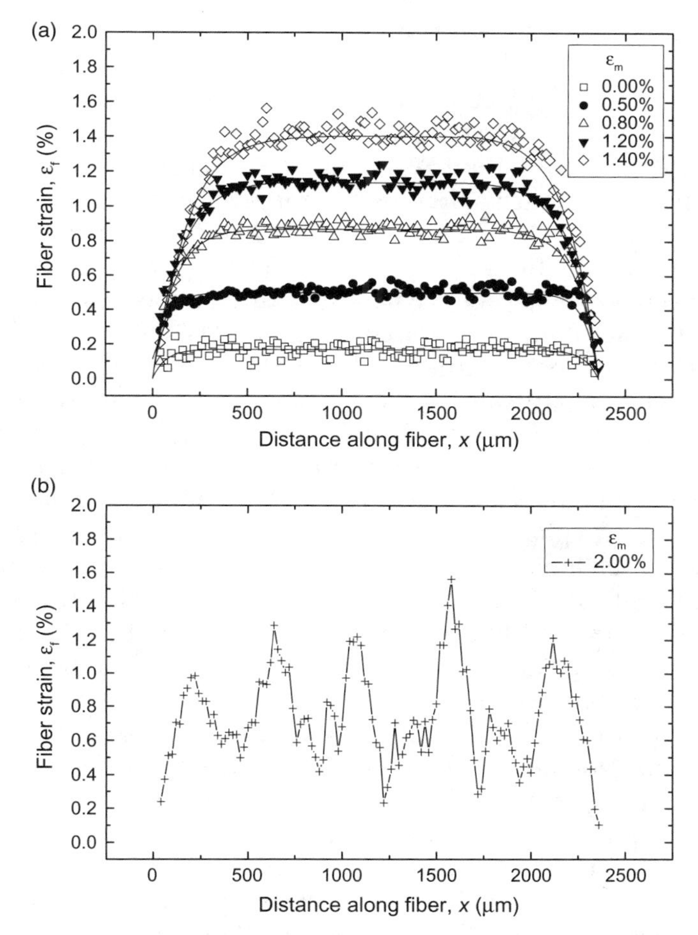

Figure 4.11 Distribution of fiber strain along a single PIPD M5 fiber embedded in an epoxy resin matrix at different levels of matrix strain. (a) Elastic stress transfer up to 1.4% matrix strain. (b) Fragmentation if the fiber at 2.0% matrix strain. (Data replotted from So et al. [104].)

however, the failure strain of the fiber is exceeded and the fiber undergoes multiple fragmentation. Further analysis of the fiber strain distributions in Figure 4.11 allows the distribution of shear stress along the fiber-matrix interface to be determined [104].

Raman spectroscopy has also been used to follow composite micromechanics in a number of different single-fiber geometries such as pullout [105] or microbond [106]. It has also been employed in the study of fiber–fiber interactions for higher levels of fiber loading [107] and even in woven polymer fiber composites to map the complex distributions of fiber strain [108].

4.5.2 Molecular and Fibrous Nanocomposites

The concept of a molecular composite, containing diacetylenes, was described explicitly by Angkaew et al. [109], along with attempts to characterize its deformation. A full understanding of the local deformation mechanisms in such materials has not yet been achieved, and hence techniques for developing this area are likely to receive significant interest in the near future.

It is now possible to produce nanostructured fibers using electrospinning by the addition of hierarchical pores [110], carbon nanotubes (CNTs) [111], or montomorillonite [112]. The use of CNTs is reminiscent of the ideas first put forward by Takayanagi and his collaborators in 1980, that of a molecular composite [113]. This type of system, where a molecular hard component reinforces a soft polymeric phase, was realized by Takayanagi and others a few years later [114], but only in the form of a film of nylon with phase-separated and dispersed chains of PpPTA. The principle of such molecular composites (or as has become customary of late, nanocomposites), as outlined by Takayanagi et al. [113], is that the aspect ratio of the rod-like molecules (in this case PpPTA [114]) should have a high enough aspect ratio, with the covalent bond between reinforcement and the "matrix" material in an ideal state.

It is entirely conceivable, based on the concepts of Takayanagi et al. [113,114], that polymer fibers could be structured in terms of their local orientation to generate desired mechanical properties by the addition of a second phase. However, one hurdle to overcome is the development of a satisfactory interface, and one way to characterize the efficacy of that property is to use Raman spectroscopy. The other is adequate dispersion, or as stated by Takayanagi et al. [114], phase separation. This degree of dispersion can also be ascertained using Raman spectroscopy, and this will now be discussed.

It is worth noting though that the Takayanagi model [10] has been shown to readily apply to urethane-diacetylene segmented block copolymers [99]. Block copolymers are rapidly becoming useful for nanostructuring materials, the progress of which has been reported elsewhere [115].

After the initial work on a lap-jointed polydiacetylene fiber composite [95], other papers were published that showed that local deformation could be followed [116–118] in various forms of molecular composite. The study of the local deformation of two-phase nanocomposites using Raman spectroscopy has been undertaken recently for CNT systems as reviewed by Wagner [119]. An example of this is shown in Figure 4.12 for a poly(vinyl alcohol) (PVA) nanofiber containing

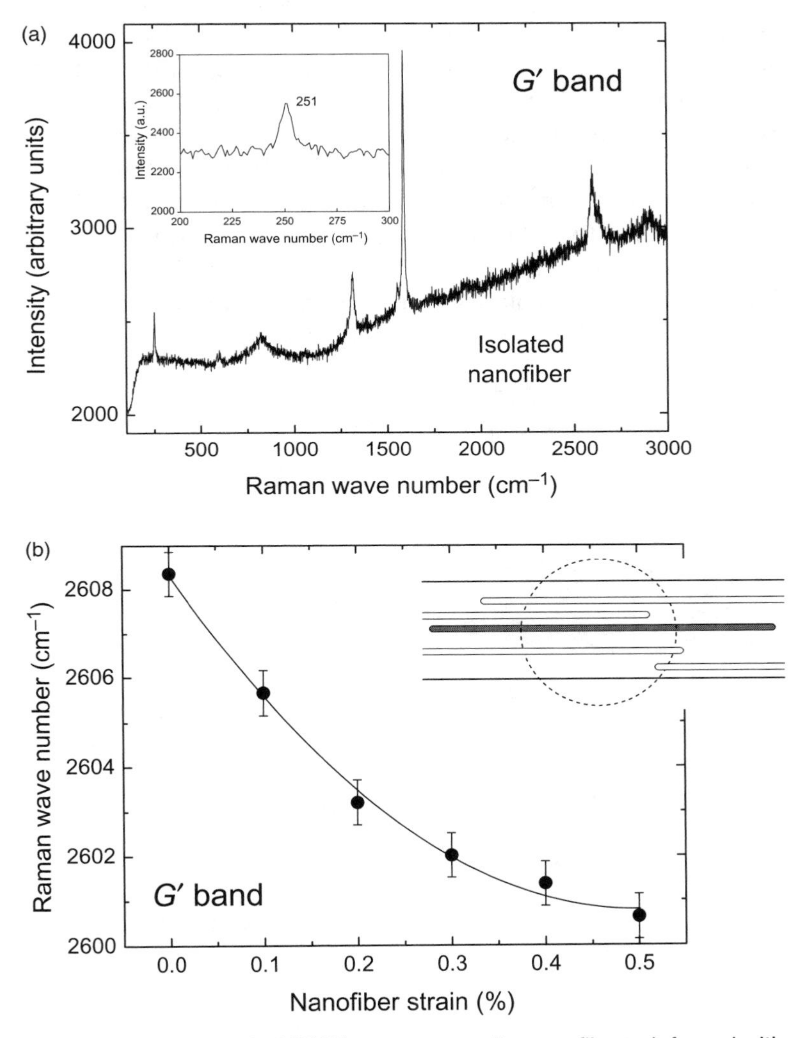

Figure 4.12 Polyvinyl alcohol (PVA) nanocomposite nanofiber reinforced with single wall carbon nanotubes (SWNTs). (a) Raman spectrum of the nanofiber with the radial breathing mode (RBM) inset. (b) Shift of the *G′* band with nanofiber strain. (Data replotted from Kannan et al. [120].)

isolated and aligned single wall carbon nanotubes (SWNTs) [120]. The Raman spectrum of the nanofiber shows that even though it contains only 0.04% of SWNTs, the spectrum is dominated by the resonance Raman scattering from the SWNTs. The single radial breathing mode at 251 cm^{-1} also implies that the nanotubes are debundled and isolated in the nanofiber. The shift of the *G′* Raman band of the nanotubes during the deformation of the nanofiber is shown in Figure 4.12b and this demonstrates efficient stress transfer from the PVA matrix to the reinforcement in the nanocomposite. The possible microstructure of the nanofiber is also shown (inset in Figure 4.12).

In recent times, the use of Raman spectroscopy to follow the deformation of natural-based nanocomposites has been reported [121]. This work involved the deformation of dispersed microcrystalline cellulose in an epoxy resin matrix. The resin was deformed in four-point bending. The 1095-cm^{-1} Raman peak from the microcrystalline cellulose was found to be resolvable from a composite spectrum of the matrix material and the reinforcing phase (see Figure 4.13a). The position of the

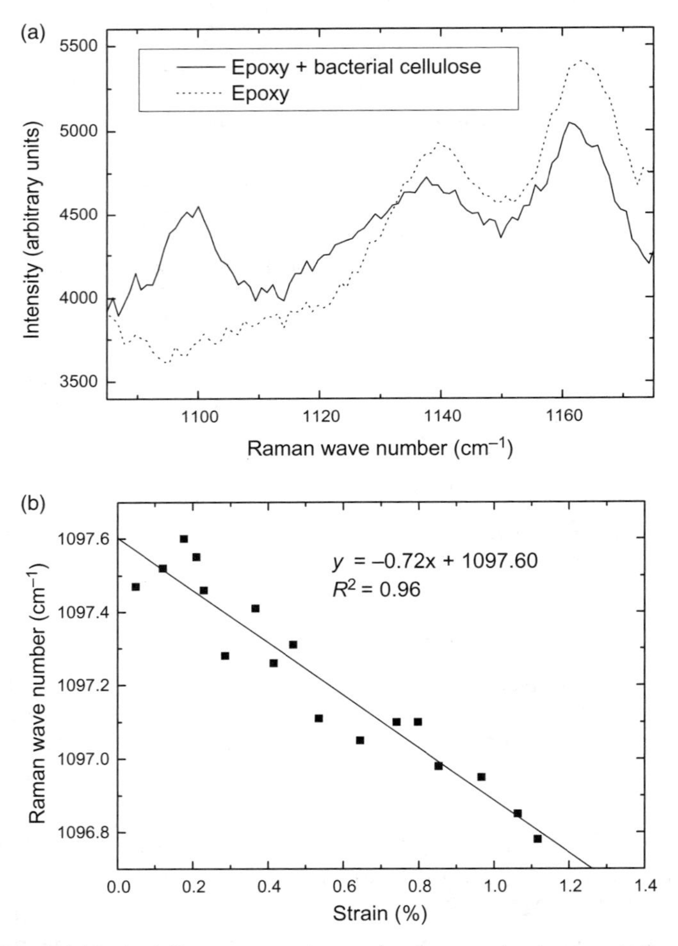

Figure 4.13 (a) Typical Raman spectrum of epoxy and microcrystalline cellulose composite and (b) shift of the 1095-cm^{-1} Raman peak for a fibrous fragment of microcrystalline cellulose [121].

1095-cm^{-1} peak was followed as a function of strain (see Figure 4.13b) which enabled a prediction of the modulus of a fragment of microcrystalline cellulose (25 ± 4 GPa) [121]. Subsequent work has used similar methods to obtain the modulus of nanowhiskers of cellulose (141 GPa) [94] and of filaments of bacterial cellulose (114 GPa) [122].

The first report of the mechanical properties of a cellulose-based nanocomposite was published in 1995 by Favier et al. [123]. A great deal of work has been done to generate a number of cellulose nanocomposites by Dufresne et al. [124,125] and Kvien et al. [126] using a variety of sources for the base reinforcement material. Another important development in the field has been the use of bacterial cellulose for the manufacture of transparent and high-performance composite materials [127,128].

Other major developments in the use of natural fibers for composite materials have been the production of all-cellulose composites. The concept was first suggested by Nishino and coworkers [129]. The principle of these materials is that you dissolve other cellulose fibers and use this as a matrix that is then reinforced by the intact material. Quite remarkable properties have been obtained for these materials, wherein twofold increases in the modulus compared with epoxycellulose have been observed [130]. All-cellulose nanocomposites have also been produced [131] which also show enhanced properties compared with a composite of two chemically distinct phases.

4.6 CONCLUSIONS

The development of high-performance fibers for a variety of applications has invigorated research into the understanding of molecular and crystalline orientation, local deformation, and morphology. It has been shown that Raman spectroscopy has been a critical technique for our understanding of these concepts. Early developments in these areas showed that it is possible to predict properties of semicrystalline polymers on the basis of stiff crystals embedded in an amorphous matrix, such as in a nanocomposite. However, it is the morphology of this composite structure that will ultimately determine the mechanical properties. Moreover, the location of the amorphous material in relation to the crystals affects whether the structure responds under uniform stress or strain. It is clear though that a distinction between natural biopolymer-based fibers and those produced by synthetic routes can be made, wherein the former generally appear to have uniform strain-type structures and the latter uniform stress. It has been shown that general relationships can be made between physical properties such as Raman band shift and applied stress and strain that give a clear signature of whether the structure is a uniform stress or strain type. Some materials, such as cellulose, have been shown to have more of a hybrid structure, whereas it is clear that measurements upon silks may have been misinterpreted, with most experimental evidence pointing toward a uniform strain-type microstructure.

It is worthwhile noting at this point that the relationships derived for molecular deformation shown in Table 4.1 really only apply to the elastic and often initial response of the polymer fibers to deformation. Other key factors play a role at higher levels of deformation, such as hydrogen bonding for instance. It has been shown that hydrogen bonding deformation can be detected using spectroscopic techniques. Nonlinearity in the band shift profiles seen with Raman is often the manifestation of such effects, but much more work is required to understand how this relates to the mechanical properties of the fibers, particularly in the presence of water and at elevated temperatures.

Finally, nanocomposites are a subject of wide interest. With the advent of new processing capabilities, such as electrospinning, wherein fibers can be produced with nano-reinforcements built into polymeric structures, it is now possible to engineer new fibers with controlled architectures. Specific examples of where spectroscopic techniques might become critical are in following the properties of the interfaces between the matrix and crystals in a nanocomposite fiber. Little is understood about how the nature of an interface in a nanocomposite might differ from that in mesoscale composites, particularly when surface-to-volume ratios increase significantly. Perhaps the next challenge is to take what we know about the effects seen in conventional fibers and not only to apply the techniques used to understand their deformation, but to test the concepts arising from studying microstructure for nanocomposite fibers.

ACKNOWLEDGMENTS

The authors wish to thank the Engineering and Physical Sciences Research Council for funding research reported here under grant numbers GR/S44471/01, GR/M82219/01, GR/J68878/01, GR/G09047/01, and GR/F06050/01.

REFERENCES

1. Treloar, L.R.G. (1941). *Trans. Faraday Soc.*, 37, 84–97.
2. Kolsky, H., Shearman, A.C. (1943). *Proc. Phys. Soc.*, 55, 383–395.
3. Crawford, S.M., Kolsky, H. (1951). *Proc. Phys. Soc. B*, 64, 119–125.
4. Crawford, S.M., Kolsky, H. (1951). *Proc. Phys. Soc. A.*, 64, 215–215.
5. Hermans, P.H. (1949). *Physics and Chemistry of Cellulose Fibers*. London: Elsevier Publishing Company.
6. Ward, I.M. (1962). *Proc. Phys. Soc.*, 80, 1176–1188.
7. Halpin, J.C., Kardos, J.L. (1972). *J. Appl. Phys.*, 43, 2235–2241.
8. Harris, B. (1999). *Engineering Composite Materials*. London: IOM Communications Ltd.
9. Northolt, M.G., van der Hout, R. (1985). *Polymer*, 26, 310–316.
10. Takayanagi, M., Imada, K., Kajiyama, T. (1966). *J. Polym. Sci. C*, 15, 263.
11. Young, R.J., Eichhorn, S.J. (2007). *Polymer*, 48, 2–18.
12. Young, R.J. (1995). *J. Text. Inst.*, 86, 360–381.
13. Raman, C.V., Krishnan, K.S. (1928). *Nature*, 121, 501–502.
14. Steinfeld, J.I. (1979). *Molecules and Radiation: An Introduction to Molecular Spectroscopy*. Cambridge, MA and London: MIT Press.
15. Liang, C.Y., Krimm, S., Sutherland, G.B.B.M. (1956). *J. Chem. Phys.*, 25, 543–548.
16. Liang, C.Y., Krimm, S., Sutherland, G.B.B.M. (1956). *J. Chem. Phys.*, 25, 549–562.
17. Liang, C.Y., Krimm, S. (1956). *J. Chem. Phys.*, 25, 563–571.
18. Krimm, S., Liang, C.Y. (1956). *J. Polym. Sci.*, 22, 95–112.
19. Krimm, S., Liang, C.Y., Sutherland, G.B.B.M. (1956). *J. Polym. Sci.*, 22, 227–247.
20. Palm, A. (1951). *J. Phys. Chem.*, 55, 1320–1324.
21. Flaugh, P.L., O'Donnell, S.E., Asher, S.A. (1984). *Appl. Spectrosc.*, 38, 847–850.
22. Delhaye, M., Dhamelincourt, P. (1975). *J. Raman Spectrosc.*, 3, 33–43.
23. Zhurkov, S.N., Novak, I.I., Vettegren, V.I. (1964). *Dokl. Akad. Nauk SSSR*, 157, 1431–1435.
24. Zhurkov, S.N., Korsukov, V.I., Novak, I.I. (1969). Infrared spectroscopic study of the chemical bonds in stressed polymers, Fracture 1969. Proceedings of the 2nd International Conference on Fracture, 545–550.
25. Roylance, D.K., deVries, K.L. (1971). *J. Polym. Sci.—Polym. Lett. Edn.*, 9, 443–447.
26. Evans, R.A., Hallam, H.E. (1976). *Polymer*, 17, 838–839.

27. Cunningham, A., Ward, I.M., Willis, H.A., Zichy, V. (1974). *Polymer*, 15, 749–756.
28. Wool, R.P. (1975). *J. Appl. Polym. Sci.—Polym. Phys. Edn.*, 13, 1795–1808.
29. Wool, R.P. (1980). *Polym. Eng. Sci.*, 20, 805–815.
30. Mitra, V.K., Risen, W.M., Baughman, R.H. (1977). *J. Chem. Phys.*, 66, 2731–2736.
31. Badger, R.M. (1934). *J. Chem. Phys.*, 2, 128–131.
32. Batchelder, D.N., Bloor, D. (1979). *J. Polym. Sci.—Polym. Phys. Edn.*, 17, 569–581.
33. Young, R.J., Bloor, D., Batchelder, D.N., Hubble, C.L. (1978). *J. Mater. Sci.*, 13, 62–71.
34. Galiotis, C. (1982). Polydiacetylene single crystal fibres. PhD Thesis, University of London.
35. Penn, L., Milanovich, F. (1979). *Polymer*, 20, 31–36.
36. Galiotis, C., Robinson, I.M., Young, R.J., Smith, B.J.E., Batchelder, D.N. (1985). *Polym. Comm.*, 26, 354–355.
37. Edwards, H.G.M., Hakiki, S. (1989). *Brit. Polym. J.*, 21, 505–512.
38. Young, R.J., Lu, D., Day, R.J. (1991). *Polym. Int.*, 24, 71–76.
39. Day, R.J., Robinson, I.M., Zakikhani, M., Young, R.J. (1987). *Polymer*, 28, 1833–1840.
40. Shen, D.Y., Hsu, S.L. (1982). *Polymer*, 23, 969–973.
41. Prasad, K., Grubb, D.T. (1989). *J. Polym. Sci. B-Polym. Phys. Edn.*, 27, 381–403.
42. Kip, B.J., van Eijk, C.P., Meier, R.J. (1991). *J. Polym. Sci. B-Polym. Phys. Edn.*, 29, 99–108.
43. Young, R.J., Day, R.J., Zakikhani, M. (1990). *J. Mater. Sci.*, 25, 127–136.
44. Yeh, W-Y., Young, R.J. (1999). *Polymer*, 40, 857–870.
45. Yeh, W-Y., Young, R.J. (1998). *J. Macromol. Sci. Phys.* B37, 83–118.
46. Northolt, M.G., Baltussen, J.J.M., Schaffers-Korff, B. (1995). *Polymer*, 36, 3485–3492.
47. Zhang, H., Davies, G.R., Ward, I.M. (1992). *Polymer*, 33, 2651–2658.
48. Kong, K., Eichhorn, S.J. (2005). *J. Macromol. Sci. Phys.*, 44, 1123–1126.
49. Kitagawa, T., Yabuki, K., Young, R.J. (2001). *Polymer*, 42, 2101–2112.
50. Kitagawa, T., Yabuki, K., Young, R.J. (2002). *J. Macromol. Sci. Phys.*, 42, 61–76.
51. Young, R.J., Lu, D., Day, R.J., Knoff, W.F., Davis, H.A. (1992). *J. Mater. Sci.*, 27, 5431–5440.
52. Bollas, D., Parthenios, J., Galiotis, C. (2006). *Phys. Rev. B*, 73, 094103.
53. Blackwell, J., Vasko, P.D., Koenig, J.L. (1970). *J. Appl. Phys.*, 41, 4375–4379.
54. Ellis, J., Bath, J. (1940). *J. Am. Chem. Soc.*, 62, 2859–2861.

55. Liang, C.Y., Marchessault, R.H. (1959). *J. Polym. Sci.*, 37, 385–395.
56. Liang, C.Y., Marchessault, R.H. (1959). *J. Polym. Sci.*, 39, 269–278.
57. Marchessault, R.H., Liang, C.Y. (1960). *J. Polym. Sci.*, 43, 71–84.
58. Lin, V.J.C., Koenig, J.L. (1975). *Bull. Am. Phys. Soc.*, 20, 372.
59. Wiley, J.H., Atalla, R.H. (1987). *Carbohydr. Res.*, 160, 113–129.
60. Pemberton, J.E., Sobocinski, R.L. (1989). *J. Am. Chem. Soc.*, 11, 432–434.
61. Edwards, H.G.M., Farwell, D.W., Webster, D. (1997). *Spectrochim Acta A*, 53, 2383–2392.
62. Edwards, H.G.M., Farwell, D.W. (1995). *J. Raman Spectrosc.*, 26, 901–909.
63. Hogg, L.J., Edwards, H.G.M., Farwell, D.W., Peters, A.T. (1994). *J. Soc. Dyers Colour*, 110, 196–199.
64. Akhtar, W., Edwards, H.G.M., Farwell, D.W., Nutbrown, M. (1997). *Spectrochim. Acta A*, 53, 1021–1031.
65. Hamad, W.Y., Eichhorn, S. (1997). *ASME J. Eng. Mat. Tech.* 119, 309–313.
66. Eichhorn, S.J., Hughes, M., Snell, R., Mott, L. (2000). *J. Mater. Sci. Lett.*, 19, 721–723.
67. Eichhorn, S.J., Sirichaisit, J., Young, R.J. (2001). *J. Mater. Sci.*, 36, 3129–3135.
68. Eichhorn, S.J., Young, R.J. (2001). *Cellulose*, 8, 197–207.
69. Fischer, S., Schenzel, K., Fischer, K., Diepenbrock, W. (2005). *Macromol. Symp.*, 223, 41–56.
70. Peetla, P., Schenzel, K., Diepenbrock, W. (2006). *Appl. Spectrosc.*, 60, 682–691.
71. Gierlinger, N., Schwanniger, M., Reinecke, A., Burgert, I. (2006). *Biomacromolecules*, 7, 2077–2081.
72. Hinterstoisser, B., Salmén, L. (1999). *Cellulose*, 6, 251–263.
73. Åkerholm, M., Hinterstoisser, B., Salmén, L. (2004). *Carbohydr. Res.*, 339, 569–578.
74. Langan, P., Nishiyama, Y., Chanzy, H. (1999). *J. Am. Chem. Soc.*, 121, 9940–9946.
75. Langan, P., Nishiyama, Y., Chanzy, H. (2001). *Biomacromolecules*, 2, 410–416.
76. Nishiyama, Y., Langan, P., Chanzy, H. (2002). *J. Am. Chem. Soc.*, 124, 9074–9082.
77. Nishiyama, Y., Sugiyama, J., Chanzy, H., Langan, P. (2003). *J. Am. Chem. Soc.*, 125, 14300–14306.
78. Jarvis, M. (2003). *Nature*, 426, 611–612.
79. Wang, Y.N., Galiotis, C., Bader, D.L. (2000). *J. Biomech.*, 33, 483–486.
80. Sirichaisit, J., Young, R.J., Vollrath, F. (2000). *Polymer*, 41, 1223–1227.

81. Sirichaisit, J., Brookes, V.L., Young, R.J., Vollrath, F. (2003). *Biomacromolecules*, 4, 387–394.
82. Termonia, Y. (1994). *Macromolecules*, 27, 7378–7381.
83. O'Brien, J.P., Fahnestock, S.R., Termonia, Y., Gardner, K.H. (1998). *Adv. Mat.*, 10, 1185–1195.
84. Stephens, J.S., Fahnestock, S.R., Farmer, R.S., Kiick, K.L., Chase, D.B., Rabolt, J.F. (2005). *Biomacromolecules*, 6, 1405–1413.
85. Brookes, V.L. (2005). Study of deformation behaviour and structure analysis of natural protein fibres by Raman Spectroscopy. PhD Thesis, University of Manchester Institute of Science and Technology.
86. Sinsawat, A., Putthanarat, S., Magoshi, Y., Pachter, R., Eby, R.K. (2002). *Polymer*, 43, 1323–1330.
87. Sinsawat, A., Putthanarat, S., Magoshi, Y., Pachter, R., Eby, R.K. (2003). *Polymer*, 44, 909–910.
88. Tashiro, K. (1993). *Prog. Polym. Sci.*, 18, 377–435.
89. Brookes, V.L., Young, R.J., Vollrath, F. (2008). *J. Mater. Sci.*, 43, 3728–3732.
90. Eichhorn, S.J., Young, R.J., Davies, G.R. (2005). *Biomacromolecules*, 6, 507–513.
91. Kitagawa, T., Tashiro, K., Yabuki, K. (2002). *J. Polym. Sci. B*, 40, 1269–1280.
92. Kitagawa, T., Tashiro, K., Yabuki, K. (2002). *J. Polym. Sci. B*, 40, 1281–1287.
93. Cael, J.J., Gardner, K.H., Koenig, J.L., Blackwell, J. (1975). *J. Chem. Phys.*, 62, 1145–1153.
94. Sturcova, A., Davies, G.R., Eichhorn, S.J. (2005). *Biomacromolecules*, 6, 1055–1061.
95. Galiotis, C., Young, R.J., Batchelder, D.N. (1983). *J. Mater. Sci. Lett.*, 2, 263–266.
96. Galiotis, C., Young, R.J., Yeung, P.H.J., Batchelder, D.N. (1984). *J. Mater. Sci.*, 19, 3640–3648.
97. Schadler, L.S., Galiotis, C. (1995). *Int. Mater. Rev.*, 40, 116–134.
98. Andrews, M.C., Young, R.J. (1993). *J. Raman Spectrosc.*, 24, 539–544.
99. Stanford, J.L., Young, R.J., Day, R.J. (1991). *Polymer*, 32, 1713.
100. Cox, H.L. (1952). *Brit. J. Appl. Phys.*, 3, 72–79.
101. Andrews, M.C., Young, R.J. (1993). *J. Raman Spectrosc.*, 24, 539–544.
102. Andrews, M.C., Young, R.J., Mahy, J., Schaap, A.A., Grabandt, O. (1998). *J. Comp. Mater.*, 32, 893–908.
103. de Lange, P.J., Mäder, E., Mai, K., Young, R.J., Ahmad, I. (2001). *Composites A Appl. Sci. Manufacturing*, 32, 331–342.
104. So, C.L., Bennett, J.A., Sirichaisit, J., Young, R.J. (2003). *Plastics Rubber and Composites*, 32, 199–205.

105. So, C.L., Young, R.J. (2001). *Composites Part A Appl. Sci. Manufacturing*, 32, 445–455.

106. Mottershead, B., Eichhorn, S.J. (2007). *Composites Sci. Technol.*, 67, 2150–2159.

107. Wagner, H.D., Amer, M.S., Schadler, L.S. (1996). *J. Mater. Sci.*, 31, 1165–1173.

108. Lei, S.Y., Young, R.J. (2001). *Composites A Appl. Sci. Manufacturing*, 32, 499–509.

109. Angkaew, S., Wang, H-Y., Lando, J.B. (1994). *Chem. Mater.*, 6, 1444–1451.

110. Bognitzki, M., Czado, W., Frese, T., Schaper, A., Hellwig, M., Steinhart, M., Greiner, A., Wendorff, J.H. (2001). *Adv. Mater.*, 13, 70–72.

111. Ko, F., Gogotsi, Y., Ali, A., Naguib, N., Ye, H., Yang, G., Li, C., Willis, P. (2003). *Adv. Mater.*, 15, 1161–1165.

112. Fong, H., Liu, W.D., Wang, C.S., Vaia, R.A. (2002). *Polymer*, 43, 775–780.

113. Takayanagi, M., Ogata, T., Morikawa, M., Kai, T. (1980). *J. Macromol. Sci. Phys.*, 17, 591–615.

114. Ueta, S., Sakamoto, T., Takayanagi, M. (1993). *Polym. J.*, 25, 31–40.

115. Hamley, I.W. (2003). *Nanotechnology*, 14, 39.

116. Hu, X., Stanford, J.L., Young, R.J. (1994). *Polymer*, 35, 80–85.

117. Casado, R., Lovell, P.A., Stanford, J.L., Young, R.J. (1997). *Macromol. Symp.*, 118, 395–400.

118. Young, R.J., Day, R.J., Ang, P.P. (1990). *Polym. Comm.*, 31, 47–49.

119. Zhao, Q., Wagner, H.D. (2004). *Philos. Trans. R. Soc. A*, 362, 2407–2424.

120. Kannan, P., Eichhorn, S.J., Young, R.J. (2007). *Nanotechnology*, 18, 235707.

121. Eichhorn, S.J., Young, R.J. (2001). *Cellulose*, 8, 197–207.

122. Hsieh, Y.C., Yano, H., Nogi, M., Eichhorn, S.J. (2008). *Cellulose*, 15, 1059–1065.

123. Favier, V., Chanzy, H., Cavaille, J.Y. (1995). *Macromolecules*, 28, 6365–6367.

124. Eichhorn, S.J., Baillie, C.A., Zafeiropoulos, N., Mwaikambo, L.Y., Ansell, M.P., Dufresne, A., Entwistle, K.M., Herrera-Franco, P.J., Escamilla, G.C., Groom, L., Hughes, M., Hill, C., Rials, T.G., Wild, P.M. (2001). *J. Mater. Sci.*, 36, 2107–2131.

125. Samir, M.A.S.A., Alloin, F., Dufresne, A. (2005). *Biomacromolecules*, 6, 612–626.

126. Kvien, I., Tanem, B.S., Oksman, K. (2005). *Biomacromolecules*, 6, 3160–3165.

127. Yano, H., Sugiyama, J., Nakagaito, A.N., Nogi, M., Matsuura, T., Hikita, M., Handa, K. (2005). *Adv. Mat.*, 17, 153–155.

128. Dufresne, A. (2000). *Composite Interfaces*, 7, 53–67.
129. Nishino, T., Matsuda, I., Hirao, K. (2004). *Macromolecules*, 37, 7683–7687.
130. Gindl, W., Schoberl, T., Keckes, J. (2006). *Appl. Phys. A*, 83, 19–22.
131. Gindl, W., Keckes, J. (2005). *Polymer*, 46, 10221–10225.

5

RAMAN APPLICATION IN EMULSION POLYMERIZATION SYSTEMS

Oihana Elizalde, Ph.D.
Polymer Research, Ludwigshafen, BASF SE

Jose Ramon Leiza, Ph.D.
Institute for Polymer Materials (Polymat), Department of Applied Chemistry, University of the Basque Country

5.1 INTRODUCTION

Raman spectroscopy has been widely used for the characterization of polymeric materials (films, colloids, foams, fibers, etc.) as it has been already addressed in the preceding chapters. An important amount of such polymers is produced by synthetic routes. Among them, emulsion polymers represent approximately 13% of the worldwide synthetic polymer production. In addition, emulsion polymers are "products-by-process" whose main properties are determined during the polymerization process. Therefore, in order to achieve an efficient production of high-quality materials in a consistent, safe, and environmentally friendly way, knowledge-based strategies [1] that use the polymer microstructure (including molecular weight distribution [MWD], polymer composition, branching and cross-linking density, particle morphology, and particle size distribution [PSD]) as a link between the reaction variables and the application properties are required. The implementation of such strategies requires, among other fundamental and scientifically challenging issues that are out of the scope of this chapter, accurate online monitoring, optimization, and control of the emulsion polymerization process.

In the last decade and due to the advances in fiber-optic technology, Raman spectroscopy has emerged as a robust, noninvasive, and accurate tool to monitor emulsion polymerization reactors. In this chapter, the term "emulsion polymerization" is used in a broad sense and includes other related processes that will be briefly described next. The chapter will cover in detail the monitoring of emulsion polymerization processes by means of Raman spectroscopy analyzing the most common latex properties (free monomer concentration and polymer composition) as well other properties that are more difficult or not well established as the measurement of the particle size. In addition, other applications of Raman spectroscopy to determine aspects of the microstructure of emulsion polymers that can also be extracted from the spectra will be addressed. The application of Raman spectroscopy to monitor suspension polymerization process is also considered in this chapter because of the importance of several products produced by this heterogeneous polymerization technique. Finally, future challenges and trends on the use of Raman spectroscopy in emulsion systems and especially its industrial implementation will be discussed.

5.2 POLYMERIZATION IN DISPERSED MEDIA

The yearly world production of synthetic polymers exceeds 200 million metric tons and about 13% of this amount (26 million) is produced in dispersed media [2]. The polymeric dispersions are typically commercialized as dry polymer and waterborne dispersions. Half of the total polymeric dispersions are commercialized as waterborne dispersions. Carboxylated styrene-butadiene copolymers, acrylic and styrene-acrylic latexes, and vinyl acetate homopolymer and copolymers are the main polymer classes. The main markets for these dispersions are paints and coatings (26%), paper coating and paperboard (23%), adhesives and sealants (22%), and carpet backing (11%) [3]. The other half of the polymeric dispersions are commercialized as dry products. These include styrene-butadiene rubber (SBR) for tires, nitrile rubbers, about 10% of the total polyvinyl chloride (PVC) production, 75% of the total acrylonitrile-butadiene-styrene (ABS), and redispersible powders for construction materials.

Different polymerization techniques can be used to produce polymeric dispersions, but among them emulsion polymerization is the most common because of its versatility and the possibility to synthesize a great variety of polymers with a broad property spectra that ensure meeting current market needs. In addition to emulsion polymerization, other related techniques also allow the production of polymeric dispersions that cannot be achieved by emulsion polymerization. These include suspension polymerization, miniemulsion polymerization, microemulsion polymerization, and dispersion polymerization. These techniques will be briefly described next.

5.2.1 Polymerization Techniques and Commercial Products

5.2.1.1 Emulsion Polymerization Table 5.1 presents a typical formulation for emulsion polymerization. The polymer is mainly made out of a mixture of "hard" (leading to high glass transition temperature, T_g, polymers, e.g. styrene) and "soft" (low T_g, e.g. butyl acrylate) monomers of low water solubility. In addition, small amounts of functional monomers such as acrylic and methacrylic acids are included in the formulation as they provide some special characteristics, such as improved adhesion and stability. Cross-linking agents and chain transfer agents are used to control the degree of cross-linking and MWD of the polymer.

Table 5.1 Typical formulation for emulsion polymerization

Ingredient		Content (wt %)
Hard monomer(s)	Styrene (S)	50–55
	Methyl methacrylate (MMA)	
	Vinyl chloride (VC)	
	Vinyl acetate (VAc)	
Soft monomer(s)	Butadiene (Bd)	
	Butyl acrylate (BA)	
	2-ethyl hexyl acrylate (2EHA)	
	Veova 10	
	Ethylene	
Functional monomer(s)	(Meth)acrylic acid	
	Cross-linking monomers	
Deionized water		45
Initiators		0.5
Emulsifiers		0.5–3.0
Chain transfer agents		0–1

Typically, emulsion polymerization is carried out in stirred-tank reactors, which commonly operate in a semicontinuous mode, although both batch and continuous operations are also used.

In a batch emulsion polymerization, the mixture of monomers is dispersed in water using emulsifiers. The monomer droplets are stabilized by the surfactant adsorbed on their surface. In principle, any type of surfactant may be used, but in practice anionic surfactants, nonionic surfactants, and mixtures thereof account for the vast majority of the systems used. The available surfactant partitions between the surface of the monomer droplets and the aqueous phase, and in most formulations, the amount of surfactant exceeds that needed to cover completely the monomer droplets and saturates the aqueous phase. The excess of surfactant forms micelles that are swollen with monomer (Figure 5.1a). Polymerization is commonly initiated by water-soluble initiators (both thermal—e.g. potassium persulfate—and redox—e.g. tert butyl hydroperoxide/ascorbic acid), although oil-soluble initiators (e.g. azo isobutiro nitrile [AIBN]) may also be used. When a water-soluble initiator such as potassium persulfate is added to the monomer dispersion, radicals are formed, and as these radicals are too hydrophilic to enter into the organic phases of the systems, they react with the monomer dissolved in the aqueous phase, forming oligoradicals. The growth rate of the oligoradicals is generally modest because of the low concentration of monomer in the aqueous phase. After adding some monomer units, the oligoradicals become hydrophobic enough to be able to enter into the micelles (entry into the monomer droplets is not likely because their total surface area is about three orders of magnitude smaller than that of the micelles). Because of the high concentration of monomer in the micelle, the oligoradical that has entered into the micelle grows fast, forming a polymer chain. The new species formed upon entry of a radical into a micelle is considered to be a polymer particle. The process of formation of polymer particles by entry of radicals into micelles is called *heterogeneous nucleation* [4]. The oligoradicals that do not enter into micelles will continue growing in the aqueous phase, and upon reaching some critical length, they become too hydrophobic and precipitate. The emulsifier present in the system will adsorb onto the newly formed interface, stabilizing the polymer particle. Then, monomer will diffuse into the new polymer particle. The process of formation of polymer particles by precipitation of oligoradicals is called *homogeneous nucleation* [5]. Both homogeneous and heterogeneous nucleation may be operative in a given system.

In general, homogeneous nucleation is predominant for monomers of relatively high water solubility (e.g. methyl methacrylate, 1.5 g/100 g

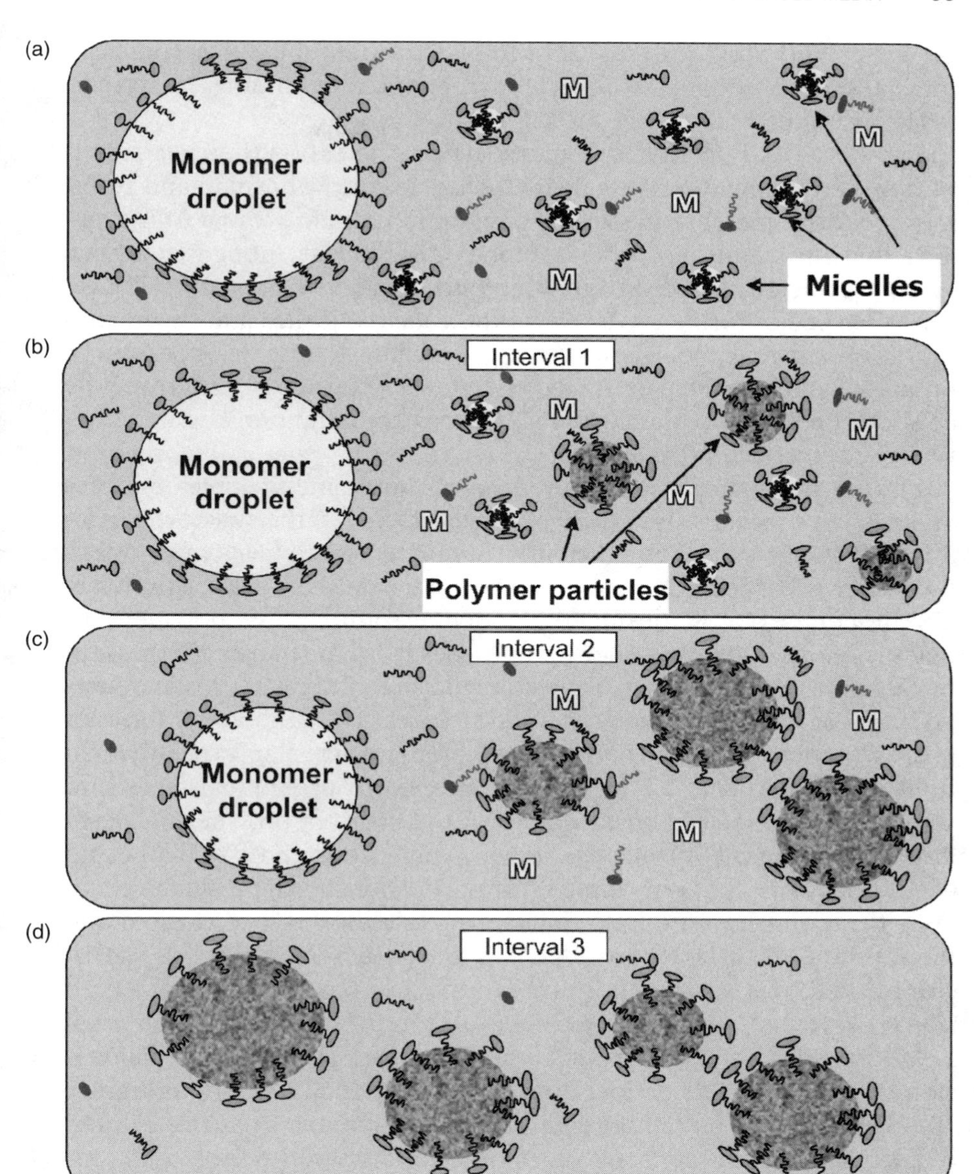

Figure 5.1 Intervals of the batch emulsion polymerization. Surfactant; Oligoradical; initiator fragment; M, monomer in aqueous phase. See text for details.

of water; and vinyl acetate, 2.5 g/100 g of water) and heterogeneous nucleation is predominant for water-insoluble monomers (e.g. styrene, 0.045 g/100 g of water).

Irrespective of the mechanism of particle nucleation (heterogeneous or homogeneous), the newly formed particles are very small and suffer a tremendous increase in surface area upon particle growth. It is arguable that the emulsifier molecules may diffuse fast enough to adsorb on the surface of these fast-growing particles, stabilizing them. Therefore, the species formed by entry of radicals in micelles and by precipitation of growing radicals in the aqueous phase may be regarded as precursor particles that only become stable particles upon growth by coagulation and polymerization [6–8]. This combined process is sometimes called *coagulative nucleation*.

During nucleation, monomer droplets, monomer-swollen micelles and monomer-swollen polymer particles coexist in the reactor (Figure 5.1b). Polymer particles efficiently compete for radicals, and hence monomer is consumed by polymerization inside the polymer particles. The monomer that is consumed by polymerization in the polymer particles is replaced by monomer that diffuses from the monomer droplets through the aqueous phase. Therefore, the size of the particles increases and that of the monomer droplets decreases. The number of micelles decreases because they become polymer particles upon entry of a radical and also because they are destroyed to provide surfactant to stabilize the increasing surface area of the growing polymer particles. After some time, all micelles disappear. This is considered to be the end of the nucleation and only limited formation of new particles may occur after this point because heterogeneous nucleation is not possible and there is no free surfactant available in the system to stabilize the particles formed by homogeneous nucleation. The stage of the batch emulsion polymerization in which particle nucleation occurs is called Interval 1. At the end of Interval 1, which typically occurs at a monomer conversion of about 5–10% (depending on the surfactant/monomer ratio), 10^{17}–10^{18} particles/L are formed. Unless coagulation occurs, the number of particles remains constant during the rest of the process.

In Interval 2, the system is composed of monomer droplets and polymer particles (Figure 5.1c). The monomer consumed by polymerization in the polymer particles is replaced by monomer that diffuses from the monomer droplets through the aqueous phase. The mass transfer rate of monomers with water solubility equal or greater than that of styrene (0.045 g/100 g of water) is substantially higher than the polymerization rate, and hence monomer partitions between the different phases of the system according to thermodynamic equilibrium.

Therefore, in the presence of monomer droplets, the concentration of the monomer in the polymer particles reaches a maximum value that is roughly constant during Interval 2.

Because of the polymerization and monomer transport, the polymer particles grow in size, and after some time the monomer droplets disappear. This marks the end of Interval 2. The monomer conversion at which Interval 2 ends depends on the capability of the polymer particle to be swollen by monomer. Thus, the transition from Interval 2 to Interval 3 occurs at about 40% conversion for styrene and at about 15% conversion for vinyl acetate. This means that most monomer polymerizes in Interval 3 (Figure 5.1d). In this interval, monomer concentration in the polymer particles decreases continuously.

In a semicontinuous reactor in which monomers, surfactant, initiator, and water may be continuously fed into the reactor, emulsion polymerization does not follow the sequence of events described. In a typical semibatch process used in industry, monomers and surfactant are fed slowly into the reactor to allow good temperature and copolymer composition control (so-called starved-feed) and hence only monomer-swollen polymer particles are present in the system. However, a fast monomer feed and a low surfactant feed will lead to a system containing monomer droplets (flooded system) and polymer particles.

The final product is a waterborne concentrated (50–60 wt % solids) dispersion of tiny (80–300 nm in diameter) polymer particles called latex. The microstructural properties of the polymer in the particles define the application properties of the latexes. The most important microstructural properties are copolymer composition, monomer sequence distribution, molar mass distribution, polymer architecture (branching, grafting, cross-linking, and gel content), particle surface functionality, particle morphology, and PSD. These properties directly affect the mechanical, adhesive, and rheological properties of the latex.

Typical commercial products include rubbery materials (styrene-butadiene copolymers, acrylonitrile-butadiene copolymers, chloroprene polymers, and carboxylated styrene-butadiene copolymers), architectural coatings (polyvinyl acetate homopolymers and copolymers, styrene-acrylic copolymers), adhesives, sealants and caulks (all acrylics and acrylic copolymers, vinyl acetate and its copolymers), and textile applications (vinyl acetate homopolymers, vinyl acrylics, all acrylics, styrene-acrylics, styrene-butadiene, and ethylene-vinyl acetate copolymers).

Dispersed polymers with submicron polymer particles are also produced by inverse emulsion polymerization, miniemulsion polymerization, dispersion polymerization, and microemulsion polymerization. Table 5.2 presents in a schematic way the main features and applications of this related techniques. On the other hand, polymeric dispersions with polymer particles with sizes well above one micron are prepared by suspension polymerization.

5.2.1.2 Suspension Polymerization The suspension polymerization process is typically carried out in well-stirred batch reactors. A typical suspension polymerization recipe is shown in Table 5.3. The monomer is initially dispersed in the continuous phase (commonly water) by the combined action of surface-active agents (inorganic and/or water-soluble polymers) and agitation. All the reactants (monomers, initiators, etc.) reside in the organic phase. The polymerization occurs in the monomer droplets that are progressively transformed into sticky, viscoelastic monomer-polymer droplets and finally into rigid, spherical polymer particles in the size range of 50–500 μm [16–18]. The solids content in the fully converted suspension is typically 30–50 wt %.

In general, the suspension polymerization can be distinguished into two types, namely the "bead" and "powder" suspension polymerization. In the former process, the polymer is soluble in the monomer and smooth spherical particles are produced. In the latter process, the polymer is insoluble in the monomer and, thus, precipitates out, yielding to irregular grains or particles. The most important thermoplastic produced by the bead process is polystyrene. In the presence of volatile hydrocarbons (C4-C6) foamable beads, the so-called expandable polystyrenes (EPS) are produced. On the other hand, PVC, which is the second largest thermoplastic manufactured in the world, is an example of the powder-type suspension polymerization that is used for the production of plastisols.

One of the most important issues in the suspension polymerization process is the control of the PSD [17]. In general, the initial monomer droplet size distribution (DSD) as well as the polymer PSD depends on the type and concentration of the surface-active agent, the quality of agitation (e.g. reactor geometry, impeller type, and power input), and the physical properties (e.g. densities, viscosities, and interfacial tension) of the continuous and dispersed phases. The dynamic evolution of the droplet/PSD is controlled by the rates of two physical processes, namely, the drop/particle breakage and coalescence. The former mainly occurs in regions of high shear stress (near the agitator) or as a result of turbulent velocity and pressure fluctuations along the droplet surface. Drop/particle coalescence can either be increased or decreased by the

Table 5.2 Description of the polymerization techniques related to the conventional emulsion polymerization

Polymerization technique	Description/main features	Type of emulsion	Initiator/surfactant system	Products/ applications
Inverse emulsion polymerization [9]	Aqueous solution of a water-soluble monomer (e.g. acrylamide) dispersed in an organic continuous phase using an excess of surfactant. Stabilization of polymer particles only by steric means.	Water-in-oil	Water- and oil-soluble initiators Thermal and redox systems Nonionic surfactants	Polymers and copolymers of acrylamide. Thickeners in paints. Pushing fluids in tertiary oil recovery; drilling fluids. Flocculants.
Miniemulsion polymerization [10–12]	Droplet-size reduced (d_d = 50–1000 nm) by combining a suitable emulsifier and an efficient emulsification apparatus. Droplet degradation controlled by using costabilizer (hydrophobe of low molar mass). Droplet nucleation. Enhanced incorporation of hydrophobic materials in polymer particles.	Oil-in-water	Water- and oil-soluble initiators Thermal and redox systems Surfactants as in conventional emulsion Hydrophobe (nonreactive; hexadecane, or reactive; long-chain acrylates)	Commercial products not known yet. Potential applications include hybrid polymers, polymer/clay nanocomposites, and controlled radical polymerization products.
Microemulsion polymerization [9,13]	Thermodynamically stable dispersion achieved by appropriate type and amount of surfactant (high). Smaller particle sizes than in other techniques. Very high molar masses.	Oil-in-water or water-in-oil (so-called inverse microemulsion polymerization)	Water- and oil-soluble initiators Anionic surfactants	Acrylamide and copolymers. Cationic and anionic polymers. Flocculants.
Dispersion polymerization [14,15]	Monomer, initiator, and stabilizer dissolved in solvent that is not a solvent of the polymer. Nucleation occurs by precipitation and stabilization of the nuclei by stabilizer added or formed *in situ* during the polymerization. Polymerization starts in a homogeneous phase and moves to a heterogeneous system.	Monomer-in-solvents (water/alcohol)	Oil-soluble initiators Stabilizers key for the stability of the dispersion. Block and graft copolymers commonly used.	Micron-sized particles. Surface coatings for metal Chromatographic separation.

Table 5.3 Typical formulation for suspension polymerization

Ingredient	Amount (wt %)
Monomers	30–50
Water	50–70
Hydrocarbon for EPS (e.g. *m*-pentane)	2–4
Initiators (e.g. BPO)	0.1–0.2
Stabilizer (PVA, barium sulfate, etc.)	0.3–0.6

EPS, expandable polystyrene; BPO, benzoyl peroxide; PVA, polyvinyl alcohol.

turbulent flow field. At sufficiently high concentration of surface-active agents, it can be assumed to be negligible for very dilute dispersions.

5.2.2 Benefits of Raman Spectroscopy in Emulsion Polymerization

As described in the previous section, the characteristics of emulsion polymers are defined during the polymerization process and there is almost no room for modification upon finishing the polymerization. Therefore, monitoring of polymerization variables is necessary to fully implement advanced closed–loop control strategies and to ensure the consistent, safe, and optimal production of polymeric dispersions with the required quality. Even when advanced control strategies are not implemented, online monitoring of polymerization processes is a must, as it generates an enormous amount of useful information that can be used for modeling and optimization purposes in the long term and to modify reaction formulations in the short term. For instance, online Raman spectroscopy is used to monitor free monomer in industrial semibatch emulsion polymerization reactors, and the decision on reactor shutdown to start a new reaction is taken based on the Raman measurements of the free monomer. Furthermore, it may allow for significant reduction of time-consuming off-line analyses performed in the lab.

Sensors used for monitoring of polymerization reactors can be classified into two large categories: (1) sensors for monitoring of *reactor operation conditions or process variables*, and (2) sensors for monitoring of the *trajectory of polymer properties* during polymerization. Temperature, pressure, flow rates, and level are measurements in the first group that are routinely performed at the plant site. On the other hand, sensors used for monitoring of trajectories of structural polymer properties are very difficult to develop, but provide much more useful information. Raman spectroscopy and other spectroscopic techniques (near-infrared

[NIR] and mid-range [MIR] spectroscopies) are sensors in the second category that have been used to monitor polymerization reactors. The advantage of these spectroscopic techniques relies on the noninvasive nature of the measurement (the reactor content is not handled outside of the reaction vessel). All these spectroscopic techniques have been used to monitor emulsion polymerization and other related systems (see earlier discussion) [19–30] and each presents advantages with respect to other invasive (e.g. gas chromatographs and densitometers) or noninvasive (e.g. reaction calorimetry and ultrasounds) techniques that have been used so far [31]. The multiple advantages that the Fourier transform (FT) spectrometers offer, combined with the excitation in the NIR region, have rediscovered Raman spectroscopy for polymer analysis. In recent years, its use in the area of polymeric materials has grown considerably, in part because all the problems regarding the existence of fluorescence (the main limitation of the conventional Raman for polymer analysis) were solved. Furthermore, Raman spectroscopy has shown to be better suited than absorption techniques (mainly MIR and NIR) to monitor emulsion polymerization systems [20,32]. Some of these advantages are related to the spectroscopic information while others to the instrumentation itself.

5.2.2.1 Spectroscopic Advantages Water is an ideal medium because its Raman spectrum is very weak. It does not absorb the laser radiation, it does not heat up due to the radiation, and it does not attenuate the emitted Raman radiation (which would cause a distortion of the relative intensities of the spectrum, affecting the quantification). Figure 5.2 shows the Raman spectra of water and poly(methyl meth-

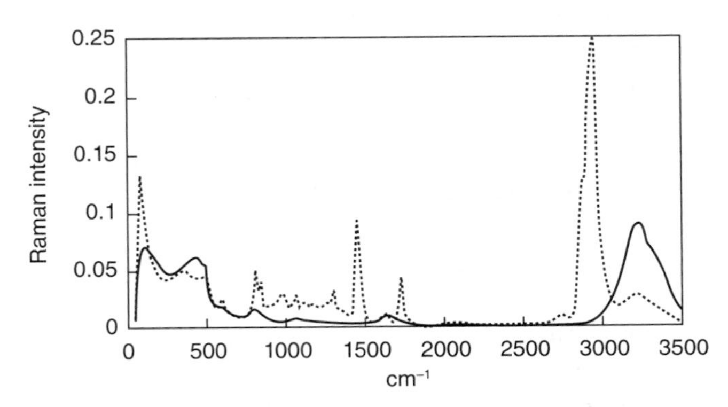

Figure 5.2 Fourier transform (FT)-Raman spectra of water (solid line) and a PMMA/PBA latex (dashed line) taken at room T, using a near-infrared (NIR) laser, with a power of 500 mW, 8 cm^{-1} resolution and accumulating 50 scans.

acrylate-co-butyl acrylate) latex synthesized by emulsion polymerization. Water presents two broad bands, one above 3000 cm^{-1} corresponding to the stress vibration of the (O–H) bond, and the other below 500 cm^{-1} due to the deformation vibration in the plane of the (O–H) bond. As can be observed, the zone of interest for emulsion polymers, where the characteristic bands of most of the compounds appear, between 3200 and 600 cm^{-1}, is relatively clean of water signals.

Another important advantage of Raman spectroscopy is that the functional groups that usually are inactive or give weak IR absorptions give rise to intense Raman bands. Raman spectroscopy offers a higher sensitivity toward the carbon–carbon double bonds. Therefore, it is an ideal tool to monitor polymerizations, because the disappearance of the double (C=C) bond is the most significative change in these processes. Raman spectroscopy also presents a high potential for the monitoring of triple bonds (monitoring of isocyanates in urethane reactions).

5.2.2.2 Instrumental Advantages The main advantages are:

(1) FT-Raman offers the possibility of using silica fibers (that have a low cost), with a length up to 200 m, to transmit the radiation to and from the sample. This allows the implementation of Raman probes in industrial polymerization reactors for remote analysis and control of the process.
(2) With the bidimensional detectors, several sampling points can be multiplexed using different fibers that transmit the radiation from each point.
(3) The band shapes are not affected when opaque and light-scattering samples are analyzed, as it happens, for example, in MIR. This overcomes the sampling problems existent in the MIR spectrometers.
(4) Little or no sample preparation or sample extraction is required.

Raman spectroscopy also presents several disadvantages. It is a very weak effect, typically between four to six orders of magnitude weaker than fluorescence. Therefore, Raman sensors require quite sophisticated instrumentation in order to obtain the desired signal levels, especially at low concentrations. Because of the poor Raman signal, the technique is extremely exposed to interferences due to the presence of small amounts of fluorescent impurities. In the conventional dispersive Raman spectroscopy, which uses a laser in the visible region, almost 90% of the samples contain fluorescent impurities that interfere with the Raman spectra. Since the fluorescence is usually several orders of magnitude stronger than the Raman signal, the information contained in the Raman spectrum is totally masked. Fluorescence can be avoided by moving to the NIR region as in the newly available dispersive equipment (cheaper than the FT-Raman ones) that can work at wavelengths as high as 830 nm (typically 785 nm) and hence exploit the advantages of dispersive Raman equipment that offers higher sensitivity and resolution. However, hydrogen-containing compounds (such as water) have absorption bands in the NIR due to overtones and combinations of the normal vibrations. This is the so-called *self-absorption effect*, defined as the absorption of Raman-scattered light by the sample itself before the light is collected back [33]. This perturbation of the spectrum will occur when the sample has an absorption band at the same frequency as the Raman scatter, and as a consequence, the corresponding Raman bands will be *suppressed* relative to other bands in the spectrum.

Another drawback of the technique is that the light transmitted through the optic fibers can excite some molecules of the fiber itself, and therefore, generate an intense background that must be reduced using filters in order to obtain satisfactory results.

5.3 MONITORING OF HETEROGENEOUS POLYMERIZATION REACTIONS

5.3.1 Introduction

Raman spectroscopy has been used in a broad range of applications in the field of polymeric materials, starting from the evaluation of polymer and copolymer blend compositions, evaluation of monomer and polymer compositions during polymerizations, determination of additive content in polymer samples, and estimation of end-use properties of polymer materials. Although fewer publications have been reported for the online monitoring and control of polymerization processes, Raman spectroscopy presents a high potential in this field. There are several references related to the use of Raman spectroscopy to follow the evolution of several processes, including polymerizations. Gulari et al. [26] used Raman spectroscopy to monitor methyl methacrylate (MMA) and styrene (Sty) bulk homopolymerizations. Clarkson et al. [24] studied the bulk homopolymerizations of several commercial acrylic monomers. Raman scattering was also used to monitor solution homo [25,34] and copolymerizations [35]: homopolymerizations of MMA [25], butyl acrylate (BA) [34] and copolymerizations of styrene and BA [35].

In the case of heterogeneous systems, there are relatively fewer published works due to the higher complexity of the systems. Spectroscopic measurements are influenced by the heterogeneity of the medium; therefore, monitoring heterogeneous polymerization reactions is a much more challenging task in comparison with the homogeneous or bulk processes. As in any other polymerization process, in the heterogeneous ones there is a continuous change of composition, since the monomer amount decreases during the reaction and polymer content increases. In addition, process parameters such as the polymerization temperature have an effect on the Raman signal as well. On top of all, one needs to consider that there are reactions that change from being homogeneous at the beginning to a heterogeneous end product, and that medium heterogeneity can therefore change extremely during the reaction. Since medium heterogeneity causes changes in the Raman signal [20,27,36–38], it must be taken into account in order to extract accurate quantitative information from the acquired spectra. In all heterogeneous processes, different phases are present in the reaction mixture (monomer droplets, polymer particles, micelles, etc.) and the concentration and size of each of these phases change during the polymerization reaction. (Raman intensity is proportional to the number of vibrating centers present in the scattering volume.) There is also a risk of having light-scattering effects, either because the incoming (unshifted) photons scatter off their path into the sampling volume, causing a lower effective photon density in the sampled volume, or because the already scattered radiation can also be scattered by the latex particles before it is collected back. Let us take as an example the prediction of monomer consumption: in this case, the overall monomer signal is the sum of the contributions of the monomer in each of the individual phases (aqueous, monomer droplets and monomer-swollen polymer particles), each contribution depending on the interactions between the molecules in that phase.

5.3.2 Quantitative Methods

A Raman spectrum is formed from different band positions that contain information about the appearance of the chemical compounds present in the reaction mixture. Each of these bands has a defined intensity, which is directly related to the amount of the compound responsible for that band. The simplest way to construct a calibration model that relates two properties is by *univariate regression*, where only one measured variable is related to the property that we want to predict. In the case of spectral calibration for example, this would be to relate the concentration of a compound to the intensity at a certain wavelength or to the ratio between the areas of two peaks. Using this method, only the information from very few spectral variables is used, neglecting the information from the rest of the spectrum. Unfortunately, the mixture of different compounds often gives spectra with overlapping bands, and the intensities are not directly related to the content of a single compound. This is the typical scenario when trying to monitor many copolymerization reactions. An example of such a system is the copolymerization of acrylic monomers, such as BA/MMA, an industrially very common combination of (co)monomers in latex used for paint and coating applications, or the system VAc/BA. (Spectra of both examples can be found in Figure 5.3.)

Because unique and isolated bands corresponding to each monomer cannot be identified for those systems (the monomers present overlapped bands for the double bonds and carbonyl groups due to the similarities of the chemical structure), univariate calibration techniques cannot be used. Consequently, it is necessary to use *chemometric* [39–43] methods to extract the property of interest from the information contained in the Raman spectra. By applying chemometric methods, the complete spectrum, or parts of it (multivariate information) can be related to the value of the property of interest.

Figure 5.4 presents a schematic representation of the calibration procedures employed using chemometric or multivariate methods. In the calibration process, a model must be built using known properties of the latex (measured by reference techniques such as gas chromatog-

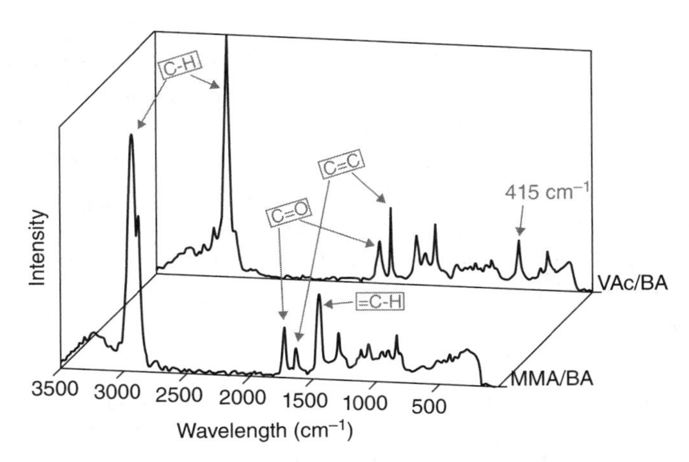

Figure 5.3 Spectra of two copolymer latexes collected during polymerization: methyl methacrylate/butyl acrylate (MMA/BA, front) and VAc/BA (back) comonomer systems.

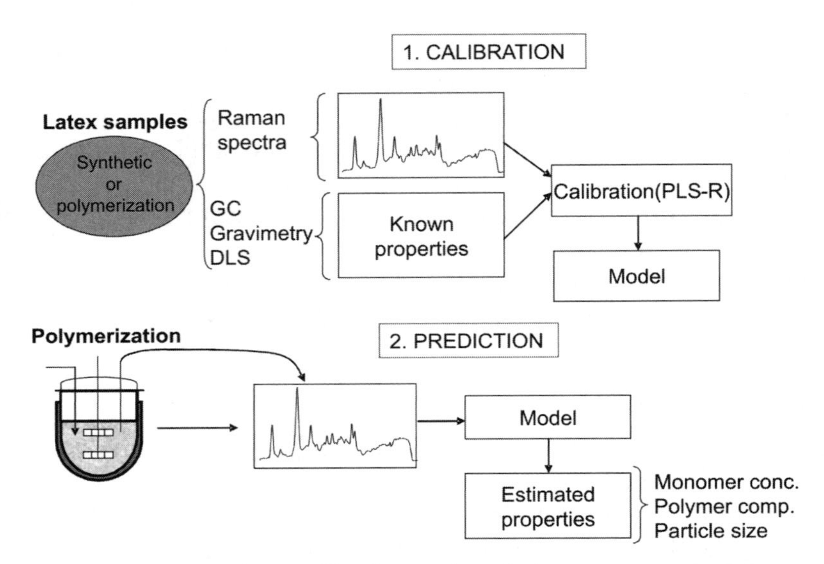

Figure 5.4 Schematic of the calibration and prediction procedure for latex properties using Raman spectra. GC, gas chromatography; DLS, dynamic light scattering; PLS-R, partial least-squares regression.

raphy for monomer concentration and gravimetry for solids content) and the spectra acquired for the latex. With this information and a given calibration procedure (partial least-square regression [PLS-R], principal component analysis [PCA], and others), a calibration model can be developed. Upon building the model, one can determine the unknown properties of a latex by taking the Raman spectra of the latex.

Most of the works published in the open literature employ multiple linear regression (MLR) [44] or methods based on factor or principal component extraction to extract quantitative information from spectroscopic data. Among the methods based on principal component extraction [43,45], principal component regression (PCR) and PLS-R [46] are the most used ones. In these data reduction methods, the spectral data are separated into independent spectral variations called *factors, principal components*, or *latent variables* that are directly related to the property of interest (such as solids content, copolymer composition, or monomer concentration).

5.3.2.1 Calibration Data The calibration data set is very important in order to obtain good prediction results. Even when the selected multivariate calibration method is good, when no appropriate data are available to build the calibration models the results will not be good. In order to assure good prediction results, the calibration data set must fulfill certain requirements. On the one hand, the data set must be representative of the future data from which we would like to obtain property predictions, and the measurement conditions should also be as similar as possible. On the other hand, special attention should be paid to the property value range of the calibration data set. Therefore, the sample set used should be as similar as possible to the one contained in the reactor during the polymerization process, and include variations in factors that change during the reactions and have an influence on the Raman measurements. For model construction, real process samples or synthetically prepared laboratory samples (see Figure 5.4) can be employed. Most of the works reported in the literature used real process samples to monitor emulsion polymerizations [20,23,46–48].

5.3.3 Monitoring Emulsion Polymerization Reactions

The first reported works on monitoring of polymerization reactions via Raman spectroscopy were for bulk and solution processes, where the polymerization medium is homogeneous. It was only some years later that the first works dealing with emulsion polymerization systems were published. In comparison with the homogeneous systems, the publications related to emulsion polymerization are still much fewer, due to the difficulties related to medium heterogeneity.

One of the already mentioned advantages of Raman spectroscopy, and of spectroscopic methods in general, is their ability to give information about different properties, such as polymer amount or composition. Therefore, it is an ideal tool to simultaneously monitor the evolution of several properties during emulsion polymerization. In the open literature, the evolution of monomer conversion, copolymer composition, and particle size in emulsion polymerization by means of Raman spectroscopy has been reported. The next subsections summarize the most relevant results for the three properties.

5.3.3.1 Monomer Conversions Figure 5.5 presents the Raman spectra taken during a batch emulsion polymerization of MMA/BA carried out at 75 °C. It can be clearly seen that the double-bond peak at around 1635 cm^{-1} continuously decreased, which was an indication that the overall monomer concentration (both MMA and BA) decreased. However from the spectra it is also possible to extract the individual monomer concentrations with the help of appropriate calibration models. Table 5.4 presents examples of monitoring monomer concentration in emulsion polymerization systems with indications of the monomer system (and whether homo- or copolymerization reac-

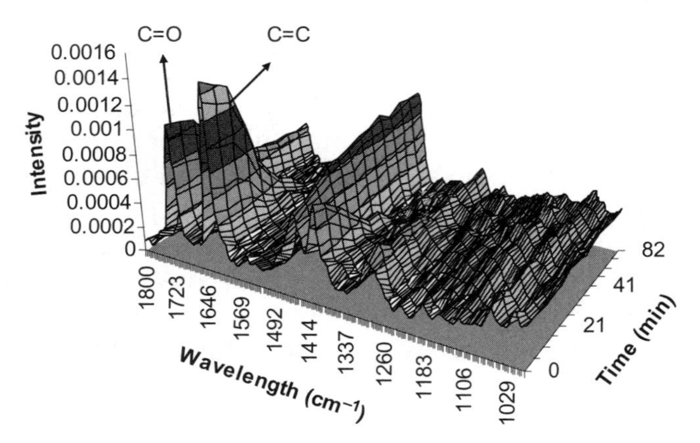

Figure 5.5 Online evolution of Raman spectra taken during a batch emulsion copolymerization of methyl methacrylate/butyl acrylate (MMA/BA) = 50/50.

tions were monitored), Raman equipment employed, calibration data and method employed, and any additional detail that was found noticeable.

The potential of Raman spectroscopy to monitor complex monomer systems (e.g. MMA/BA polymerization) with the aid of PLS-R was shown by Elizalde et al. [30,48] and is briefly reproduced here. The authors also compared Raman spectroscopy with another noninvasive monitoring technique used in polymerization reactors: reaction calorimetry. The performance of both techniques is shown in Figures 5.6 and 5.7 and can be followed in detail elsewhere [29].

It is noteworthy that during the semibatch emulsion polymerizations of MMA/BA, the instantaneous conversion was high during the whole process. This means that monomer concentrations were relatively low, especially the concentration of MMA, which is the more reactive monomer in this system. The Raman predictions were challenging under these conditions because of the low monomer concentrations present in the reactor. Figure 5.6 presents the comparison of the overall conversion measured by gravimetry and the two online techniques for several copolymerization experiments. As it can be seen, in general, there was not much difference between both techniques, and they compared well with gravimetry, which was used as the reference technique.

Figures 5.7 shows the comparison of the free amounts of MMA and BA monomer calculated by means of gravimetry and gas chromatography (reference off-line techniques) and by Raman spectroscopy and

Table 5.4 Examples of monitoring monomer concentration (conversion) in emulsion polymerization systems

Polymerization system	Raman equipment	Acquisition and spectra treatment	Calibration samples	Calibration method	Comments
Styrene	FT-Raman (1064 nm) [20,37,49] Dispersive Raman (532 nm) [38] Dispersive Raman (514.5 nm) [50] Dispersive low-resolution Raman spectrometer (785 nm) [51]	Laser power (LP) = 450 mW, resolution (R) = 8 cm^{-1}, 32 scans [20,37,49] LP = 200 mW, R = 4.5 cm^{-1}, normalized spectra [38] LP = 140 mW, acquisition with a probe [50] R = 30 cm^{-1}, acquisition with a probe, spectra smoothing [51]	Synthetic samples and real process samples [20,37,50] Synthetic samples [38,51]	Multivariate linear (PLS) and nonlinear approaches [20,37,50] Univariate [38,50] Univariate (internal standard) [51]	Estimation of particle size [20,37] Comparison of several chemometric techniques [49] Correlation with particle size [38] Monomer consumption [50] Monomer consumption in batch and semibatch reactions [51]
Butyl acrylate (BA)	FT-Raman (1064 nm) [20,52] Dispersive Raman Low-resolution Raman spectrometer (785 nm) [51]	LP = 450–510 mW, R = 8 cm^{-1}, 32 scans, acquisition with a probe, normalized spectra [20,52] R = 30 cm^{-1}, acquisition with a probe, spectra smoothing [51]	Synthetic samples	Univariate model (C=C peak area), multivariate (PCR, PLS) [20,52] Univariate (internal standard) [51]	Monitoring of monomer concentration (univariate versus multivariate) [20,52] BA homopolymerization using a polystyrene seed [51]

Table 5.4 *Continued*

Polymerization system	Raman equipment	Acquisition and spectra treatment	Calibration samples	Calibration method	Comments
Methyl methacrylate (MMA)	Dispersive Raman (514.5 nm) [50] Dispersive Raman (785 nm) [53]	LP = 550 mW, acquisition with a probe, spectra smoothing [50] LP = 250 mW, R = 4 cm^{-1}, acquisition with a probe (through reactor window) [53]	Synthetic samples [50] Real process samples [53]	Multivariate (simple least-squares fit) [50] Univariate (peak intensities), multivariate (PCA, PLS) [53]	Monitoring of monomer consumption
VAc	FT-Raman (1064 nm) [20,52,54,55]	LP = 450–510 mW, R = 8 cm^{-1}, 32 scans, acquisition with a probe, normalized sectra [20,52] LP = 300 mW, R = 4 cm^{-1}, 36 scans [54]	Synthetic samples	Univariate model (C=C peak area), multivariate (PCR, PLS) [20,52] Univariate (peak height, peak area, Beers' Law) [54] Univariate (relative peak intensities) [55]	Monitoring of monomer concentration (univariate versus multivariate) [20,52]
St/Bd	FT-Raman (1064 nm) [23]	R = 8 cm^{-1}, 200 scans, mean centered spectra, first and second derivatives and multiplicative scatter correction	Real process samples	Multivariate (PLS)	Monitoring of solids content and styrene concentration

VAc/BA	FT-Raman (1064 nm) [20,27,29]	LP = 450 mW, R = 8 cm^{-1}, 8 scans, acquisition with and without a probe (through reactor window), spectra normalization, smoothing and filtering	Synthetic samples and real process samples	Multivariate (PLS)	Effect of increase in noise level on the estimation of BA and VAc concentration. Comparison of model prediction using synthetic samples and process samples [20,27] Comparison with reaction calorimetry [29]
MMA/BA	Dispersive Raman (532 nm) [56]	LP = 400 mW, R = 1.5 cm^{-1}, acquisition with a probe (through reactor window), normalized spectra [56]	Synthetic samples [56] Real process samples [29,30,48,57]	Classical least-squares regression (using an internal standard) [56] Multivariate (PLS) [29,30,48,57]	Prediction of individual monomer concentrations using the polystyrene seed as standard [56] Prediction of individual monomer concentrations, solids content and copolymer composition [29,30,48,57] Comparison with reaction calorimetry [29] Calibration model maintenance strategies [57]
	FT-Raman (1064 nm) [29,30,48,57]	LP = 1 W, R = 4 cm^{-1}, 200 scans, acquisition with a probe, normalized spectra, first and second derivatives [29,30,48,57]			
St/BA	Dispersive Raman (785 nm) [53,58]	LP = 250 mW, R = 4 cm^{-1}, acquisition with a probe (through reactor window), normalized spectra [53] Normalized spectra [58]	Real process samples	Univariate (peak intensities), multivariate (PCA, PLS) [53] Univariate [58]	Individual monomer concentration

Table 5.4 ***Continued***

Polymerization system	Raman equipment	Acquisition and spectra treatment	Calibration samples	Calibration method	Comments
MMA/BA/AMA	FT-Raman (1064 nm) [59,60]	LP = 300 mW, R = 2–4 cm^{-1}, normalized spectra [59,60]		Univariate	Monitoring of overall monomer conversion (solids content)
VeoVa9/BA	Dispersive Raman (532 nm) [28]	LP = 400 mW, R = 1.5 cm^{-1}, acquisition using a probe (through a glass window), spectra smoothing, baseline correction and normalization	Synthetic samples	Multivariate (classical least squares)	Monitoring of individual monomer concentrations during batch and semibatch EP
S/MA	Dispersive Raman (532 nm) [56]	LP = 400 mW, R = 1.5 cm^{-1}, acquisition with a probe (through reactor window) normalized spectra	Synthetic samples	Univariate	Individual monomer concentrations

AMA, Allyl methacrylate; S/Bd, styrene/butadiene; VAc/BA, vinyl acetate/butyl acrylate; S/BA, styrene/butyl acrylate; S/MA, styrene/methyl acrylate.

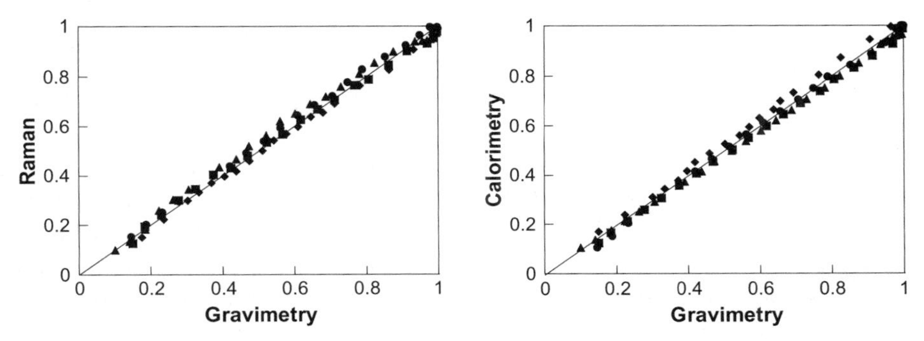

Figure 5.6 Comparison of the overall conversions measured for the semibatch emulsion copolymerization of methyl methacrylate/butyl acrylate (MMA/BA) with gravimetry: Raman spectroscopy (left) and calorimetry (right). Symbols refer to different runs.

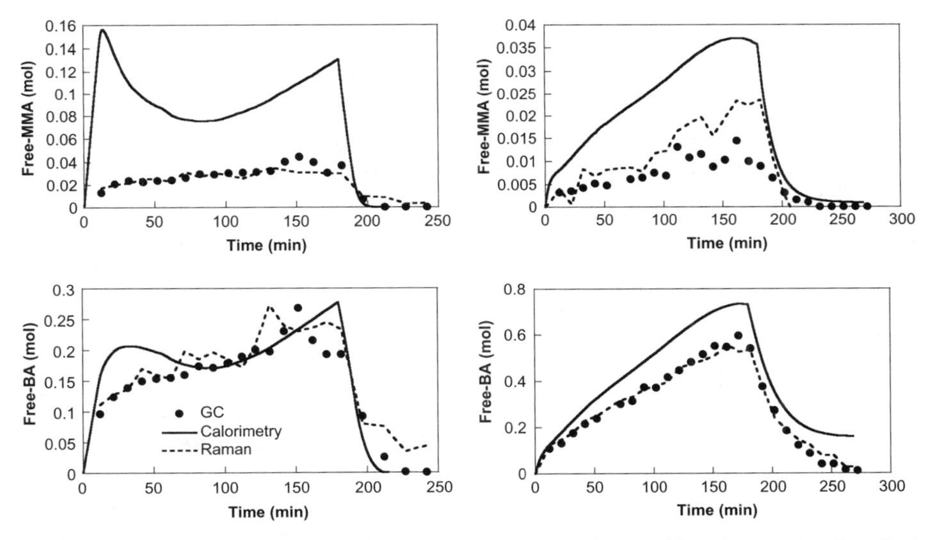

Figure 5.7 Comparison of the instantaneous conversion and free amounts of methyl methacrylate (MMA) and butyl acrylate (BA) as measured by the three techniques for a reaction with a molar feed composition of MMA/BA = 50/50 (left panel) and MMA/BA = 10/90 (right panel). Dots: gas chromatography (GC) data. Continuous line: reaction calorimetry. Dashed line: Raman spectroscopy.

calorimetry. The left panels of Figure 5.7 correspond to a semibatch experiment with feed compositions 50/50, and the right panels to a feed composition 10/90 on MMA/BA. Predictions by Raman spectroscopy (dashed line) are better than for reaction calorimetry in both reactions. Furthermore, the prediction of the BA concentration is more accurate than that of MMA, because of the very low concentration of MMA in this starved-feed reactions and especially for the MMA/BA = 10/90 composition (see right panels of Figure 5.7).

The inaccuracies of reaction calorimetry are inherent to the measurement, because reaction calorimetry directly provides the overall conversion but not the free amounts of each monomer that are inferred with the help of a model [68]. Note that the higher the instantaneous conversion, the higher is the error associated with the estimation of free monomer from reaction calorimetry. This is why the predictions for emulsion polymerizations that proceeded at lower instantaneous conversions (e.g. VAc/BA system) were significantly better [29].

5.3.2.2 Polymer Composition One of the most often studied polymer properties is the composition of different components. Raman spectroscopy has been reported for the prediction of copolymer composition, the determination of the amount of additives, etc. [62,63]. Also it presents important applications in the field of paint analysis [59,64].

Considering how exactly Raman spectroscopy helps to predict the unreacted monomer amounts, even in the case of complex monomer systems under starved conditions, the online monitoring of copolymer composition during polymerization is possible and has been reported in the literature [30,48]. Most of the published works, though, show only the prediction of monomer concentrations, since the copolymer composition can be easily inferred from those values (and material balances of the reactor) without the need to construct a new calibration model. Nevertheless, as can be seen in Figure 5.8, the Raman spectra contain enough information to build calibration models that directly predict the evolution of copolymer composition during emulsion polymerization, independently from the unreacted monomer amounts. Elizalde et al. [30] monitored the evolution of copolymer composition (among other properties) for the BA/MMA systems under industrial conditions (high solids content, starved conditions). Due to the complexity of the system, PLS-R was used to build the calibration models, using data acquired during seeded semibatch BA/MMA emulsion polymerizations, at three different molar compositions: 50/50, 70/30, and 90/10. Details on the polymerization conditions can be found elsewhere [30,48].

In order to check the model performance experimentally, a reaction, totally independent from the calibration data set, was carried out with a MMA/BA = 50/50 molar composition. Figure 5.9 shows the performance of the PLS-R model for the cumulative copolymer composition. The model predicted well the evolution of the cumulative molar copolymer composition (based on n-BA) with slight deviations at the

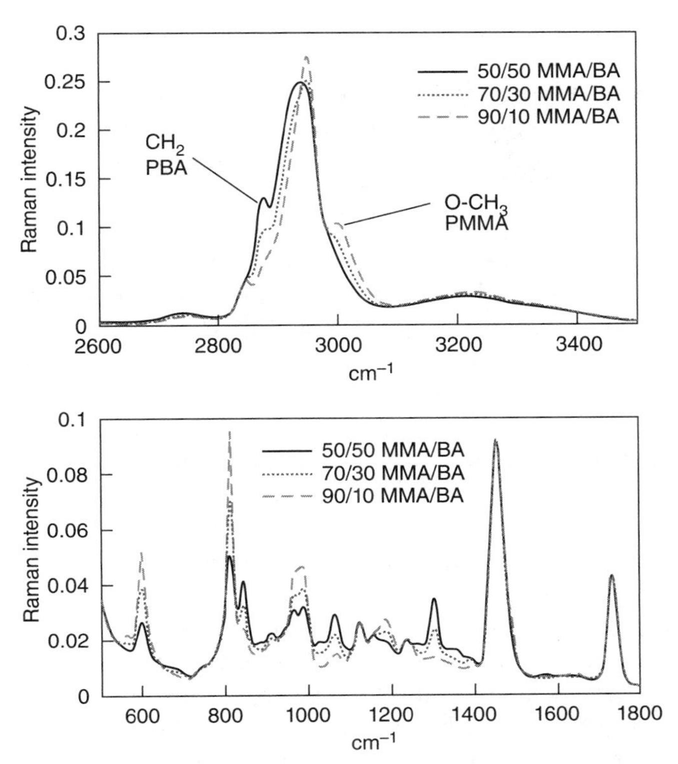

Figure 5.8 Spectra of three latexes having different methyl methacrylate/butyl acrylate (MMA/BA) composition. Top: detail of the region between 3400 and 2600 cm^{-1}. Bottom: detail of the region between 1800 and 500 cm^{-1}.

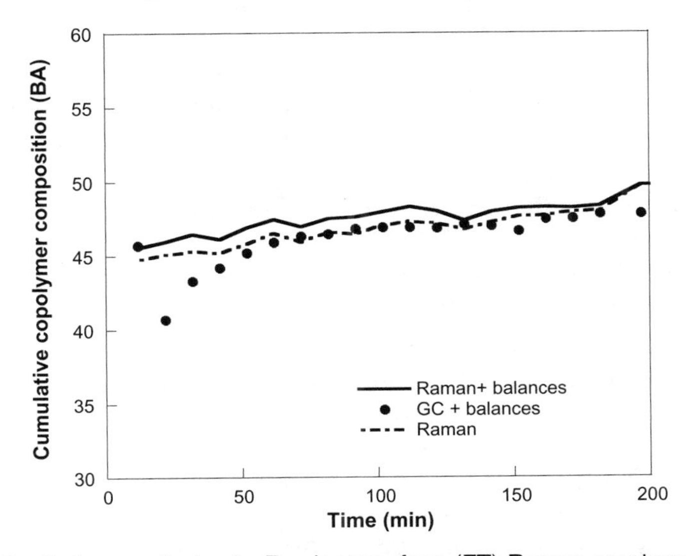

Figure 5.9 Online monitoring by Fourier transform (FT)-Raman spectroscopy of the evolution of the copolymer composition during a semibatch emulsion copolymerization of methyl methacrylate/butyl acrylate (MMA/BA) = 50/50. GC, gas chromatography.

beginning and at the end of the polymerization. For comparison purposes, the prediction based on the monomer concentration (also obtained by Raman) and the material balances of the reactor is included.

5.3.3.3 Particle Size The rheology, stability, and film-forming process of a latex are strongly dependent on its particle size (PS) and PSD. The PSD of a latex is governed by the mechanisms of nucleation, particle growth, and particle aggregation/coagulation. As explained in Section 5.2, particle nucleation in emulsion polymerization occurs by means of homogeneous, micellar (heterogeneous), droplet, and coagulative nucleations. During a polymerization reaction, it is rare that only one of these mechanisms is responsible for the particle nucleation since usually more than one mechanism takes place in parallel (although one is generally dominant). Parameters such as the water solubility of the monomer(s) and the amount and type of stabilizer (emulsifier, protective colloid) are key parameters affecting particle nucleation. Despite the large amount of work already devoted to the study of particle nucleation, there are still many unanswered questions.

Once the particles have been formed, they will grow due to propagation reactions taking place inside the polymer particles. In addition, stability issues during the polymerization reaction can lead to particle agglomeration, which in some cases will massively influence the final PSD. It would be extremely helpful, therefore, to have a tool to monitor the evolution of the PSD online during the reaction. It would, on the one hand, help to answer many open questions related to the particle nucleation stage. On the other hand, it would help to identify early particle agglomeration and to apply closed-loop control strategies for the PSD.

The techniques that are able to perform the online evaluation of PSDs include fiber-optic dynamic light scattering (FODLS), turbidimetry, size fractionation techniques (such as capillary hydrodynamic fractionation chromatography [CHDF] and field-flow fractionation [FFF]), and ultrasound spectroscopic techniques [31]. However, the only technique that can be considered mature is FODLS, which still presents several fully unsolved issues (as multiple scattering in concentrated dispersions).

In an attempt to improve this situation, some authors have taken advantage of the noninvasive measurement provided by the Raman spectroscopy and they have tried to correlate the Raman spectra with the particle size of polymer latexes. The number of works in this direction is still scarce [37,38,65] and there is not an agreement in the existence of a direct relationship of the Raman spectra and the particle size.

Van den Brink et al. [38] showed that for latex with limited scattering properties (small particle sizes, <140 nm, and low solids content), turbidity laws are valid; but with stronger light-scattering latexes (larger particles and higher solids), the turbidity laws fail. However, they found that a correlation of particle size and the focal depth could be made. They correlated the intensity ratio at two different focal depths to particle size and/or particle concentration. They did not estimate the evolution of the particle size during emulsion polymerization. Ito et al. [65] analyzed mixtures of commercial latexes (SBR, PMMA, polyacrylonitrile) varying only the amount of each polymer, keeping always a constant solids content. They performed PLS-R based on particle size measurements obtained by light scattering. According to the calibration model, the important wave numbers correlated with the particle size were mainly due to lattice vibrations in molecular crystals observed at wavelengths below 400 cm^{-1} and others corresponding to the SBR. However, no online monitoring of the particle size during an emulsion polymerization reaction was shown and the calibrated particle size region was rather narrow (137 ± 28 nm). Reis et al. [37] performed analysis with samples acquired during emulsion polymerizations of styrene, which represent real-life samples with varying monomer and polymer amounts, and not only polymer composition and particle size. The authors claim that particle size can be measured, but they did not clearly understand the relationship between the polymer particle size and the Raman spectra. Besides, when the particle size was estimated, other properties such as monomer concentration and solids content were poorly estimated and an explanation for such behavior was not offered. Section 5.5.2 addresses future challenges of Raman spectroscopy to monitor particle size in emulsion polymerization systems.

5.3.4 Monitoring Other Heterogeneous Polymerization Systems

Raman spectroscopy has also been used to monitor other heterogeneous polymerization processes such as those described in Section 5.2. In most of the cases, the evolution of the monomer conversion was attempted. Interestingly, in the suspension polymerization examples that yield dispersions of polymer particles with sizes spanning from few microns to several thousand microns, in addition to the monomer concentration, particle size was correlated with the Raman spectra. On the other hand, monitoring processes that yielded an encapsulated epoxy

Table 5.5 Monitoring other polymerization processes

Process/Goal	Monitored property	Instrument/ Calibration details	Reference no.
Suspension polymerization/PSD has a strong impact on properties and must be controlled during polymerization	Monomer concentration (homopolymerization) Particle size (micron range)	FT-Raman, NIR excitation (1064 nm) Spectral region below 600 cm^{-1}	36
Microemulsion polymerization/Very compartmentalized system, very fast kinetics and difficult to monitor	Monomer concentration (homopolymerization)	Dispersive Raman (514 nm) Univariate	66
Anionic dispersion polymerization	Monomer concentration (block copolymers of styrene-butadiene) Polymer composition (polybutadiene)	FT-Raman, NIR excitation (1064 nm) In-house least-square method	67
Encapsulation/ Production of microcapsules that contain gas, solid, or liquid in different shell materials	Encapsulation efficiency (quantification of encapsulated epoxy resin amount)	FT-Raman, NIR excitation (1064 nm) Univariate (relative peak intensities)	68 and 69

PSD, particle size distribution; FT, Fourier transform; NIR, near-infrared.

resin in a urea-formaldehyde shell have also been reported. Additional details about these processes are given in Table 5.5.

5.4 OTHER APPLICATIONS

5.4.1 Monomer Partitioning

A very important feature in heterogeneous systems is the monomer partitioning between the different phases (water phase, monomer droplets, and polymer particles). The polymerization rate, initiator efficiency, radical transfer, and copolymer composition are directly affected by the monomer concentration in the polymerization loci. Therefore, understanding and measuring monomer partitioning is of great interest in understanding and modeling kinetics of emulsion polymerization.

Several works can be found in the open literature showing theoretical models that predict monomer partitioning and their experimental validation [75]. These experimental validations are usually based on time-consuming separation techniques that cannot be applied for *in situ* monitoring monomer partitioning. The *in situ* monitoring of monomer partitioning based on scattering techniques [73,76], vapor pressure measurements [7], and Raman spectroscopy [77] has been reported by different authors, and especially the last reference by Hergeth and Codella [77] is of much interest since it is useful for quantifying the selective partitioning of monomer mixtures. In this work, Hergeth and Codella [77] measured via Raman spectroscopy the partition of acrylonitrile between polymer particles and the aqueous phase of polybutadiene lattices. This was possible due to the features that differentiate acrylonitrile from other monomers: its relatively high water solubility and the lack of solubility of polyacrylonitrile in its monomer. The use of Raman spectroscopy for such application provides clear advantages since it is a noninvasive technique, and therefore no separation step of the dispersed polymer phase is required, greatly reducing the measurement time.

It is well known from the literature that the frequency of the nitrile (–C≡N) stretching vibration is affected by its molecular environment (such as the solute/solvent interactions). Nyquist [78] observed that in the case of acetonitrile and benzonitrile, the (–C≡N) stretching band shifted to higher wave numbers when increasing the mole fraction of water. The explanation they gave to this shift was the formation of water oligomers and the strong preferential solvation in those systems. Taking into account this work, Hergeth and Codella [77] observed that the stretching vibration of the (–C≡N) bond in acrylonitrile is as well sensitive to its molecular environment. While the C=C band does not shift in the Raman spectrum, the (–C≡N) band is shifted from that observed for pure AN, due to interactions with solvent (in this case water). This shift makes it possible to differentiate between the AN in the polymer and the aqueous phases, and therefore, to monitor AN partitioning *in situ*. Unfortunately, this feature has not been demonstrated for other monomers used in emulsion polymerization and hence Raman spectroscopy cannot be generalized as a technique to determine monomer partitioning.

5.4.2 Polymer Morphology

In order to develop a product with the desired final properties, it is important first to understand the relationship between the final application properties and the polymer architecture, and second, to be able to produce in a controlled and reproducible way the desired polymer architecture. The polymer architecture is defined by a combination of several variables such as the particle size and PSD of a latex, its molecular weight and MWD, the gel content, copolymer composition, and particle morphology, to mention some of the most relevant ones. A critical point in order to achieve the desired architecture at the end of the polymerization process is to be able to monitor the evolution of the aforementioned variables during the reaction. The potential of Raman spectroscopy to monitor the copolymer composition and particle size has already been discussed earlier in this chapter.

This section focuses on the particle morphology, a very important property, since it is directly related to the film morphology and this has a direct effect on the application properties. This is especially interesting in the case of adhesives, where adhesive properties such as tack, resistance to shear and to peel are strongly affected by the particle and film morphology. There are plenty of publications in the open literature where different strategies to achieve a certain final particle morphology are discussed. Among all the works, the majority of them are devoted to the production of core-shell particles, usually achieved by employing a two-stage polymerization process. Typically the core is produced during the first stage, and in the second one, the monomer(s) forming the shell are added into the reactor. Despite the large number of publications describing different methods to yield core-shell particles, the online monitoring and detection of this kind of morphology still remains a challenge. A couple of recent publications relating the particle morphology with spectroscopic data, Raman in one case [79] and NIR on the other [80], are worth mentioning in detail here, since they are the only published works where spectroscopic techniques have been used to detect particle morphology.

Lascelles et al. [79] focused on the production of polypyrrol-coated polystyrene particles with sizes in the μm range and used several techniques to characterize the surface composition and nanomorphology of these core-shell particles. In their work they employed an FT-Raman spectrometer (1064 nm, resolution of 4 cm^{-1} and 2000 scans per measurement) to characterize both uncoated and coated PS particles. The spectrum of the uncoated PS particles matched perfectly the typical spectrum of polystyrene, but when analyzing the polypyrrol-coated particles, the authors achieved a highly surprising and interesting result. Four samples of PS particles coated with different amounts of polypyrrol (1.0%, 4.6%, 6.5%, and 8.9%) gave Raman spectra that were identical to the spectrum of pure polypyrrol and no signals attributable to

PS could be detected. The authors then carried out control experiments where Raman spectra of simple PS and polypyrrol mixtures were taken, and in this case the peaks corresponding to PS were easy to identify. Therefore, they concluded that the core-shell morphology was responsible to the changes observed in the spectra and that the polypyrrol layer must be responsible for the complete attenuation of the signal coming from the PS core (even at very low polypyrrol contents, which resulted in an extremely thin shell of 2 nm). This result is most probably not universal and applicable to any core-shell type of particle, since the observed phenomenon depends on the composition of both core and shell polymers, but it provides a rapid and reliable way to detect the formation of this kind of morphology. Unfortunately, the authors just analyzed the end dispersions and did not follow the spectral changes during the shell formation.

Lenzi et al. [80] reported that the NIR spectra collected during the polymerization process with a probe attached to the reactor window may contain useful information about the structure of core-shell polystyrene beads produced through simultaneous semibatch emulsion/suspension polymerizations. In their work, both core and shell had the same composition (polystyrene) but each of them was formed via a different polymerization technique. The core was formed in the first stage via suspension polymerization and the shell in the second via emulsion polymerization. In this second stage, nanometer-size particles are formed which attach to the much larger core that has a smooth surface, forming a rough and porous shell around it. The appearance of the core-shell structure always led to changes in the NIR spectra (appearance of three bands at 1729, 1765, and 1810 nm) that could not be obtained with polymer suspensions, polymer emulsions, or mixtures of polymer suspensions and emulsions. Due to the complexity of the system, the authors were not able to assign the three new peaks, but their results indicate that NIR spectra can be useful to detect the modification of the particle surface. After the spectral changes attributed to the particle morphology were detected, the Raman signal remained constant, regardless of the thickness of the shell.

5.4.3 Physical Properties: Latex Viscosity

As already shown in previous sections, Raman spectroscopy is a powerful tool to acquire chemical information about polymer dispersions and to monitor the polymerization process. Unfortunately, very little has been published relating Raman spectra and physical properties of dispersions. An interesting study by Ito et al. [81] shows how Raman spectroscopy coupled with multivariate calibration techniques (PLS-R) can be used to predict the viscosity of waterborne automotive paints. The authors studied automotive paint systems based on mixtures of water-soluble acrylic resins, styrene-acrylic lattices, and melamine formaldehyde resins. The authors measured the viscosity of different formulations (mixtures) at 22 °C using a conventional viscosimeter and acquired as well the Raman spectra of the formulations at the same temperature. Afterward, PLS-R models were built to correlate the measured viscosity data with the acquired Raman spectra (as depicted in Figure 5.4). According to Ito et al. [81], the resulting PLS-R models were able to properly predict viscosity from Raman data (with a correlation coefficient > 0.935).

In our opinion, in this particular example, the authors are using the PLS-R model as some sort of "black box" approach. It is clear that different formulations will result in different viscosities, since they have a different chemical composition and different amounts of water-soluble resin. As a consequence, the Raman spectra will vary as well. But the changes in the spectra will be caused by the difference in the chemical composition, and not directly by the viscosity. This could be confirmed by, for example, measuring the spectra of two samples of same chemical composition but different viscosity (e.g. by measuring the viscosity of the same sample at a higher temperature): no changes in the Raman spectra should be observed.

But despite the aforementioned points, we think that the approach described by Ito et al. is a smart and fast way to obtain the viscosity of mixtures from their Raman spectra (since spectra acquisition takes a short time). This approach of course will only be valid if the effect of temperature is not considered, namely, when the viscosity measurements are always carried out under the same conditions. This is the case in industrial environments, where an approach of this type could be extremely interesting to reduce the measurement time. This work shows the potential of Raman spectroscopy to simultaneously monitor chemical and physical information of formulations of high industrial interest. This is of great importance in order to establish quantitative relationships between the chemical/physical properties and final application properties that will lead to high-performance products.

5.4.4 Monitoring Film Formation and Characterization of Films

Film formation from aqueous latex dispersions is very complex. The current understanding of the film formation process can be described by a mechanism that includes four consecutive stages (see schematic representation in Figure 5.10): (1) the concentration of the latex dispersion; (2) particle contact; (3) particle deformation; and (4) the interdiffusion of the polymer chains across the particle boundaries, which is necessary to form a mechanically stable film [82,83]. End-use properties of the latex (mechanical and optical properties to mention a few) are strongly influenced by the film formation process. One of the main concerns during film formation is the stratification of the polymer dispersions. For instance, it has been shown that the mobility of low-molar-mass species, in particular surfactant molecules, may be affected by latex glass transition temperature (free volume of the polymer matrix), surface tension at the film–air and film–substrate interfaces, compatibility, and coalescence times. To gain insight into the film formation and develop models for the process that would help in producing better coatings and paints, different experimental techniques have been used. These include small-angle neutron scattering (SANS), attenuated total reflectance Fourier transform infrared (ATR-FTIR), photoacoustic measurements, high-resolution cryogenic scanning electro microscopy, and nuclear magnetic resonance, among others. Film drying is conventionally measured by gravimetric techniques, although spectroscopic techniques have also been employed.

In the last years, confocal Raman spectroscopy has emerged as an accurate analytical technique for quantitative monitoring film formation in waterborne latexes prepared by emulsion polymerization or by other means. The main advantage of the Raman spectroscopic technique with respect to the ATR-FTIR is that Raman allows measuring

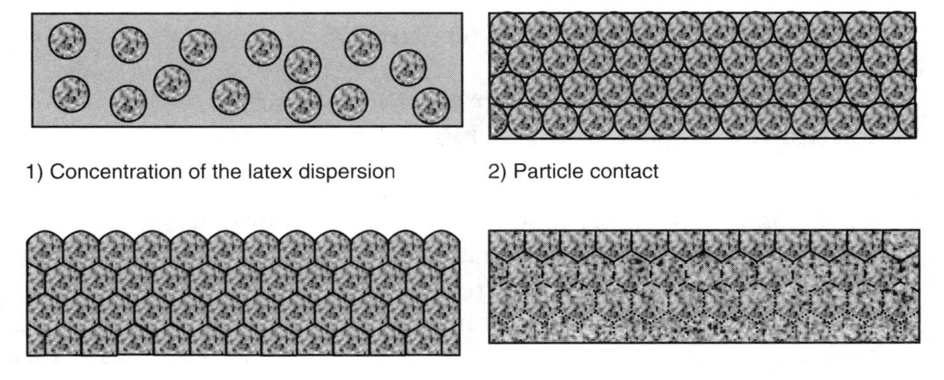

Figure 5.10 Schematic representation of the different stages during the film-forming process.

drying because the water peaks (or peaks due to interactions of water with other molecules) can be accurately monitored in the wet state of the film, something which is not straightforward with infrared techniques. Thus, confocal Raman spectroscopy has been applied to measure the distribution (depth profiling) of water-soluble (initiator moieties such as sulfates) and surface-active species (surfactants) in acrylic latex systems [84,85] and to analyze phase separation and surfactant stratification in styrene/butyl acrylate copolymer and latex blend films [86]. More recently, an excellent work on drying and film formation of latexes in both vertical and horizontal directions has been presented by Ludwig et al. [87] using an inverse micro-Raman spectroscopy by combining an inverse microscope with a confocal Raman spectrometer. The equipment allows scanning a film of 150 μm in 30 seconds at slices of 5 μm and the same scans can be performed horizontally starting from the edge to the center at every 3 mm. In the same way, oxidation of waterborne alkyd resins [88] during film formation was followed by tracking the oxidation reaction of the double bonds and the subsequent reactions of the radicals formed. It was observed for model compounds (methyl oleate, ethyl linoleate, and methyl linoleate) that oxidation proceeded first converting the *cis* C=C to *trans* C=C and then to conjugated double bonds, and upon reaching a maximum level, the double bonds tend to disappear by addition of radicals to these double bonds. The confocal Raman spectroscopy has shown to be a complementary tool to ATR-FTIR and atomic force microscopy in understanding drying and film formation processes.

5.5 FUTURE CHALLENGES AND TRENDS

5.5.1 Challenges of the Implementation of Raman Spectroscopy in Industrial Polymerization Reactors

As it has been shown throughout the chapter, Raman spectroscopy is a robust, noninvasive, and accurate sensor to monitor emulsion polymerization processes and polymer latex characteristics. However, most of the applications shown earlier are related to monitoring and control of laboratory reactors. It is nevertheless desirable to implement this technique in large-scale industrial polymerization plants. There are already several emulsion polymer-producing companies that use Raman probes in polymerization reactors. The implementation of Raman spectroscopy in harsh environments, such as those of emulsion polymer plants, has been possible because of the possibility of applying fiber-optic technology to transfer the signal to the reactor and back to the detector. This means that the most sensitive part of the Raman equipment (the laser source and the detector) can be installed in the control room in a less demanding environment. In any case, there are a number of aspects that must be considered in the implementation of advanced hardware instruments in industrial environment.

5.5.1.1 Safety An important issue of implementing Raman spectroscopy in industrial reactors is the use of a laser source in a dispersion plant. Two main risks are associated with the use of lasers in working environments: (1) exposure to direct and indirect light from a laser can cause irreversible damage to the eye and also severe burning; (2) the laser, at sufficiently high power, can set fire in an explosive atmosphere (it has been shown for styrene and butadiene that the power limit is of about 200 mW [89]). Fortunately, the laser power typically measured out of a Raman probe is 100 mW (for an FT-Raman equipment with a NIR laser [1064 nm] and a power of 1.5 W) and furthermore, beam attenuation derivation shows that 8 cm of water is enough to absorb 80% of the beam power and hence significantly reduce the risk of the laser light when implemented in industrial emulsion polymerization reactors. Nevertheless, very stringent safety procedures must be followed to implement Raman spectroscopy in a safe manner in an industrial environment. Typically, procedures are necessary to avoid eye contact with the laser when the reactor is empty and installing the optical fibers into nitrogen-purged steel tubes through the plant to avoid light leakage to the environment.

5.5.1.2 Selection of the Excitation Wavelength of the Laser Source A balance between emission intensity (sensitivity) and fluorescence in technical systems which are seldom very clean, and therefore fluoresce, is necessary. As discussed in other chapters and in Section 5.2, fluorescence can be avoided by using excitation in the NIR region, but this implies the use of FT technology and more expensive detectors (e.g. liquid N_2 cooled Ge detectors) to enhance the sensitivity. Also, self-absorption of Raman-emitted light may complicate quantification and requires a constant distance between the probe and the analyzed material or constant focus depth of the light-collecting lenses in the reaction mixture, especially in transparent mixtures.

If excitation wavelengths in the visible range are selected, one has to cope with strong fluorescence, which requires spectra pretreatment. From an industrial point of view, the decision on which laser is chosen is generally made based on availability and cost criteria (i.e. replacement costs and lifetime). Sometimes it is better to install a HeNe laser at 633 nm (which is cheap and long lasting) instead of a solid-state laser at 785 nm (or 835 nm), even if the 785-nm laser would be better in terms of reducing the effect of fluorescence. Ease of maintenance, availability of spare parts, and 24-hour support by the manufacturer are very important issues to be considered for the installation of advanced instruments in polymerization plants.

5.5.1.3 Requirements of the Probe The following points must be considered: (1) in turbid mixtures, a confocal probe is a prerequisite for getting spectra (focus of the lenses should be between 0.5 and 1 mm in front of the window); (2) the probe material and the housing of the probe have to withstand the reaction ingredients and the reactor pressure and temperature for a long time; (3) the probe must be free of any electrical contacts; and (4) the probe (i.e. specially the light path within the probe) has to withstand the vibrations of an industrial reactor.

5.5.1.4 Position of the Probe in the Reactor Stirred-tank reactors are typically used to produce emulsion polymers. Lab-scale reactors and pilot-scale stirred-tank reactors can be safely assumed to be perfectly mixed reactors and hence monitoring in any location of the reactor should be representative of the state of the reactor. However, in large-scale reactors and with high solids content formulations, this is not necessarily true, as has been demonstrated by computational fluid dynamic simulations of the mixing behavior of viscous polymerization reactors [90]. Therefore, the location of the Raman probe in a large reactor is not straightforward. Figure 5.11 shows two potential locations based on common openings of industrial reactors. The first location is near the impeller, using an opening on top of the reactor and protecting tubing. This location will ensure a representative analysis of the whole reactor content. Potential difficulties with this setup include (1) difficult maintenance, as accurately adjusting the position of the probe is difficult; (2) possibility of fouling due the presence of a protruding device; and (3) high sensitivity to mechanical vibrations due to the length of the tubing and the proximity to the impeller.

The second location is at the bottom of the reactor. The advantages of this location are that it is easily accessible and hence maintenance is possible, and that it is possible to build a device smooth enough to avoid or minimize clogging. The main disadvantage of this setup is the uncertainty of whether the latex analyzed at the bottom of the reactor is representative of the whole content of the reactor. Computational fluid dynamic calculations performed in a polymerization reactor

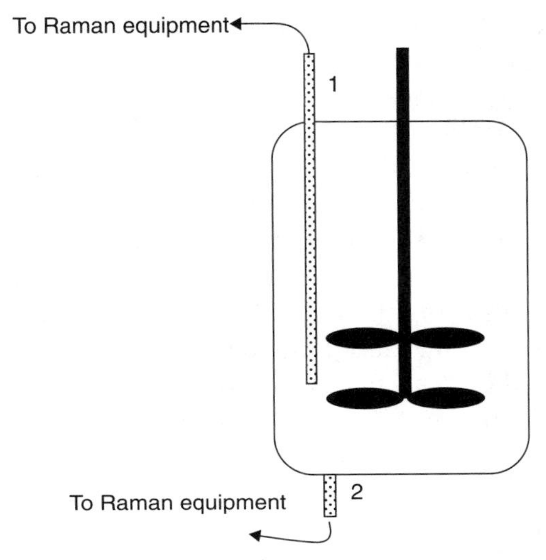

Figure 5.11 Possible locations of Raman probes in an industrial polymerization reactor.

similar to those employed in industry indicated that this position of the probe is well mixed and hence this location is well suited [89]. To our knowledge, most of the companies currently monitoring emulsion polymerization reactors by means of Raman spectroscopy have chosen the second alternative.

5.5.1.5 Fouling and Calibration Maintenance In lab reactors, fouling is not an issue because upon finishing a reaction the reactor is thoroughly cleaned, as are the different inserts of the reactor (impeller and temperature probes, as well as Raman or other spectroscopic probes if installed), and hence cleanliness is guaranteed and it would not affect spectroscopic measurements or calibrations models used to determine the properties of the latex. In industry, polymerization reactors are not cleaned after each reaction (in some cases, high-pressure water cleaning is performed between batches), and hence reactor walls can be fouled. In addition, any insert or protruding device in the reactor is prone to suffer polymer buildup. This polymer buildup in the wall causes a decrease of the overall heat transfer coefficient that might lead to a significant increase of batch time during a campaign. Besides, if a polymer layer builds up in the Raman probe, this might jeopardize the accuracy of the Raman analysis or might affect the calibration models developed to determine latex properties.

As it has been shown (see Section 5.3), PLS-R calibration models provide a good prediction of the properties of interest (solids content, unreacted monomer concentrations, and polymer composition) in the range of variation of the different variables used in the calibration set. These variables include those not directly related to the variables included in the calibration model, such as stirring speed and temperature. In addition, there are other factors related to the spectrometer (e.g. maintenance and replacement of the laser power, the optical fibers, the probe head, and alignment of the spectrometer and filters) that might make a calibration model fail in correctly predicting the property of the latex. It is therefore necessary to implement calibration maintenance strategies to ensure good performance of the spectroscopic equipment for a long time.

There are several multivariate calibration tools described in the literature that can help in the construction of more robust models. They are known as *multivariate calibration standardization methods* [91] and can be divided into two categories:

(1) Methods that help to improve the robustness of the calibration model by data preprocessing (*variable selection*), the incorporation of measurement conditions into the calibration model (*global calibration models*), and/or the application of robust multivariate calibration techniques such as *variable selection partial least squares* (VS-PLS) [92].
(2) Methods to adapt the calibration model by transforming the measured spectra, the regression parameters of the model, or the predictions by the calibration model [93].

The *global calibration models* are the ones most often used in the literature, and the references stated previously are mainly based on such models. To use these methods, the calibration samples should be measured under the previously existing conditions and under changed or new conditions. Then, all these measurements are combined into the same data set, which means that in addition to the spectral variation caused by the component concentrations, external spectral variation caused, for example, by the new measurement/instrumental conditions is also included. The objective is to try to model the effect of external spectral variation by including it in the model. It is assumed that these new sources of external variation can be modeled by additional PLS factors [94].

The drawbacks of these methods are that they require measuring the same data under different conditions that could cause mismatches between the prediction of the property and the real value. In industrial practice (and in the lab too), this is not always possible because spectra acquired during a set of polymerization reactions are used to build the calibration models. Therefore, the data cannot be analyzed again. There are few alternatives to this calibration maintenance problem. Elizalde et al. [57] compared two pragmatic alternatives: (1) building calibration models based on spectra from new experiments (in their case, a change in the probe head was made and the spectrometer was realigned, which resulted in spectral changes and mismatches in property predictions; note that other maintenance in the equipment as well as the effect of fouling on the calibration models might cause similar mismatches in the property predictions); and (2) building PLS models for the differences between the off-line data (real values) and the predictions provided by the old models. The new prediction is given by the sum of the old model plus that of the PLS model that accounts for the old model mismatch. Figure 5.12 presents the predictions of the different calibration models for the solids content and the concentration of MMA and BA in a semibatch emulsion polymerization carried out after the Raman probe was replaced in the equipment. The figure shows the prediction of the previous/old model (before the probe was changed), the prediction using a calibration model based on only two new experiments carried out with the new probe, and the predictions of the old model plus a new model for the differences (which predicts the difference between the off-line data and the old model). These results showed that the second alternative (modeling the difference and adding it to the prediction of the old model) was faster in resetting the quality of the calibrations to conditions similar to those before the instrument was repaired when a small number of new experiments is available. Notwithstanding, the authors pointed out that as more reactions are available under the new conditions, developing calibration models based only on the new data should be considered.

5.5.1.6 Other Potential Advantages of Raman Spectroscopy As mentioned in Section 5.5.1.4, the location of the probe is challenging in emulsion polymerization reactors due to the lack of perfect mixing in the large reactors employed in industrial plants. Although in most cases the bottom location might provide reasonable results, under some circumstances it might be necessary to address other alternatives. In polymerizations where low-vapor-pressure monomers (e.g. butadiene and ethylene) are employed, it is likely that an important amount of these monomers remains on the gas phase (headspace) or might form a layer on top of the latex. In those cases, using a single Raman probe

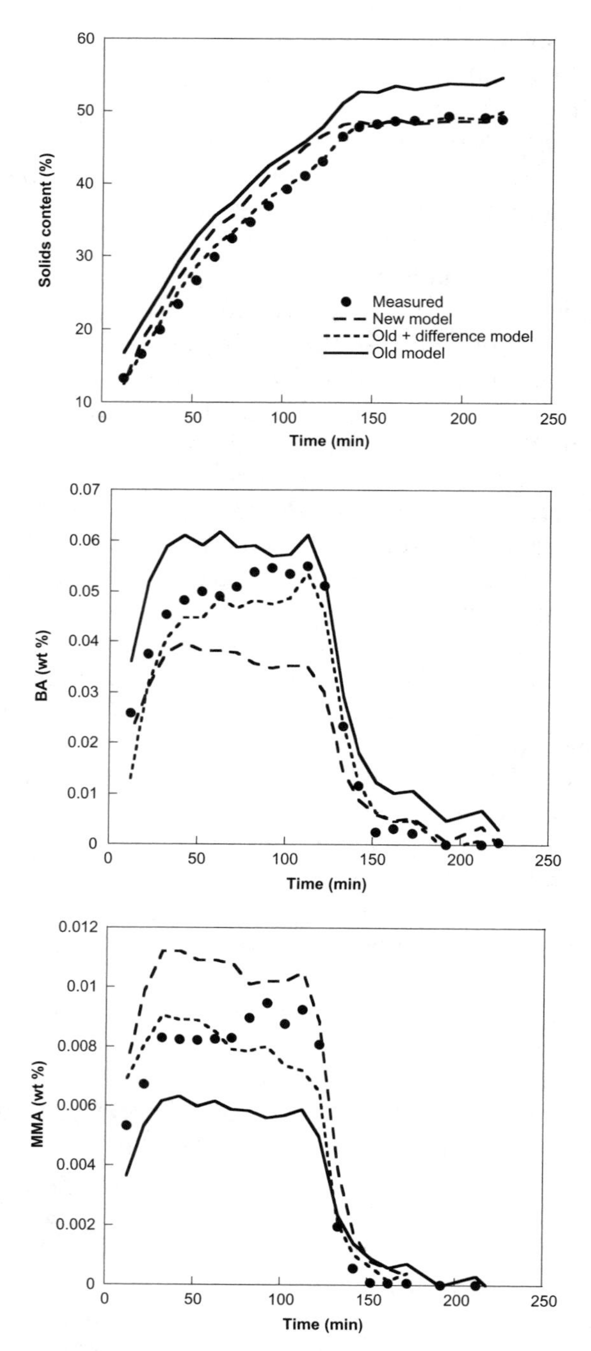

Figure 5.12 Prediction of the solids content and unreacted *n*-butyl acrylate (BA) and methyl methacrylate (MMA) amounts for an experiment using different calibration models (see text for details).

in the liquid phase (latex) might not provide an accurate picture of the state of the reactor, and although pressure of the reactor (which is always monitored) might provide a rough estimate of the concentration of the monomers in the gas phase, more accurate information might be obtained if the headspace of the reactor is also monitored by a Raman probe. The implementation of a Raman probe in the headspace is technically feasible and easier than in the latex phase (no fouling problems from a maintenance point of view).

Another potential advantage of using a Raman probe to monitor and control industrial polymerization reactors is the sensitivity of the Raman spectra to changes in stability of the latex. Industrial formulations are very robust in terms of stability, because a massive coagulation in a 50-m^3 reactor would have a significant economic loss and hence must be avoided. It has been shown that the Raman spectra are sensitive to the particle size and in general to the state of the dispersion [37,38,65]. Therefore, information about the colloidal stability of the latex is somehow included in the spectra. Therefore, monitoring of industrial emulsion polymerization reactors by Raman spectroscopy in addition to the main benefits described in this chapter might offer an early indication of the stability of the polymeric dispersion and hence allow for a shutdown of the reactor before a catastrophic incident occurs in the polymerization.

However, it should be kept in mind that all the useful information contained in the Raman spectra must be previously extracted and presented to the operator in a clear and simplified way (e.g. red and green light or simple graphs, but no Raman spectra) in order to be accepted in the plant.

5.5.2 Challenges in Particle Size Prediction

As discussed in Section 5.3, some authors claim that particle size can be determined from the spectra collected during emulsion polymerization and appropriate calibration models [37,38,65]. However, only one of the works showed the online monitoring of the particle size during the course of the polymerization. In that work, the authors claim that particle size can be measured, but they did not clearly understand the relationship between the polymer particle size and the Raman spectra. Besides, when the particle size was estimated, other properties such as monomer concentration and solids content were poorly estimated and an explanation for such behavior was not offered. Further research in this area is required to demonstrate that there exists a relationship between the particle size of a polymeric dispersion and the Raman spectra and that this relationship is strong enough to build a robust and accurate calibration model for the measurement of the particle size during an emulsion polymerization.

One of the drawbacks of the approaches discussed here and in Section 5.3 to determine the particle size of polymer latexes based on Raman spectra is that for each system, a calibration model must be built. In other words, a calibration model built from data of polystyrene latex standards to predict the particle size will not correctly predict the particle size of the latexes made of a different monomer(s) system.

Another approach that has not been exploited by the Raman spectroscopy manufacturers is to use the elastically scattered light, so-called Rayleigh scattering, which is filtered out in all the equipment to get the inelastically scattered light and analyze the Raman effect. Analysis of the elastically scattered light (which is stronger than the Raman effect) should be similar to the analysis carried out in dynamic (quasi-elastic) light-scattering equipment to measure the particle size. Dynamic light scattering is a well-established technique to measure particle size in heterogeneous systems without the requirement of any calibration model [95–97]. The main difficulty of this technique is to measure the particle size in concentrated dispersions (as the emulsion polymerization reactors that typically reach 50% at the end of the polymerization). Two problems need to be solved for concentrated dispersions: the effect of multiple scattering and particle interaction [98]. Currently, there is commercial equipment that can measure the particle size in concentrated dispersions where the aforementioned problems have been minimized.

We believe that Raman spectroscopic equipment could be adapted to analyze, in addition to the inelastically scattered light, the elastically scattered light that would provide a significantly more robust and independent information about the particle size. Furthermore, both sets of data, the one coming from the Raman spectra and that of the scattered light, could be used together to obtain a more accurate information of the emulsion polymerization processes.

5.5.3 Monitoring Polymerizations Carried Out Under Supercritical Carbon Dioxide (sc-CO_2)

Raman spectroscopy might also become an important analytical tool to monitor noninvasively polymerizations carried out under sc-CO_2 conditions; sc-CO_2 has shown advantages over other green solvents (ionic liquids, fluorous phases, and sc-H_2O) because of the relatively low temperatures and pressures required to achieve the supercritical conditions, the chemical inertness, and the highly tunable solvent behavior [99]. The commercialization by Dupont of the production of Teflon® in sc-CO_2 is an example. Both polymerizations in homogeneous phase (fluorinated acrylates) and in heterogeneous phase (most of the common high-molecular-weight polymers are not soluble in sc-CO_2) can be carried under sc-CO_2 conditions.

So far Raman spectroscopy has been used to monitor sc-CO_2 density changes. Blatchford and Wallen [100] reported the successful monitoring of the density evolution in pure sc-CO_2 conditions by integrating Raman spectroscopic peaks. Kessler et al. [101,102] studied the decomposition of several peroxide initiators at high pressures using Raman spectroscopy. After having identified the characteristic wave number of the O–O stretching vibrations of the peroxide, they have monitored the latter's intensity during the decomposition reaction. Such peroxides (butyl peroxypivalate, di-tert-butyl peroxide, and 2,2-bis(tert-butylperoxy)butane)) are important to the polymer industry since they are currently being used as free-radical initiators for the polymerization of olefins.

REFERENCES

1. Asua, J.M. (2004). Emulsion polymerization: From fundamental mechanisms to process developments. *J. Polym. Sci. A Polym. Chem.*, 42, 1025–1041.
2. Daniel, J.C. (2006). Le grand familles de produits et de procédes insutriels. In: Daniel JC, Pichot C, eds. *Les Latex Synthétiques: Élaboration, Propriétés, Applications*. Paris: Lavoisier, 319–329.
3. Schmidt-Thummes, J., Schwarzenback, E., Lee, D.I. (2002). In: Urban D, Takamura K, eds. *Polymer Dispersions and their Industrial Applications*. Weinheim: Wiley-VCH, 75–101.
4. Harkins, W.D. (1947). A general theory of the mechanism of emulsion polymerization. *J. Am. Chem. Soc.*, 69, 1428.
5. Priest, W.J. (1952). Particle growth in the aqueous polymerization of vinyl acetate. *J. Phys. Chem.*, 56, 1077–1082.
6. Feeney, P.J., Napper, D.H., Gilbert, R.G. (1987). Surfactant-free emulsion polymerizations: Predictions of the coagulative nucleation theory. *Macromolecules*, 20, 2922–2930.
7. Fitch, R.M., Tsai, C.H. (1971). In: Fitch RM, ed. *Polymer Colloids*. New York: Plenum, 73–102.
8. Ugelstad, J., Hansen, F.K. (1976). Kinetics and mechanism of emulsion polymerization. *Rubber Chem. Technol.*, 49, 536–609.

9. Candau, F. (1997). Inverse emulsion and microemulsion polymerization. In: Lovell PA, El-Aasser MS, eds. *Emulsion Polymerization and Emulsion Polymers*. New York: John Wiley & Sons, Inc., 723–742.
10. Asua, J.M. (2002). Miniemulsion polymerization. *Prog. Polym. Sci.*, 27, 1283–1346.
11. Ugelstad, J., Elaasser, M.S., Vanderhoff, J.W. (1973). Emulsion polymerization: Initiation of polymerization in monomer droplets. *J. Polym. Sci. C Polym. Lett.*, 11, 503–513.
12. El-Aasser, M.S., Miller, C.M. (1997). Preparation of latexes using miniemulsions. In: Asua, J.M., ed., *Polymeric Dispersions: Principles and Applications*. Dordrecht, The Netherlands: Kluwer Academic Publishers, 109–126.
13. Candau, F. (1997). Microemulsion polymerization. In: Asua, J.M., ed., *Polymeric Dispersions: Principles and Applications*. Dordrecht, The Netherlands: Kluwer Academic Publishers, 127–140.
14. Barrett, K.E.J. (1974). *Imperial Chemical Industries, Ltd., Paints Division, Dispersion Polymerization in Organic Media*. London: John Wiley & Sons, Ltd.
15. Sudol, E.D. (1997). Dispersion polymerization. In: Asua, J.M., ed., *Polymeric Dispersions: Principles and Applications*. Dordrecht, The Netherlands: Kluwer Academic Publishers, 141–154.
16. VivaldoLima, E., Wood, P.E., Hamielec, A.E., Penlidis, A. (1997). An updated review on suspension polymerization. *Ind. Eng. Chem. Res.*, 36, 939–965.
17. Yuan, H.G., Kalfas, G., Ray, W.H. (1991). Suspension polymerization. *J. Macromol. Sci. Rev. Macromol. Chem. Phys.*, C31, 215–299.
18. Kotoulas, C., Kiparissides, C. (2007). Suspension polymerization. In: Asua, J.M., ed., *Polymer Reaction Engineering*. Oxford: Blackwell Publishing, 209–232.
19. Chatzi, E.G., Kammona, O., Kiparissides, C. (1997). Use of a midrange infrared optical-fiber probe for the on-line monitoring of 2-ethylhexyl acrylate/styrene emulsion copolymerization. *J. Appl. Polym. Sci.*, 63, 799–809.
20. Reis, M.M., Araujo, P.H.H., Sayer, C., Giudici, R. (2004). Comparing near infrared and Raman spectroscopy for on-line monitoring of emulsion copolymerization reactions. *Macromol. Symp.*, 206, 165–178.
21. Vieira, R.A.M., Sayer, C., Lima, E.L., Pinto, J.C. (2002). In-line and in situ monitoring of semi-batch emulsion copolymerizations using near-infrared spectroscopy. *J. Appl. Polym. Sci.*, 84, 2670–2682.
22. Wu, C.C., Danielsen, J.D.S., Callis, J.B., Eaton, M., Ricker, N.L. (1996). Remote in-line monitoring of emulsion polymerization of styrene by short-wavelength near-infrared spectroscopy. 1. Performance during normal runs. *Process Control Qual.*, 8, 1–23.

23. Bauer, C., Amram, B., Agnely, M., Charmot, D., Sawatzki, J., Dupuy, N., Huvenne, J.P. (2000). On-line monitoring of a latex emulsion polymerization by fiber-optic FT-Raman spectroscopy. Part I: Calibration. *Appl. Spectrosc.*, 54, 528–535.
24. Clarkson, J., Mason, S.M., Williams, K.P.J. (1991). Bulk radical homopolymerization studies of commercial acrylate monomers using near-infrared Fourier-transform Raman spectroscopy. *Spectrochim. Acta A Mol. Biomol. Spectrosc.*, 47, 1345–1351.
25. Damoun, S., Papin, R., Ripault, G., Rousseau, M., Rabadeux, J.C., Durand, D. (1992). Radical polymerization of methyl-methacrylate in solution monitored and studied by Raman spectroscopy. *J. Raman Spectrosc.*, 23, 385–389.
26. Gulari, E., Mckeigue, K., Ng, K.Y.S. (1984). Raman and FTIR spectroscopy of polymerization: Bulk polymerization of methyl-methacrylate and styrene. *Macromolecules*, 17, 1822–1825.
27. Reis, M.M., Araujo, P.H.H., Sayer, C., Giudici, R. (2004). Development of calibration models for estimation of monomer concentration by Raman spectroscopy during emulsion polymerization: Facing the medium heterogeneity. *J. Appl. Polym. Sci.*, 93, 1136–1150.
28. Van den Brink, M., Pepers, M., van Herk, A.M., German, A.L. (2001). On-line monitoring and composition control of the emulsion copolymerization of VeoVa 9 and butyl acrylate by Raman spectroscopy. *Polym. React. Eng.*, 9, 101–133.
29. Elizalde, O., Azpeitia, M., Reis, M.M., Asua, J.M., Leiza, J.R. (2005). Monitoring emulsion polymerization reactors: Calorimetry versus Raman spectroscopy. *Ind. Eng. Chem. Res.*, 44, 7200–7207.
30. Elizalde, O., Asua, J.M., Leiza, J.R. (2005). Monitoring of high solids content starved-semi-batch emulsion copolymerization reactions by Fourier transform Raman spectroscopy. *Appl. Spectrosc.*, 59, 1270–1279.
31. Leiza, J.R., Pinto, J.C. (2007). Control of polymerization reactors. In: Asua, J.M., ed., *Polymer Reaction Engineering*. Oxford: Blackwell Publishing, 315–361.
32. Hergeth, W.D. (1997). Optical spectroscopy on polymeric dispersions. In: Asua, J.M., ed., *Polymeric Dispersions: Principles and Applications*. Dordrecht, The Netherlands, Kluwer Academic Publishers, 243–256.
33. Hergeth, W.D. (1998). Raman scattering on polymeric dispersions. *Chem. Eng. Technol.*, 21, 647–651.
34. Van den Brink, M., van Herk, A.M., German, A.L. (1999). On-line monitoring and control of the solution polymerization of n-butyl acrylate in dioxane by Raman spectroscopy. *Process Control Qual.*, 11, 265–275.
35. Van den Brink, M., Hansen, J.F., De Peinder, P., van Herk, A.M., German, A.L. (2001). Measurement of partial conversions during the solution

copolymerization of styrene and butyl acrylate using on-line Raman spectroscopy. *J. Appl. Polym. Sci.*, 79, 426–436.

36. Santos, J.C., Reis, M.M., Machado, R.A.F., Bolzan, A., Sayer, C., Giudici, R., Araujo, P.H.H. (2004). Online monitoring of suspension polymerization reactions using Raman spectroscopy. *Ind. Eng. Chem. Res.*, 43, 7282–7289.
37. Reis, M.M., Araujo, P.H.H., Sayer, C., Giudici, R. (2003). Evidences of correlation between polymer particle size and Raman scattering. *Polymer*, 44, 6123–6128.
38. Van den Brink, M., Pepers, M., van Herk, A.M. (2002). Raman spectroscopy of polymer latexes. *J. Raman Spectrosc.*, 33, 264–272.
39. Brereton, R.G. (1990). *Chemometrics Applications of Mathematics and Statistics to Laboratory Systems*. New York: Ellis Horwood.
40. Massart, D.L. (1988). *Chemometrics: A Textbook*. Amsterdam, The Netherlands: Elsevier.
41. Sharaf, M.A., Illman, D.L., Kowalski, B.R. (1986). *Chemometrics*. New York: Wiley, 1986.
42. Mark, H. (1991). *Principles and Practice of Spectroscopic Calibration*. New York: John Wiley & Sons, Inc.
43. Martens, H., Ns, T. (1989). *Multivariate Calibration*. Chichester, England: John Wiley & Sons, Ltd.
44. Beebe, K.R., Kowalski, B.R. (1987). An introduction to multivariate calibration and analysis. *Anal. Chem.*, 59, A1007.
45. Haaland, D.M., Thomas, E.V. (1988). Partial least-squares methods for spectral analyses. 1. Relation to other quantitative calibration methods and the extraction of qualitative information. *Anal. Chem.*, 60, 1193–1202.
46. Geladi, P., Kowalski, B.R. (1986). Partial least-squares regression: A tutorial. *Anal. Chim. Acta*, 185, 1–17.
47. Svensson, O., Josefson, M., Langkilde, F.W. (1999). Reaction monitoring using Raman spectroscopy and chemometrics. *Chemometrics Intelligent Lab. Syst.*, 49, 49–66.
48. Elizalde, O., Leiza, J.R., Asua, J.M. (2004). On-line monitoring of all-acrylic emulsion polymerization reactors by Raman spectroscopy. *Macromol. Symp.*, 206, 135–148.
49. Reis, M.M., Araujo, P.H.H., Sayer, C., Giudici, R. (2007). Spectroscopic on-line monitoring of reactions in dispersed medium: Chemometric challenges. *Anal. Chim. Acta*, 595, 257–265.
50. Wang, C., Vickers, T.J., Mann, C.K. (1993). Use of water as an internal standard in the direct monitoring of emulsion polymerization by fiberoptic Raman spectroscopy. *Appl. Spectrosc.*, 47, 928–932.
51. McCaffery, T.R., Durant, Y.G. (2003). Monitoring of seeded batch, semibatch, and second stage emulsion polymerization by low resolution Raman spectroscopy. *Polym. React. Eng.*, 11, 507–518.

52. Reis, M.M., Uliana, M., Sayer, C., Araujo, P.H.H., Giudici, R. (2005). Monitoring emulsion homopolymerization reactions using FT-Raman spectroscopy. *Braz. J. Chem. Eng.*, 22, 61–74.
53. Al-Khanbashi, A., Dhamdhere, M., Hansen, M. (1998). Application of in-line fiber-optic Raman spectroscopy to monitoring emulsion polymerization reactions. *Appl. Spectrosc. Rev.*, 33, 115–131.
54. Brookes, A., Dyke, J.M., Hendra, P.J., Strawn, A. (1997). The investigation of polymerisation reactions in situ using FT-Raman spectroscopy. *Spectrochim. Acta A Mol. Biomol. Spectrosc.*, 53, 2303–2311.
55. Ozpozan, T., Schrader, B., Keller, S. (1997). Monitoring of the polymerization of vinylacetate by near IR FT Raman spectroscopy. *Spectrochim. Acta A Mol. Biomol. Spectrosc.*, 53, 1–7.
56. Pepers, M. (2004). Online monitoring and control of copolymerizations. PhD Thesis, Eindhoven University of Technology.
57. Elizalde, O., Asua, J.M., Leiza, J.R. (2005). Monitoring of emulsion polymerization reactors by Raman spectroscopy: Calibration model maintenance. *Appl. Spectrosc.*, 59, 1280–1285.
58. Van den Brink, M., Pepers, M., van Herk, A.M., German, A.L. (2000). Emulsion (co) polymerization of styrene and butyl acrylate monitored by on-line Raman spectroscopy. *Macromol. Symp.*, 150, 121–126.
59. Claybourn, M., Agbenyega, J.K., Hendra, P.J., Ellis, G. (1993). Fourier-transform Raman spectroscopy in the study of paints. *Adv. Chem. Ser.*, 443–482.
60. Claybourn, M., Massey, T., Highcock, J., Gogna, D. (1994). Analysis of processes in latex systems by Fourier-transform Raman spectroscopy. *J. Raman Spectrosc.*, 25, 123–129.
61. de Buruaga, I.S., Leiza, J.R., Asua, J.M. (2000). Model-based control of emulsion terpolymers based on calorimetric measurements. *Polym. React. Eng.*, 8, 39–75.
62. Everall, N., King, B. (1999). Raman spectroscopy for polymer characterization in an industrial environment. *Macromol. Symp.*, 141, 103–116.
63. Williams, K.P.J., Mason, S.M. (1990). Future directions for Fourier-transform Raman spectroscopy in industrial analysis. *Spectrochim. Acta A Mol. Biomol. Spectrosc.*, 46, 187–196.
64. Ellis, G., Claybourn, M., Richards, S.E. (1990). The application of Fourier-transform Raman spectroscopy to the study of paint systems. *Spectrochim. Acta A Mol. Biomol. Spectrosc.*, 46, 227–241.
65. Ito, K., Kato, T., Ona, T. (2002). Non-destructive method for the quantification of the average particle diameter of latex as water-based emulsions by near-infrared Fourier transform Raman spectroscopy. *J. Raman Spectrosc.*, 33, 466–470.
66. Feng, L.Z., Ng, K.Y.S. (1991). Characterization of styrene polymerization in microemulsions by Raman spectroscopy. *Colloids Surf.*, 53, 349–361.

67. Bandermann, F., Tausendfreund, I., Sasic, S., Ozaki, Y., Kleimann, M., Westerhuis, J.A., Siesler, H.W. (2001). Fourier-transform Raman spectroscopic on-line monitoring of the anionic dispersion block copolymerization of styrene and 1,3-butadiene. *Macromol. Rapid Commun.*, 22, 690–693.

68. Cosco, S., Ambrogi, V., Musto, P., Carfagna, C. (2006). Urea-formaldehyde microcapsules containing an epoxy resin: Influence of reaction parameters on the encapsulation yield. *Macromol. Symp.*, 234, 184–192.

69. Cosco, S., Ambrogi, V., Musto, P., Carfagna, C. (2007). Properties of poly(urea-formaldheyde) microcapsules containing an epoxy resin. *J. Appl. Polym. Sci.*, 105, 1400–1411.

70. Gardon, J.L. (1968). Emulsion polymerization. 6. Concentration of monomers in latex particles. *J. Polym. Sci. A Polym. Chem.*, 6, 2859.

71. Mathey, P., Guillot, J. (1991). Swelling of polybutadiene Aas latex particles and cast flms by styrene and acrylonitrile monomers. *Polymer*, 32, 934–941.

72. Maxwell, I.A., Kurja, J., Vandoremaele, G.H.J., German, A.L., Morrison, B.R., Partial swelling of latex particles with monomers. *Makromol. Chem. Macromol. Chem. Phys., 193*, 2049–2063.

73. Morton, M., Kaizerman, S., Altier, M.W. (1954). Swelling of latex particles. *J. Colloid Sci.*, 9, 300–312.

74. Omi, S. (1985). A generalized computer modeling of semibatch, n-component emulsion polymerization systems and its applications. *Zayro Gijutsu*, 3, 426.

75. Vanzo, E., Marchess, R.H., Stannett, V. (1965). Solubility and swelling of latex particles. *J. Colloid Sci.*, 20, 62.

76. Goodwin, J.W., Ottewill, R.H., Harris, N.M., Tabony, J. (1980). A study by small-angle neutron-scattering of the swelling of polystyrene latex particles by monomer. *J. Colloid Interface Sci.*, 78, 253–256.

77. Hergeth, W.D., Codella, P.J. (1994). Monomers in polymer dispersions. 4. Partition of acrylonitrile in rubber latex as studied by Raman spectroscopy. *Appl. Spectrosc.*, 48, 900–903.

78. Nyquist, R.A. (1990). Solvent-induced nitrile frequency shifts: acetonitrile and benzonitril. *Appl. Spectrosc.*, 44, 1405–1407.

79. Lascelles, S.F., Armes, S.P., Zhdan, P.A., Greaves, S.J., Brown, A.M., Watts, J.F., Leadley, S.R., Luk, S.Y. (1997). Surface characterization of micrometre-sized, polypyrrole-coated polystyrene latexes: Verification of a "core-shell" morphology. *J. Mater. Chem.*, 7, 1349–1355.

80. Lenzi, M.K., Lima, E.L., Pinto, J.C. (2006). Detecting core-shell structure formation using near infrared spectroscopy. *J. Near Infrared Spectrosc.*, 14, 179–187.

81. Ito, K., Kato, T., Ona, T. (2004). Rapid viscosity determination of waterborne automotive paint emulsion system by FT-Raman spectroscopy. *Vibrational Spectrosc.*, 35, 159–163.
82. Keddie, J.L. (1997). Film formation of latex. *Mater. Sci. Eng.*, R21, 101–170.
83. Winnik, M.A. (1997). Latex film formation. *Curr. Opin. Colloid Interface Sci.*, 2, 192–199.
84. Belaroui, F., Grohens, Y., Boyer, H., Holl, Y. (2000). Depth profiling of small molecules in dry latex films by confocal Raman spectroscopy. *Polymer*, 41, 7641–7645.
85. Belaroui, F., Hirn, M.P., Grohens, H., Marie, P., Holl, Y. (2003). Distribution of water-soluble and surface-active low-molecular-weight species in acrylic latex films. *J. Colloid Interface Sci.*, 261, 336–348.
86. Zhao, Y.Q., Urban, M.W. (2000). Phase separation and surfactant stratification in styrene/n-butyl acrylate copolymer and latex blend films. 17. A spectroscopic study. *Macromolecules*, 33, 2184–2191.
87. Ludwig, I., Schabel, W., Kind, M., Castaing, J.C., Ferlin, P. (2007). Drying and film formation of industrial waterborne lattices. *AIChE J.*, 53, 549–560.
88. Oyman, Z.O., Ming, W., van der Linde, R. (2003). Oxidation of model compound emulsions for alkyd paints under the influence of cobalt drier. *Prog. Org. Coatings*, 48, 80–91.
89. Agnely, M. (2000). *Reliable multidetection sensor for advanced control in latex production (Remap)*. Final report. EU Project BRPR CT96 0275.
90. Arevalillo, A. (2008). Modelado de reactores de polimerizacion en emulsion mediante fluidodinámica computacional. PhD Thesis, University of The Basque Country.
91. Denoord, O.E. (1994). Multivariate calibration standardization. *Chemometrics Intelligent Lab. Syst.*, 25, 85–97.
92. Lindgren, F., Geladi, P., Rannar, S., Wold, S. (1994). Interactive variable selection (Ivs) for Pls. 1. Theory and algorithms. *J. Chemometrics*, 8, 349–363.
93. Swierenga, H., Haanstra, W.G., de Weijer, A.P., Buydens, L.M.C. (1998). Comparison of two different approaches toward model transferability in NIR spectroscopy. *Appl. Spectrosc.*, 52, 7–16.
94. Wulfert, F., Kok, W.T., Smilde, A.K. (1998). Influence of temperature on vibrational spectra and consequences for the predictive ability of multivariate models. *Anal. Chem.*, 70, 1761–1767.
95. Brown, W. (1993). *Dynamic Light Scattering the Method and Some Applications*. Oxford: Clarendon Press.
96. Pecora, R. (1985). *Dynamic Light Scattering Applications of Photon Correlation Spectroscopy*. New York: Plenum.

97. Schmitz, K.S. (1990). *An Introduction to Dynamic Light Scattering by Macromolecules*. Boston: Academic Press.
98. Finsy, R. (1994). Particle sizing by quasi-elastic light-scattering. *Adv. Colloid Interface Sci.*, 52, 79–143.
99. Kemmere, M. (2005). Recent developments in polymer processes. In: Meyer, T., Keurentjes, J., eds., *Handbook of Polymer Reaction Engineering*. Weinheim: Wiley-VCH Verlag GmbH.
100. Blatchford, M.A., Wallen, S.L. (2002). Development and validation of spectroscopic methods for monitoring density changes in pressurized gaseous and supercritical fluid systems. *Anal. Chem.*, 74, 1922–1927.
101. Kessler, W., Luft, G., Zeiss, W. (1997). In situ Raman spectroscopy for the study of high-pressure reactions. *Berichte der Bunsen Ges. Phys. Chem. Chem. Phys.*, 101, 698–702.
102. Luft, G., Kessler, W., Zeiss, W. (1999). Activation of the decomposition of some peroxides with trialkylaluminium compounds under high pressure. *Angew. Makromol. Chem.*, 269, 36–41.

6

RAMAN APPLICATIONS IN LIQUID CRYSTALS

Naoki Hayashi, PhD
Research Manager, Analysis Technology Center, FUJIFIL Corporation

6.1 INTRODUCTION

Liquid crystals are the substances that exhibit intermediate properties between the "fluidity" of conventional liquid and the "order" of solid crystal. Reinitzer [1], an Austrian botanist, synthesized the derivative of a cholesteryl benzoate (Figure 6.1). He found that it shows a viscous and clouded state at a temperature range between those of crystalline and liquid phases. Lehmann [2] observed this material by polarizing an optical microscope and found an optical anisotropy as well as fluidity. He named this state "flüssige Kristalle" (liquid crystal).

6.2 STRUCTURES AND PHASES OF LIQUID CRYSTALS

Liquid crystals are categorized mainly into two categories: *lyotropic liquid crystals* and *thermotropic liquid crystals*. A lyotropic liquid crystal contains two components: a solute and a solvent. None of the components show liquid crystal phase solely. Depending on the ratio of the two components, a lyotropic liquid crystal phase could be formed.

Thermotropic liquid crystals are composed of one or more constituents each of which shows liquid crystal behavior. Their structure is

Raman Spectroscopy for Soft Matter Applications, Edited by Maher S. Amer

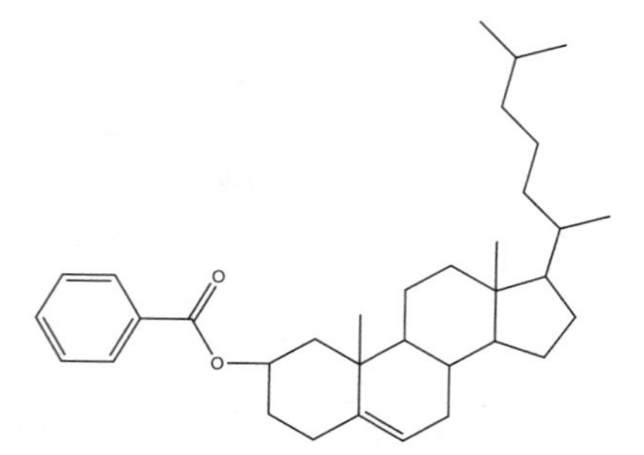

Figure 6.1 The chemical structure of the derivative of cholesteryl benzoate synthesized by Reinitzer.

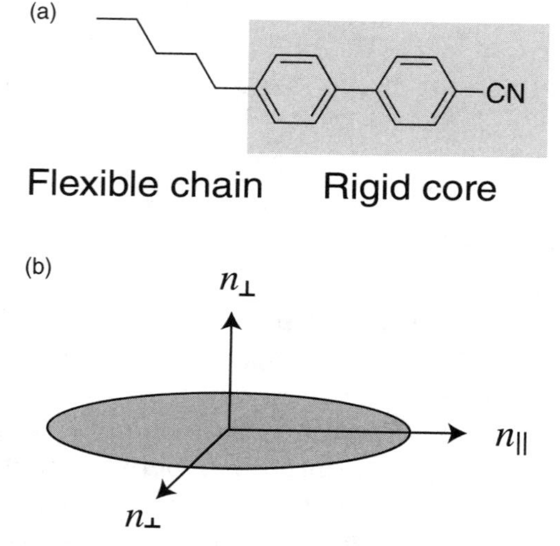

Figure 6.2 (a) Rod-like shape of the nematic liquid crystal. The chemical structure of 4′-*n*-4-pentyl-cyanobiphenyl (5CB) is shown. (b) Refractive index ellipsoid of nematic (N) or smectic *A* (Sm*A*) liquid crystals.

based upon a rigid core called "mesogen" and a flexible chain attached to the core (Figure 6.2). The rigid core causes anisotropy in the system, and the flexible chains cause the system's fluidity.

Based on polarized light microscopy observation, Friedel [3] classified the liquid crystal phases into three types: nematic (N), cholesteric (Ch), and smectic (Sm) (Figure 6.3).

In the nematic phase, the molecules are oriented into a particular direction, but there is no order in the molecular position (Figure 6.3b). The unit vector pointing toward the average direction of the molecular long axes is defined as $\boldsymbol{n}$, which is called the $\boldsymbol{n}$-director. There exists

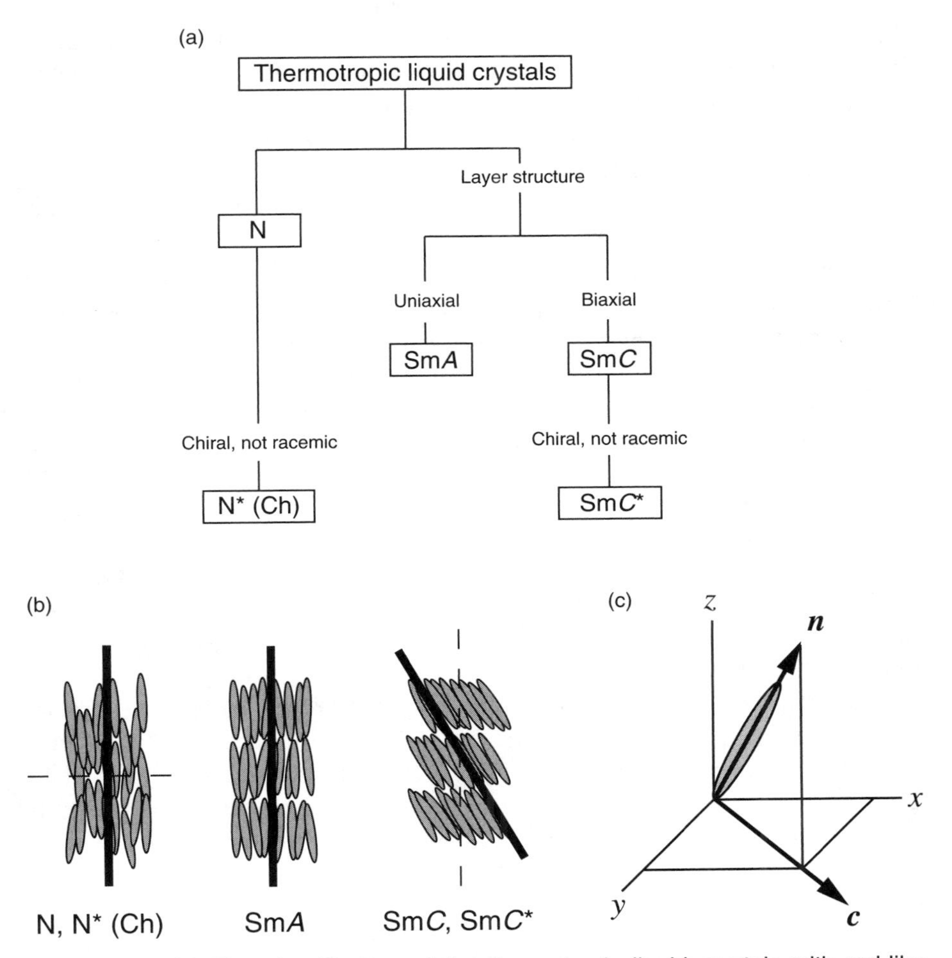

Figure 6.3 (a) The classification of the thermotropic liquid crystals with rod-like shape and low molecular weight. (b) The structures of the thermotropic liquid crystals. The liquid crystal molecules are represented by ellipsoids. The solid lines show the averages of the orientation of the molecular long axes, which are parallel to the ***n***-directors. The broken lines represent the helical axes when they have helical structures. (c) The ***c***-director in smectic *C* (Sm*C*) phase. The *xy*-plane is parallel to the smectic layer, and the *z*-axis is perpendicular to the layer. Sm*A*, smectic *A*; N*, chiral nematic or cholesteric phase; Ch, cholesteric; Sm*C**, chiral smectic *C*.

rotational symmetry of the molecular distribution about its long axis. The system has uniaxial symmetry with respect to the ***n***-director.

When the nematic phase is composed of chiral molecules, the phase, then, has chirality and is referred to as *chiral nematic* or *cholesteric* phase (N*). The molecules are oriented in a plane, but the adjacent planes are twisted along a helical axis. Cholesteric phase is the same as the nematic phase in terms of their thermodynamics. The ***n***-director rotates around a helical axis that is perpendicular to the ***n***-director. The twist direction is determined by the chirality of the molecules. When the size of the helical pitch is comparable to the visible light wavelength, 400–700 nm, they reflect light with a wavelength matching their pitch size more efficiently.

The *smectic phase* has an additional level of order, that is, the positional order along the molecular orientation. The smectic phase has a layer structure. The phase is further classified into smectic *A* (Sm*A*) and smectic *C* (Sm*C*) phases. In the Sm*A* phase, the ***n***-director is parallel to the layer normal, and the phase has uniaxial symmetry. In the Sm*C* phase, however, the molecules or ***n***-director tilts with respect to the layer normal. In this phase, the order about tilting direction has to be considered, and thus, the phase has biaxial symmetry. The ***c***-director is useful to describe this biaxial phase (Figure 6.3c). The ***c***-director is the unit vector that is parallel to the projection of the ***n***-director on the smectic layer. When a liquid crystal molecule has a chiral structure, the phase (Sm*C**) shows a helical structure in the bulk without any external field. In this case, the helical pitch is very long ($\sim 10^{-1}$–$\sim 10^{2}\,\mu m$) compared with the layer thickness that is comparable to the molecular length (several nanometers). The helical axis is perpendicular to the smectic layer, and the ***c***-director rotates around the helical axis. The Sm*C* phase has C_{2h} symmetry; however, the mirror and inversion symmetry is lost in the Sm*C** phase, so that the symmetry of the Sm*C** phase is reduced to C_2 symmetry. Meyer et al. [4] synthesized *p*-decyloxybenzylidene *p′*-amino 2-methyl butyl cinnamate (DOBAMBC) and reported ferroelectricity in the Sm*C** phase (Figure 6.4). There are subphases of Sm*C** that show antiferro- and ferrielectricity. Chandani et al. [5,6] discovered 4-(1-methyl-heptyloxycarbonyl)phenyl 4′-octyloxybiphenyl-4-carboxylate (MHPOBC) showing the ferro-, ferri-, and antiferroelectricity in Sm*C**-like phases (Figure 6.5).

In the case of liquid crystal showing all phases described previously, it shows the phase sequence isotropic crystal—N–Sm*A*–Sm*C* (including its subphases)—as temperature decreases.

Some polymers also shows liquid crystal phases. They are characterized by a typical polymer structure, i.e. main-chain, side-chain, and hybrid types (Figure 6.6). The phase structures of high molecular weight liquid crystals (polymer) are the same as those with low molecular weight.

Other than the aforementioned liquid crystal phases, various types of molecules (e.g. disk-like molecule [7], bent-core molecule [8]) create

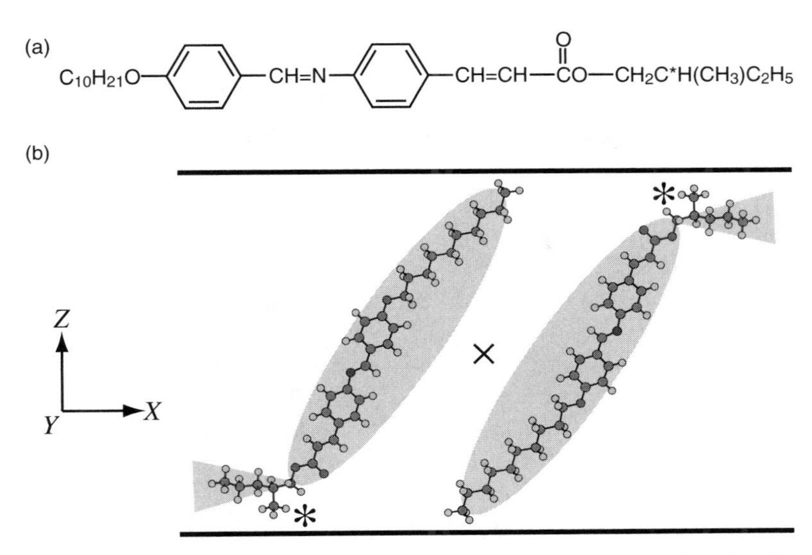

Figure 6.4 (a) The chemical structure of *p*-decyloxybenzylidene *p*′-amino 2-methyl butyl cinnamate (DOBAMBC). (b) The C_2 symmetry within a layer. The *z*-axis is parallel to the smectic layer normal, and the *x*-axis is parallel to the ***c***-director. The mark "X" at the center of the layer indicates the C_2 symmetry axis. The spontaneous polarization appears parallel to this axis.

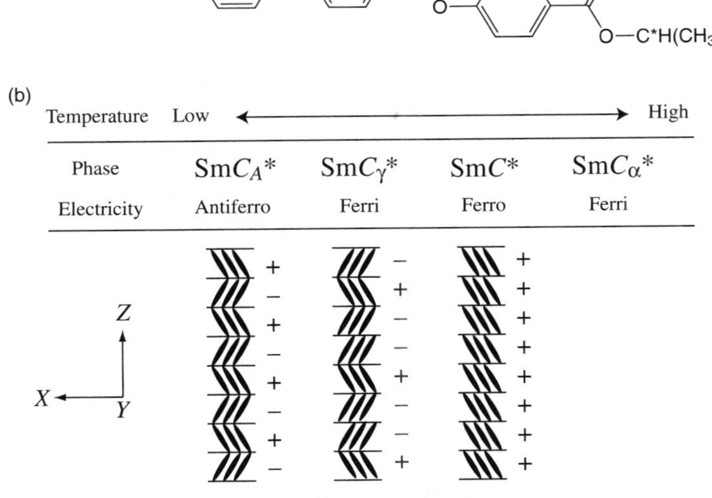

Figure 6.5 (a) The chemical structure of 4-(1-methyl-heptyloxycarbonyl)phenyl 4′-octyloxybiphenyl-4-carboxylate (MHPOBC). (b) The structure of Sm*C** and the subphases. The sign on the right of the layers represent the directions of spontaneous polarization within the layer. Sm*C**α phase has no stable structure for molecular tilting direction in a layer but has a very short helical pitch [47].

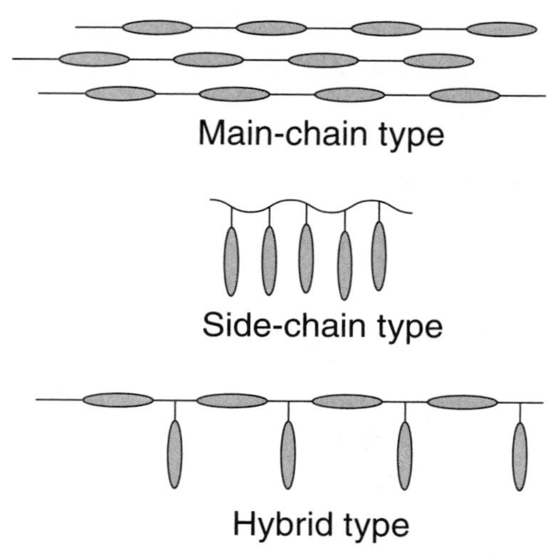

Figure 6.6 The structures of liquid crystal polymers. The mesogens are represented by ellipsoids.

specific liquid crystal structures. Such structures will not be discussed in here. For a detailed description of such specific structures, the reader is referred to Demus et al. [9].

6.3 LIQUID CRYSTAL DISPLAYS (LCDs)

Liquid crystals have a set of unique properties that can be summarized as follows:

(1) unique anisotropic optical, mechanical, electrical, and magnetic properties;
(2) fluidity resulting from easy to produce molecular reorientation;
(3) easy to obtain uniform long-range repetition with a long correlation length resulting from an easy to self-assemble structure;
(4) easy to control molecular orientation at the surface, i.e. a small external field can induce the change.

In liquid crystals, the refractive indices in the direction parallel to the molecular long axis and short axis, in rod-like molecule for example, are different (Figure 6.2b). Therefore, in oriented liquid crystals, a phase delay occurs between the light polarized parallel and perpendicular to the molecular orientation, causing a "birefringence." The orientation of liquid crystal surface molecules depends on their interaction with the surface they contact. Such interaction, and hence the orientation of the surface molecules, can be easily controlled by coating applied to the contact surface of a substrate. When amphiphilic molecules or silane couplers are used as coatings, the long axes of the molecules align perpendicular to the substrate plate. This orientational state is called "homeotropic molecular alignment" (Figure 6.7a). When other coatings, such as polyvinyl alcohol, nylon, polyimide, etc., are used, the molecular long axes are oriented parallel to the substrate plates. This state is called "homogeneous molecular alignment" (Figure 6.7b).

A "rubbing" procedure is frequently used to orient molecules in one direction. The "rubbing" makes grooves or scratches on the surface of the coating material to induce the anisotropy of the orientation of chains of the coating material molecules.

The most important application of liquid crystal is LCDs. The twisted nematic (TN)-LCD structure is illustrated in Figure 6.8. The liquid crystals are sandwiched between two glass plates with transparent electrodes. The alignment agent is coated on the plates, and the molecular orientation on the surface is fixed by the rubbing procedure. The rubbing direction of upper and lower plates is crossed so that the orientations of the liquid crystal molecules are twisted through its thickness. The polarizers are placed outside the glass plates so that the polarization axes are crossed. In the white mode, the initial state shown in Figure 6.8a, the liquid crystal molecules are oriented parallel to the glass plate. When the incident light is polarized parallel to the molecular orientation at the lower plate, the polarization of the transmitted light through the liquid crystal layer is rotated along the twisted orientation of liquid crystals, and the light can pass through the second polarizer. When electric field is applied to the cell, molecular reorientation is induced,

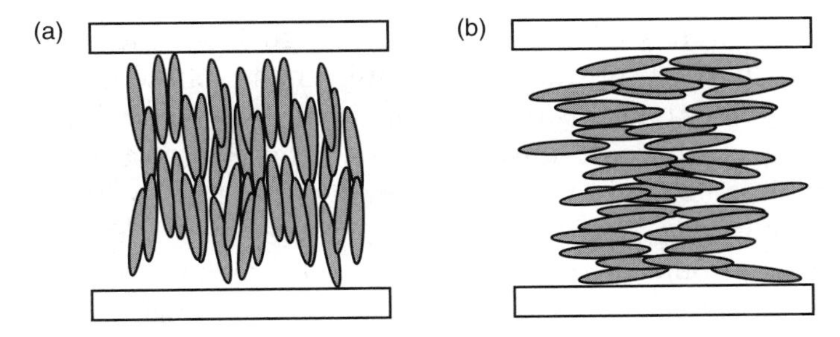

Figure 6.7 (a) Homeotropic molecular alignment and (b) homogeneous molecular alignment in nematic phase.

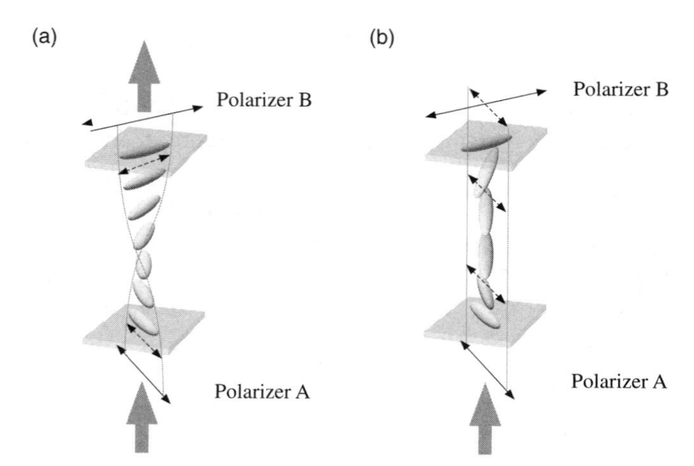

Figure 6.8 Principle of twisted nematic-liquid crystal display (TN-LCD) structure. (a) White mode; no electric field is applied. The light polarization (dashed arrows) is rotated as passing through the twisted liquid crystal layer. The polarized light can pass through polarizer B which polarization is perpendicular to polarizer A. (b) Black mode; electric field is applied to the cell. The liquid crystal molecules are aligned along the electric field vertically. The light polarization is kept parallel to polarizer A, and no light can be passed through polarizer B.

and the molecules are reoriented in a direction perpendicular to the electric field due to the dielectric anisotropy (Figure 6.8b). In this case, the polarization of the incident light is not rotated through the liquid crystal thickness, because there is no difference in the refractive index perpendicular to the molecular long axis. Thus, the liquid crystal cell acts as a light bulb.

6.4 THE ORIENTATIONAL ORDER PARAMETERS

Unique characters of liquid crystals emerge from the alignment of molecules. The physical state of the liquid crystal phase is described by "order." Nematic phase is distinguished from isotropic liquid by "orientational order" of molecules. In the Sm*A* phase, the "positional order" of molecules is needed to describe the physical state in addition to the orientational order.

The orientational order of liquid crystal molecules is given by the *molecular distribution function*. In the nematic phase, the system is uniaxial, and thus, the thermodynamic state of the phase can be represented by the molecular orientational distribution function, $f(\beta)$, where

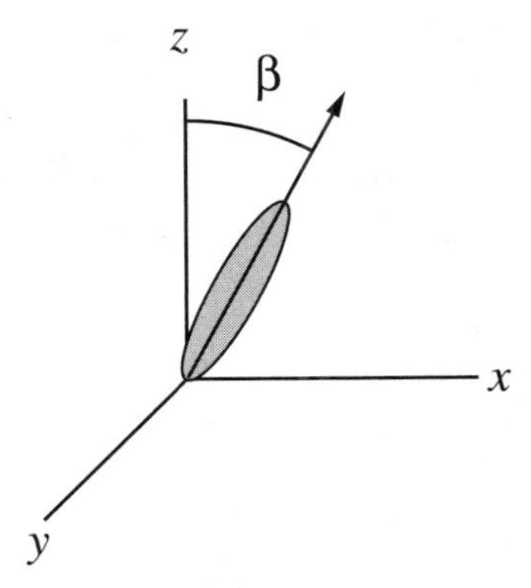

Figure 6.9 The coordinate system of the nematic phase. The ellipsoid represents the liquid crystal molecule. The *z*-axis is parallel to the ***n***-director.

β is the tilting angle between the individual molecular long axis and the ***n***-director (Figure 6.9). If the liquid crystal molecule is assumed as a rigid rod, the molecular orientational distribution function $f(\beta)$ can be expanded in terms of Legendre polynomials $P_L(\cos\beta)$as follows [10–13]:

$$f(\beta) = \sum_{L=\text{even}} \frac{2L+1}{2} \langle P_L \rangle P_L(\cos\beta). \tag{6.1}$$

Here, $\langle P_L \rangle$ is the L-th order orientational order parameter and is defined by

$$\langle P_L \rangle = \int_0^{\pi} \sin\beta d\beta P_L(\cos\beta) f(\beta). \tag{6.2}$$

The ⟨⟩ denotes the thermal average. The second and fourth order of orientational order parameter, $\langle P_2 \rangle$ and $\langle P_4 \rangle$, are given by

$$\langle P_2 \rangle = \frac{1}{2} \langle 3\cos^2\beta - 1 \rangle \tag{6.3}$$

and

$$\langle P_4 \rangle = \frac{1}{8} \langle 35\cos^4\beta - 30\cos^2\beta + 3 \rangle. \tag{6.4}$$

The orientational order parameters are zero when the orientational order is lost as in the isotropic phase. On the other hand, when the molecules are perfectly aligned without any fluctuation as in the single crystal, the order parameters are equal to 1.

The orientational order in the nematic phase was discussed theoretically with the mean-field molecular theory by Maier and Saupe [10,11,14]. Later, McMillan [15] extended the theory to the smectic phase.

6.5 RAMAN SPECTROSCOPY IN LIQUID CRYSTALS

A vibrational Raman spectrum directly provides information regarding molecular structure and molecular interaction represented in Raman shifts, intensities, and bandwidth. Furthermore, polarized Raman measurements allow the evaluation of molecular orientational order.

6.5.1 Investigating Molecular Changes with the Phase Transition

The vibrational Raman spectrum of liquid crystals changes with phase transitions because of the change in the molecular local structure and the molecular interaction.

Galbiati and Zerbi [16] measured the Raman spectra for dodecylcyanobiphenyl (12-CB) (Figure 6.10). This compound shows a phase sequence as crystal-48 °C-Sm*A*-58.5 °C-isotropic. The spectrum pattern of the Raman spectra in C–H stretching and the C–C skeletal stretching regions showed an all-trans, planer structure in crystalline phase. The Raman spectra in Sm*A* and isotropic phase showed that the structure in crystalline phase collapses and a gauche structure appears, indicating a conformational disordered state of the alkyl chains.

6.5.2 Measuring Spinning and Tumbling Diffusion Coefficients

The analysis of a polarized Raman spectrum band shape enables the measurement of the spinning and tumbling diffusion coefficients of liquid crystal molecules [17,18].

Fontana et al. measured the tumbling diffusion constant change as a function of temperature (Figure 6.11). It was reported that the diffusion constant decreases with increasing temperature. It was shown, however, that the behavior of tumbling diffusion is a normal thermally activated process.

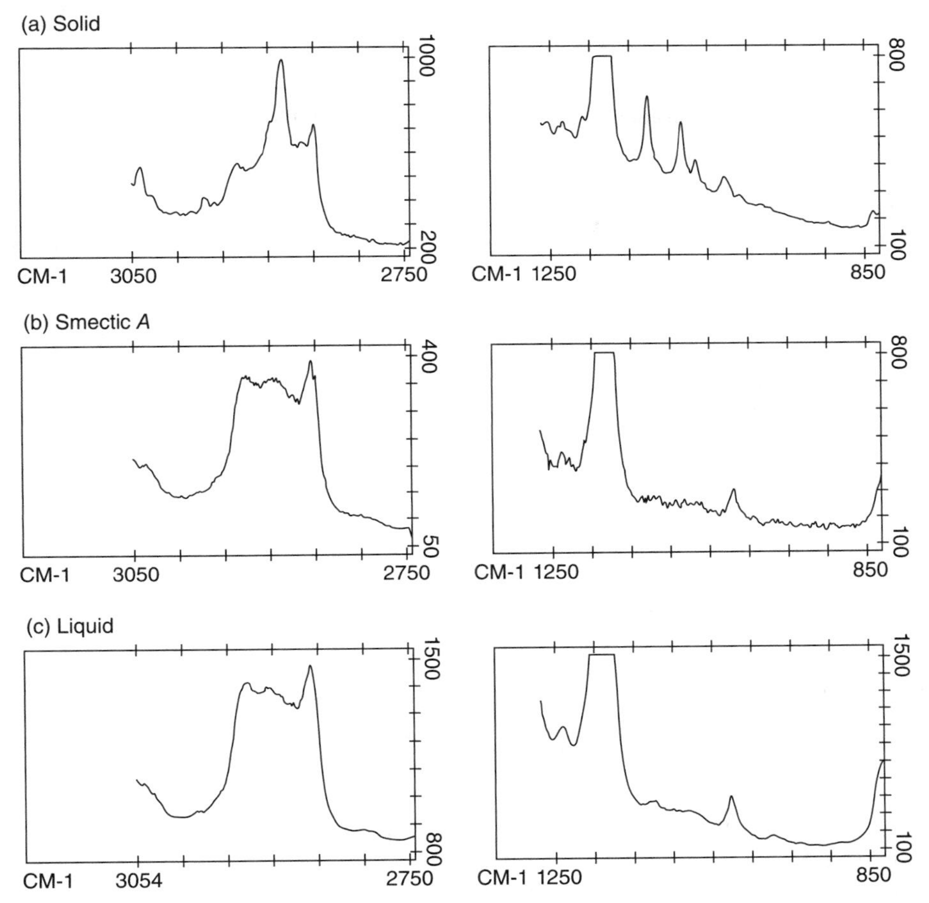

Figure 6.10 Temperature change of the Raman spectra of 12-CB in the C–H stretching region (left) and C–C skeletal stretching region (right). Measured at the solid state at 25 °C (a), smectic *A* (Sm*A*) phase at 52 °C (b), and isotropic phase at 60 °C (c). From Galbiati and Zerbi [16] with permission.

6.6 RAMAN SPECTROSCOPY MEASUREMENTS OF ORIENTATIONAL ORDER PARAMETERS

It is widely considered in the liquid crystal community that the most important parameter to consider for understanding the properties and behavior of liquid crystals is the orientational order parameter. Hence, the rest of this chapter will be devoted to discussing the theoretical and experimental aspects of this subject. The orientational order parameters can be evaluated by experimental methods [19] such as polarized

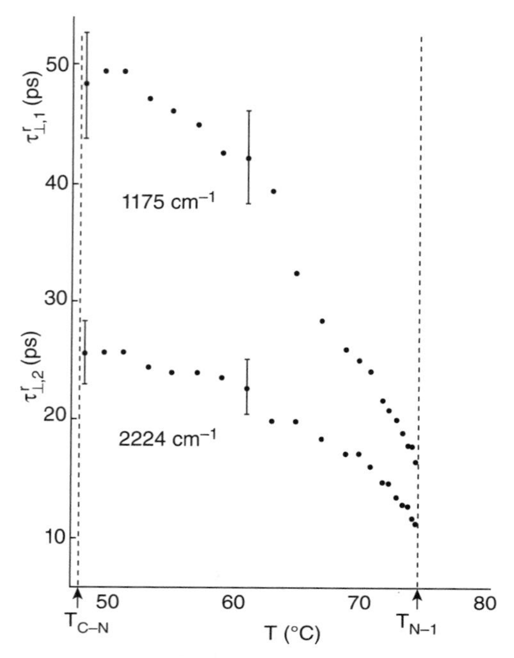

Figure 6.11 Temperature dependence of tumbling diffusion constant of 4-(*n*-octyl)-4′-ethoxytolane. $\tau^r_{\perp,1}$ and $\tau^r_{\perp,2}$ are obtained by infrared absorption (1175 cm^{-1}, benzene ring in plane deformation) and Raman scattering (2224 cm^{-1}, tolane C≡C stretching) measurements, respectively. From Fontana et al. [18] with permission.

vibrational Raman spectroscopy (e.g. Jen et al. [20]; Constant and Decoster [21]), electron paramagnetic resonance (e.g. Luckhurst and Yeates [22]), nuclear magnetic resonance (NMR) (Miyajima and Hosokawa [23]; Nakai et al. [24]), and infrared (IR) (e.g. Jang et al. [25]).

Polarized vibrational Raman scattering measurement gives not only the second-order parameter $\langle P_2 \rangle$ but also the fourth order parameter $\langle P_4 \rangle$, simultaneously. The technique has the additional advantage that a glass sample cell can be used for measurement. In this case, the texture of the sample can be observed by an optical microscope and the electro-optic response can be monitored during the polarized Raman scattering measurement.

Figure 6.12 shows a schematic illustration of macro-Raman system for the analysis of orientational order of liquid crystals. The excitation beam is incident perpendicular to the sample cell. The back-scattered light is collected by the objective lens and passes through the polarizer, which direction can be parallel or perpendicular to the polarization of the excitation light. The scattered light is depolarized by the optical

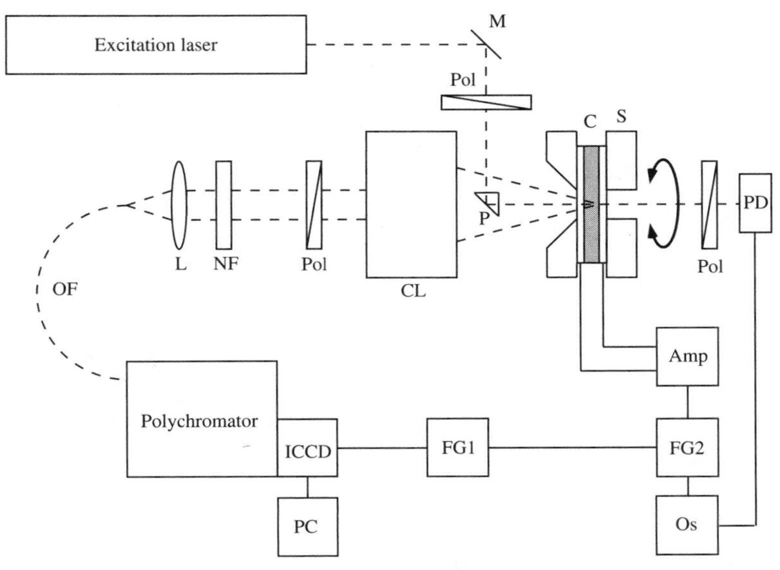

Figure 6.12 Schematic drawing of the experimental setup for macroscopic polarized Raman measurement for the analysis of orientational order. The broken lines represent light streams, and the solid lines represent electric signals. M, mirror; Pol, polarizer; P, prism; C, sample cell; S, rotating stage; ICCD, intensified CCD camera; CL, Cassegrain lens; NF, Raman notch filter; L, camera lens; OF, optical fiber; PC, personal computer; FG1, function generator for the gate pulse; FG2, function generator for applying the electric field to the sample; Amp, high-speed amplifier; Os, oscilloscope.

fiber and measured by the spectrometer. The change in orientational order can be measured by the time-resolved measurement. The optical response can be monitored by detecting the intensity of light passed through the sample cell.

A Raman microscope improves the spatial resolution of the measurement. The spatial resolution can be increased to submicron resolution by using a confocal setup. The system is shown in Figure 6.13.

6.6.1 Theoretical Analysis

In order to simplify the theoretical analysis of polarized Raman intensities and their utilization into orientational parameters measurements, the following assumption are usually made:

(1) The liquid crystal phase is restricted to nematic or Sm*A*. The system is uniaxial so that the molecular distribution is cylindri-

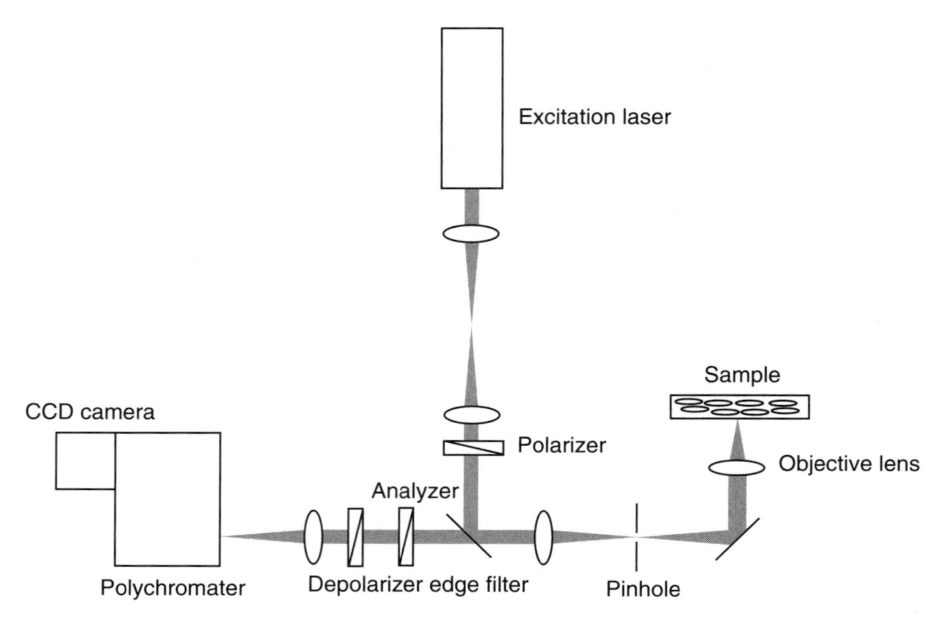

Figure 6.13 Schematic drawing of a confocal Raman microscope. The confocal system with spatial aperture improves spatial resolution up to submicron scales.

cally symmetric, and the molecular alignment is homogeneous so that the ***n***-director is parallel to the substrate plate of a sample cell.

(2) A Raman tensor of a particular vibrational mode can be treated as uniaxial:

$$\alpha^{\mathrm{Mol}} = \begin{pmatrix} \alpha_{\perp} & & \\ & \alpha_{\perp} & \\ & & \alpha_{\parallel} \end{pmatrix}. \tag{6.5}$$

(3) The principal axis of the Raman tensor with the largest value is parallel to the molecular long axis. This indicates that the distribution function of the principal axis of the Raman tensor is identical with the distribution function of the liquid crystal molecule.

Usually, the molecular fixed coordinate system is defined by x-, y-, and z-axes, where the z-axis is parallel to the molecular long axis or the longest principal axis of the Raman polarizability tensor. The polarizability tensor is given by

$$\alpha_{ij} = \hat{n}_i^M \cdot \alpha^M \cdot \hat{n}_j^M$$
$$= \left(n_{i1}^M, n_{i2}^M, n_{i2}^M\right) \cdot \begin{pmatrix} \alpha_\perp & & \\ & \alpha_\perp & \\ & & \alpha_\parallel \end{pmatrix} \cdot \begin{pmatrix} n_{j1}^M \\ n_{j2}^M \\ n_{j3}^M \end{pmatrix}, \tag{6.6}$$

where n_i and n_j are the polarization unit vector. The superscript M represents the definition in the molecular fixed coordinate system. The polarizability tensor can be rewritten in terms of spherical harmonics. The polarization vector is written by

$$\hat{n}_l^M = \sum_q (-1)^q \, n_i^{M,(q)} \hat{e}_{-q}^M, \tag{6.7}$$

where $l = i, j$, and $\hat{e}_{-q}^M$ is the irreducible spherical vector and related to the unit vector as

$$\hat{e}_0^M = \hat{z}, \quad \hat{e}_{\pm 1}^M = \left(\mp \frac{1}{\sqrt{2}}\right)(\hat{x} \pm i\hat{y}). \tag{6.8}$$

The polarizability tensor is given by

$$\alpha_{ij} = \sum_{q,q'} (-1)^{q-q'} \, n_i^{M,(q)} n_j^{M,(-q')} \alpha_{-qq'}^M \tag{6.9}$$

with

$$\alpha_{-qq'}^M = \hat{e}_{-q}^M \cdot \alpha^M \cdot \hat{e}_{q'}^M. \tag{6.10}$$

$n_i^{M,(q)}$ and $n_j^{M,(-q)}$ are the spherical tensor operators of rank-1. The linear combination of these operators can generate the spherical tensor operators of ranks-0, -1, and -2 by using the 3-j symbol [26]:

$$T_{km}^M(ij) = \sum_{q,q'} (-1)^m (2k+1)^{1/2} \begin{pmatrix} 1 & 1 & k \\ q & -q' & -m \end{pmatrix} n_i^{M,(q)} n_j^{M,(-q')} \tag{6.11}$$

Here, k is 0, 1, and 2, $|m| < k$, q-$q' = m$. The inverse function is given by

$$n_i^{M,(q)} n_j^{M,(-q')} = \sum_{k,m} (-1)^m (2k+1)^{1/2} \begin{pmatrix} 1 & 1 & k \\ q & -q' & -m \end{pmatrix} T_{km}^M(ij). \tag{6.12}$$

The following relation is derived from Equations 6.9 and 6.12

$$\alpha_{ij} = \sum_{k,m} (-1)^m \alpha^M(k,-m) T^M_{km}(ij), \tag{6.13}$$

where

$$\alpha^M(k,-m) = \sum_{q,q'} (-1)^{q-q'} (2k+1)^{1/2} \begin{pmatrix} 1 & 1 & k \\ q & -q' & -m \end{pmatrix} \alpha^M_{-qq'}. \tag{6.14}$$

$\alpha^M_{-qq'}$ is calculated as follows:

$$\begin{aligned} \alpha^M_{00} &= \hat{e}^M_0 \cdot \alpha^M \cdot \hat{e}^M_0 = \alpha^M_{zz} = \alpha_{\parallel}, \\ \alpha^M_{-11} &= \hat{e}^M_{-1} \cdot \alpha^M \cdot \hat{e}^M_1 = \frac{-1}{\sqrt{2}}(\hat{x} - i\hat{y}) \cdot \alpha^M \cdot \frac{-1}{\sqrt{2}}(\hat{x} + i\hat{y}) = -\alpha_{\perp}, \\ \alpha^M_{1-1} &= \hat{e}^M_1 \cdot \alpha^M \cdot \hat{e}^M_{-1} = -\alpha_{\perp}. \end{aligned} \tag{6.15}$$

Accordingly, $\alpha^M(k,-m)$ is given by

$$\alpha^M(0,0) = -\frac{1}{\sqrt{3}}(2\alpha_{\perp} + \alpha_{\parallel}), \quad \alpha^M(2,0) = \frac{2}{\sqrt{6}}(\alpha_{\parallel} - \alpha_{\perp}). \tag{6.16}$$

In Equation 6.13, we can find that the operator $T^M_{km}(ij)$ represents the orientational relation between the polarization of light and the principal axis of the polarizability tensor.

The Raman polarizability tensor is defined in the laboratory fixed coordinate system. The rotational transformation from the molecular fixed coordinate system into the laboratory fixed one is illustrated in Figure 6.14. The rotational operator is written by $D(\Omega)$, where Ω rep-

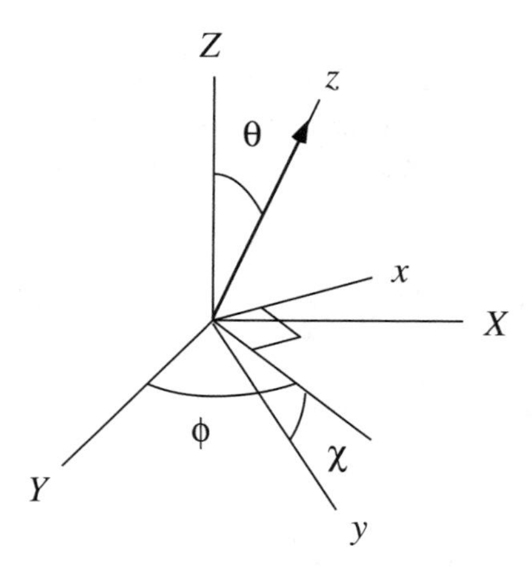

Figure 6.14 Rotational transformation from the molecular fixed coordinate system to the laboratory fixed coordinate system.

resents the Euler angle θ, ϕ, and χ. The spherical tensor operator $T^M_{km}(ij)$ is rewritten by the Wigner rotation matrix [26]:

$$T^M_{km}(ij) = D(\Omega)F^L_{km}D^{-1}(\Omega) = \sum_{m'} F^L_{km'}(ij)D^{(k)*}_{mm'}(\Omega). \tag{6.17}$$

The superscript L indicates the definition in the laboratory fixed coordinate system. By the substitution in Equation 6.13 by Equation 6.17, the polarizability tensor is obtained in the laboratory fixed coordinate system as

$$\alpha_{ij} = \sum_{k,m,m'} F^L_{km'}(ij)D^{(k)*}_{mm'}(\Omega)\alpha^M(k,-m). \tag{6.18}$$

Here, $F^L_{km'}(ij)$ is given by Equation 6.19:

$$F^L_{km'}(ij) = \sum_{q,q'}(-1)^{m'}(2k+1)^{1/2}\begin{pmatrix} 1 & 1 & k \\ q & -q' & -m \end{pmatrix} n_i^{L,(q)} n_j^{L,(-q')}. \tag{6.19}$$

The measurement system is defined in Figure 6.15. The x-, y-, and z-axes constitute the right-handed laboratory fixed coordinate frame. The z-axis is parallel to the orientation center axis or ***n***-director. The xz-plane is parallel to the substrate plane of the cell. The direction of the incident and the back-scattered light are parallel to the z-axis, and the polarization vector is represented as ***p*** and ***p′***. These polarization angles are ω and ω', respectively. The electric fields of the incident light at the depth Y in the Cartesian coordinate system is given by

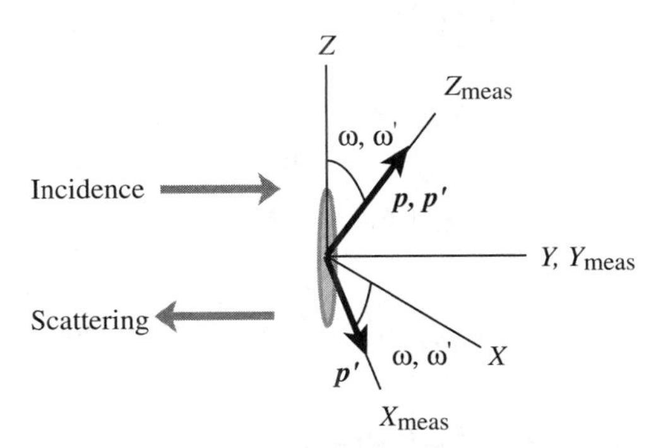

Figure 6.15 Measurement system. The z-axis is parallel to the orientation of liquid crystal molecules or ***n***-director. Incident laser polarization (***p***) is parallel to the Z_{meas}-axis, and the polarization (***p′***) of the polarized and depolarized scattering is parallel to the Z_{meas}- and X_{meas}-axes. The polarization angles of ***p*** and ***p′*** are ω and ω', respectively.

$$\hat{E}_Y(\omega) = T_X \sin\omega e^{-\delta X}\hat{X} + T_Z \cos\omega\hat{Z}. \tag{6.20}$$

Similarly, the electric field of the scattered light from the depth Y is given by

$$\hat{E}'_Y(\omega) = T_X \sin\omega e^{-\delta X'}\hat{X} + T_Z \cos\omega\hat{Z}. \tag{6.21}$$

Here, δX ($\delta X'$) is the phase difference of the X-polarized light due to the difference of the refractive index of the incident (scattered) light. T_X (T'_X) and T_Z (T'_Z) are the transmission coefficients of the incident (scattered) light with X- and Z-polarization when the lights pass the interface of the liquid crystal and the glass plate. The transmittances are given by

$$T_L = \frac{2n_g}{n_g + n_L}, \quad T'_L = \frac{2n_g}{n_g + n_L}, \tag{6.22}$$

where n_g and n_L are the refractive index of the glass plate and the liquid crystal for L-polarization ($L = X$ or Z), respectively. Equations 6.20 and 6.21 are rewritten with the spherical irreducible vector as

$$\hat{E}_Y(\omega) = \sum_q (-1)^q n_l^{L,(q)} \hat{e}_{-q}^L, \tag{6.23}$$

$$\hat{E}'_Y(\omega') = \sum_{q'} (-1)^{q'} n_l^{L,(q')} \hat{e}_{-q'}^L, \tag{6.24}$$

where $l = i, j$. The spherical irreducible vectors are defined in the laboratory fixed coordinate system as

$$\hat{e}_0^L = \hat{Z}, \quad \hat{e}_{\pm 1}^L = \frac{\mp 1}{\sqrt{2}}\left(\hat{X} - i\hat{Y}\right). \tag{6.25}$$

The polarization vectors of the incident light are calculated as

$$n_i^{L,(0)} = T_Z \cos\omega, \quad n_i^{L,(\pm 1)} = -\frac{i}{\sqrt{2}} T_X \sin\omega e^{-i\delta X}, \tag{6.26}$$

and those of the scattered light are

$$n_i^{L,(0)} = T_Z \cos\omega, \quad n_i^{L,(\pm 1)} = -\frac{i}{\sqrt{2}} T_X \sin\omega e^{-i\delta X}. \tag{6.27}$$

The Raman intensities are calculated by Equations 6.18, 6.26, and 6.27 as follows:

$$I_{\parallel}(\omega) = I(\omega, \omega) = \int_{Y=0}^{d} \left\langle \alpha_{ij,Y}^{*}(\omega)\alpha_{ij,Y}(\omega) \right\rangle, \tag{6.28}$$

$$I_{\perp}(\omega) = I\left(\omega, \omega + \frac{\pi}{2}\right) = \int_{Y=0}^{d} \langle \alpha_{ij,Y}^{*}(\omega)\alpha_{ij,Y}(\omega) \rangle, \tag{6.29}$$

where d is the thickness of the sample. Since the meaningful Raman components are $\alpha_M(0,0)$ and $\alpha_M(2,0)$ with consideration of the presumptions described before, the polarized and depolarized Raman intensities are obtained in the form

$$I_{\parallel}(\omega) = C_1(\omega) + C_2(\omega)\langle P_2 \rangle + C_3(\omega)\langle P_4 \rangle + C_4(\omega)R, \tag{6.30}$$

$$I_{\perp}(\omega) = C_5(\omega) + C_6(\omega)\langle P_2 \rangle + C_7(\omega)\langle P_4 \rangle - C_4(\omega)R. \tag{6.31}$$

Since the scattered light is collected by an objective lens, the effect of a refracting angle on the measured intensity should be taken into account when the light passes through the interface between the liquid crystal and glass plates [27]. Hence, the measured Raman intensities should be

$$I_{\parallel,\text{meas}}(\omega) = \frac{I_{\parallel}(\omega)}{n_{\parallel}(\omega)^2}, \tag{6.32}$$

$$I_{\perp,\text{meas}}(\omega) = \frac{I_{\perp}(\omega)}{n_{\perp}(\omega)^2}, \tag{6.33}$$

where $n_{\parallel}(\omega)$ and $n_{\perp}(\omega)$ are the refractive index along the Z_L- and X_L-axes, respectively:

$$n_{\parallel}(\omega) = \frac{n_Z n_X}{\sqrt{n_Z^2 \sin^2 \omega + n_X^2 \cos^2 \omega}}, \tag{6.34}$$

$$n_{\perp}(\omega) = \frac{n_Z n_X}{\sqrt{n_Z^2 \cos^2 \omega + n_X^2 \sin^2 \omega}}. \tag{6.35}$$

The coefficients, C_1 to C_7, are given as follows:

$$C_1 = T_X^2 T_X'^2 \left(a^2 + \frac{4}{45} b^2\right) + \left[-2a^2 T_X^2 T_X'^2 + \frac{b^2}{45}(-8T_X^2 T_X'^2 + 3T_Z^2 T_X'^2 + 3T_X^2 T_Z'^2)\right] \cos^2 \omega + \left[a^2 (T_X^2 T_X'^2 + T_Z^2 T_Z'^2) + \frac{b^2}{45}(4T_X^2 T_X'^2 - 3T_Z^2 T_X'^2 - 3T_X^2 T_Z'^2 + 4T_Z^2 T_Z'^2)\right] \cos^4 \omega, \tag{6.36}$$

$$C_2 = -\frac{2}{3} ab T_X^2 T_X'^2 - \frac{4}{63} b^2 T_X^2 T_X'^2 + \left[\frac{b^2}{63}(8T_X^2 T_X'^2 + 3T_Z^2 T_X'^2 + 3T_X^2 T_Z'^2)\right] \cos^2 \omega + \left[\frac{b^2}{63}(-4T_X^2 T_X'^2 - 3T_Z^2 T_X'^2 - 3T_X^2 T_Z'^2 + 8T_Z^2 T_Z'^2) + \frac{ab}{3}(-2T_X^2 T_X'^2 + 4T_Z^2 T_Z'^2)\right] \cos^4 \omega \tag{6.37}$$

$$C_3 = \frac{3}{35} b^2 T_X^2 T_X'^2 + \frac{b^2}{35}(-6T_X^2 T_X'^2 - 4T_Z^2 T_X'^2 - 4T_X^2 T_Z'^2) \cos^2 \omega + \frac{b^2}{35}(3T_X^2 T_X'^2 + 4T_Z^2 T_X'^2 + 4T_X^2 T_Z'^2 + 8T_Z^2 T_Z'^2) \cos^4 \omega. \tag{6.38}$$

$$C_4 = T_X^2 T_X'^2 T_Z^2 T_Z'^2 \cos^2 \omega - T_X^2 T_X'^2 T_Z^2 T_Z'^2 \cos^4 \omega, \tag{6.39}$$

$$C_5 = \frac{b^2 T_X^2 T_Z'^2}{15} + \left[a^2 (T_X^2 T_X'^2 + T_Z^2 T_Z'^2) + \frac{b^2}{45}(4T_X^2 T_X'^2 - 6T_X^2 T_Z'^2 + 4T_X^2 T_Z'^2)\right] \cos^2 \omega - \left[a^2 (T_X^2 T_X'^2 + T_Z^2 T_Z'^2) + \frac{b^2}{45}(4T_X^2 T_X'^2 - 3T_Z^2 T_X'^2 - 3T_X^2 T_Z'^2 + 4T_Z^2 T_Z'^2)\right] \cos^4 \omega, \tag{6.40}$$

$$C_6 = \frac{b^2 T_X^2 T_X'^2}{21} + \left[\frac{ab}{3}(-2T_X^2 T_X'^2 + 4T_Z^2 T_Z'^2) + \frac{b^2}{63}(-4T_X^2 T_X'^2 + -6T_X^2 T_Z'^2 + 8T_Z^2 T_Z'^2)\right] \cos^2 \omega + \left[\frac{b^2}{63}(4T_X^2 T_X'^2 + 3T_Z^2 T_X'^2 + 3T_X^2 T_Z'^2 - 8T_Z^2 T_Z'^2) + \frac{ab}{3}(2T_X^2 T_X'^2 - 4T_Z^2 T_Z'^2)\right] \cos^4 \omega, \tag{6.41}$$

$$C_7 = \frac{4}{35} b^2 T_X^2 T_X'^2 + \frac{b^2}{35}(3T_X^2 T_X'^2 + 8T_Z^2 T_X'^2 + 8T_X^2 T_Z'^2) \cos^2 \omega - \frac{b^2}{35}(3T_X^2 T_X'^2 + 4T_Z^2 T_X'^2 + 4T_X^2 T_Z'^2 + 8T_Z^2 T_Z'^2) \cos^4 \omega. \tag{6.42}$$

The parameters, a and b, are the average and isotropy of the Raman tensor:

$$a = (\alpha_{\parallel} + 2\alpha_{\perp})/3, \tag{6.43}$$

$$b = \alpha_{\parallel} - \alpha_{\perp}. \tag{6.44}$$

The birefringence affects only the parameter R, which is given by

$$R = c_1\left[2a^2 - \frac{4}{45}b^2 + \left(\frac{2}{3}ab - \frac{8}{63}b^2\right)\langle P_2\rangle - \frac{8}{35}b^2\langle P_4\rangle\right] + c_2 b^2\left[\frac{2}{15} + \frac{2}{21}\langle P_2\rangle - \frac{8}{35}b^2\langle P_4\rangle\right]. \tag{6.45}$$

Here, c_1 and c_2 depend on sample thickness d, birefringence Δn, incident laser light wavelength λ, and scattered light wavelength λ',

$$c_1 = \sin(K_1 d)/K_1 d, \quad c_2 = \sin(K_2 d)/K_2 d, \tag{6.46}$$

with

$$K_1 = \frac{2\pi\Delta n(\lambda + \lambda')}{\lambda\lambda'}, \quad K_2 = \frac{2\pi\Delta n(\lambda - \lambda')}{\lambda\lambda'}. \tag{6.47}$$

The Raman tensor parameter $(b/a)^2$ is determined from the depolarization ratio R_{iso} observed in the isotropic phase:

$$R_{iso} = I_X/I_Z = 3b^2/(45a^2 + 4b^2). \tag{6.48}$$

The refractive index of the liquid crystal can be evaluated using an Abbe refractometer or simply measuring the refractive angle for the wedge-shaped cell. The orientational order parameters are obtained by fitting both polarized and depolarized Raman intensities, $I_{\parallel}(\omega)$ and $I_{\perp}(\omega)$, to Equations 6.32 and 6.33, where fitting parameters are $\langle P_2\rangle$, $\langle P_4\rangle$, and R (Figure 6.16). It should be noted that R can be regarded as an independent fitting parameter since it contains the birefringence effect, $d\Delta n$, as well as $\langle P_2\rangle$ and $\langle P_4\rangle$.

When the incident and scattered light polarization are parallel ($\omega = 0$) or perpendicular ($\omega = \pi/2$) to the ***n***-director, C_4 becomes zero or the birefringence effect disappears, and Equations 6.32 and 6.33 become more simple. The orientational order parameters can be obtained by the analysis of polarized Raman intensities at $\omega = 0$ and $\pi/2$. However, the polarized Raman intensity is rather small at $\omega = \pi/2$, and the depolarized intensity is also small at $\omega = 0$ and $\pi/2$ when the orientational order is high such as in Sm*C* (cf. Figure 6.20a-ii and b-ii). The

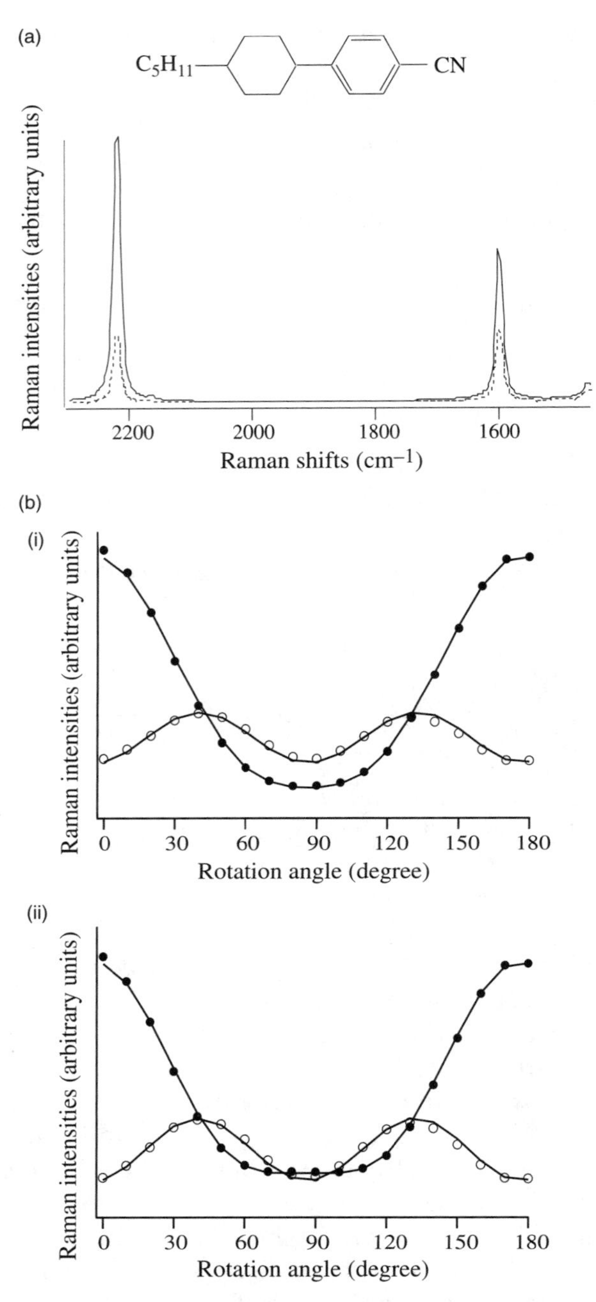

Figure 6.16 (a) Polarized Raman spectra of the nematic liquid crystal 4-(trans-4′-*n*-pentylcyclohexyl) benzonitrile in isotropic phase. The solid and broken lines show $I_{||}(\omega)$ and $I_{\perp}(\omega)$, respectively. (b) Polarized Raman intensities are plotted for the angle between the polarization direction of the incident laser light and the ***n***-director; the phenyl C–C stretching mode was at 1600 cm^{-1} (i), and the CN stretching mode was at 2220 cm^{-1} (ii). The solid and open circles are $I_{||}(\omega)$ and $I_{\perp}(\omega)$, respectively. The lines show the best fitted result by Equations 6.30 and 6.31.

measurement error becomes larger if the measurement is done for only the two configurations of $\omega = 0$ and $\pi/2$.

Since theoretical analysis is done for the Raman tensor, not for the molecule, it might be useful to consider a case where the Raman tensor tilts from the molecular long axis, or in other words, a case in which the third assumption (mentioned before) is not realized. In this case, the orientational order parameters obtained according to the above analysis are the apparent values, so that the order parameters $\langle P_2 \rangle_{app}$ and $\langle P_4 \rangle_{app}$ are used for the Raman tensor instead of $\langle P_2 \rangle$ and $\langle P_4 \rangle$, and $\langle P_2 \rangle_{mol}$ and $\langle P_4 \rangle_{mol}$ are used for molecules. $\langle P_2 \rangle_{app}$ and $\langle P_4 \rangle_{app}$ are related to $\langle P_2 \rangle_{mol}$ and $\langle P_4 \rangle_{mol}$ by the tilting angle β_0 between the Raman tensor axis and the molecular long axis [20] according to the equation

$$\langle P_L \rangle_{app} = P_L(\cos\beta_0)\langle P_L \rangle_{mol} \quad (L = 2 \text{ and } 4). \tag{6.49}$$

6.6.2 Orientational Order Parameters of Nematic Liquid Crystals

Jen et al. were the first to report $\langle P_4 \rangle$ of the nematic and smectic liquid crystal [20]. For *N*-(*p*′-butoxybenzylidene)-*p*-*n*-cyanoaniline (BBCA) in *N*-(*p*′-methoxybenzylidene)-*p*-*n*-butylaniline (MBBA), the stretching vibrational mode of cyano function group at the end of mesogen is used as the probe of the molecular orientational order (Figure 6.17). Both $\langle P_2 \rangle$ and $\langle P_4 \rangle$ are gradually decreased with increasing temperature. In the vicinity of nematic–isotropic phase transition temperature, $\langle P_2 \rangle$ is about 0.4 and $\langle P_4 \rangle$ is almost zero. In the nematic phase, it was found that the temperature dependence of the orientational order parameters differed from the Maier–Saupe theory, which is the simplest mean field molecular model for a cylindrical hard rod. The result indicates that a molecular theory, considering the detailed molecular structure, is needed when a precise molecular distribution is to be discussed.

6.6.3 The Spatial Distribution of Director and the Apparent Orientational Order

Polarized Raman spectroscopy gives information about the orientational order of microscopic molecular structure as described previously. Experimental measurements, however, is done on larger domains compared with the molecular length. Hence, the spatial distribution of the director has to be considered.

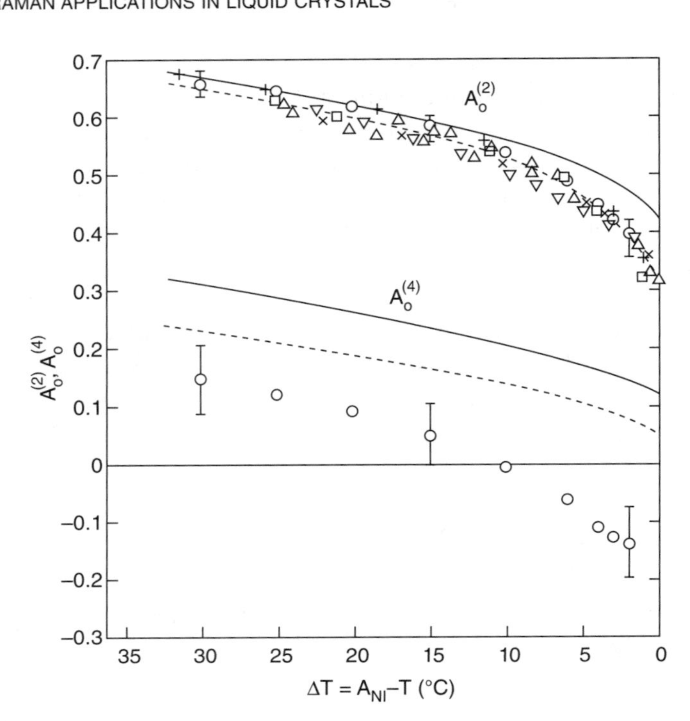

Figure 6.17 Theoretical and experimental values of the nematic orientational order parameters $\langle P_2 \rangle$ and $\langle P_4 \rangle$ of 20% *N*-(*p*′-butoxybenzylidene)-*p*-*n*-cyanoaniline (BBCA) in *N*-(*p*′-methoxybenzylidene)-*p*-*n*-butylaniline (MBBA). Solid lines, Maier–Saupe theory; dashed lines, Humphries–James–Luckhurst theory; circles, Raman measurements. The values of $\langle P_2 \rangle$ are compared with the values from other measurements. Squares, nuclear magnetic resonance (NMR) results; crosses, optical dielectric anisotropy; triangles, diamagnetic anisotropy. From Jen et al. [20] with permission.

The molecular distribution function, f, is given by the combination of the spatial distribution of director, f_{dir}, and the local molecular distribution by the thermal fluctuation, f_1:

$$f = f_{\text{dir}} \otimes f_1. \tag{6.50}$$

The operator ⊗ represents convolution operation. It should be noticed that the system might not be uniaxial any longer.

The apparent orientational order parameters can be evaluated from the experimental result by the analysis with Equations 6.32 and 6.33 for the arbitrary molecular distribution, even for the system with biaxial symmetry if the average of the molecular long axis in the measured system is defined as the center axis of the orientational order. Here,

$\langle P_2 \rangle$ and $\langle P_4 \rangle$ in Equations 6.30 and 6.31 are substituted by $\langle P_2 \rangle_{\text{app}}$ and $\langle P_4 \rangle_{\text{app}}$ to distinguish from the molecular distribution by the thermal fluctuation. The orientational order parameters describing any molecular distribution is given by [28,29]

$$\left\langle D_{m0}^{(L^*)}(\varphi,\theta,\chi)\right\rangle = \int_0^{2\Pi}\int_0^{\Pi}\int_0^{2\Pi} D_{m0}^{(L^*)}(\varphi,\theta,\chi) f_1(\varphi,\theta,\chi)\, d\varphi \sin\theta d\theta d\chi. \quad (6.51)$$

$D_{m0}^{(L^*)}(\varphi,\theta,\chi)$ is the Wigner rotation matrix and (φ,θ,χ) is the Euler angle describing the rotational transformation from the molecular fixed coordinate frame to the apparent orientational center axis. $\langle P_2 \rangle_{\text{app}}$ and $\langle P_4 \rangle_{\text{app}}$ are related to the biaxial orientational order parameters by the following equation:

$$\langle P_L \rangle_{\text{app}} = \sum_{m=-L}^{L} \left\langle D_{m0}^{(L^*)}(\varphi,\theta,\chi)\right\rangle. \quad (6.52)$$

The distribution function f_1 can be obtained by this equation when f_{dir} is given experimentally or theoretically. Conversely, f_1 can be evaluated if f_{dir} is given.

6.6.4 The V-Shaped Switching of Ferroelectric Liquid Crystal

Ferro- and antiferroelectric smectic liquid crystals (SmC^* and SmC_A^*) are widely investigated for the application to LCD. It was found that some ferroelectric or antiferroelectric smectic liquid crystals show the V-shaped electro-optic response [29–32]. They show neither threshold nor hysteresis occurring uniformly without any boundary movement (Figure 6.18). The V-shaped switching is very attractive due to the fast response to the electric field [33,34].

The "random switching model" [30,35–37] was proposed for the mechanism of the V-shaped switching (Figure 6.19a). The frustration and the competition between the ferro- and antiferroelectric behavior bring about the reduction of the interlayer molecular interaction in the particular case. Thus, the Langevin-type reorientation process of the ***c***-directors occurred. On the other hand, Takezoe et al. [38], Park et al. [39], Rudquist et al. [40], and Clark et al. [41] asserted that the charge stabilization and/or the highly collective rotation of the local in-plane directors on the SmC^* tilt cone in the macroscopic scale, and that the frustration did not play any essential role (Figure 6.19b). The molecular distributions expected by the random model and the collective model are different. Therefore, the models can be verified by the molecular distribution at the tip of V.

(a)

(i) $C_{11}H_{23}O$–[F-substituted biphenyl]–$\overset{O}{\overset{\|}{C}}O$–[difluorophenyl]–$\overset{O}{\overset{\|}{C}}O$–$\overset{*}{C}H(CF_3)(CH_2)_4OCH_3$

crystal (ca. –31°C) anti* (31°C) ferri* (36°C) Sm*A* (46°C) isotropic

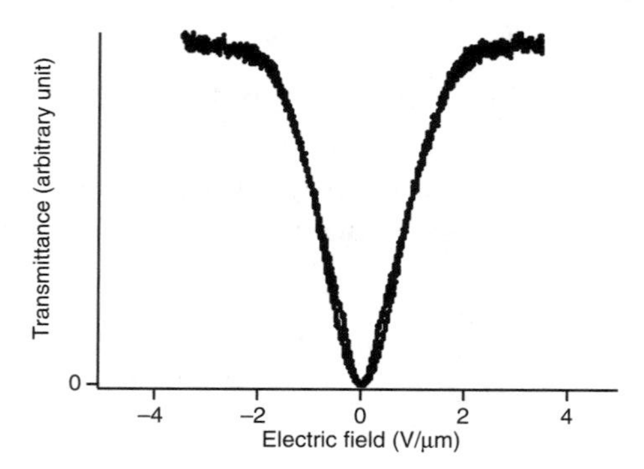

(b)

(ii) $C_{12}H_{25}O$–[biphenyl]–$\overset{O}{\overset{\|}{C}}O$–[tetrahydronaphthyl]–$\overset{O}{\overset{\|}{C}}O$–$\overset{*}{C}H(CF_3)C_6H_{13}$

(iii) $C_{10}H_{21}O$–[phenyl]–[cyclohexyl]–$\overset{O}{\overset{\|}{C}}O$–[phenyl]–$\overset{O}{\overset{\|}{C}}O$–$\overset{*}{C}H(CF_3)CH_2CO_2C_2H_5$

Mitsui mixture, d:e = 63:37 (by weight)

Free-standing film; SmI* (32°C) ferri* (34°C) AF* (56°C) FI* (59°C) Sm*C** (79°C)Sm*A* (100°C) isotropic

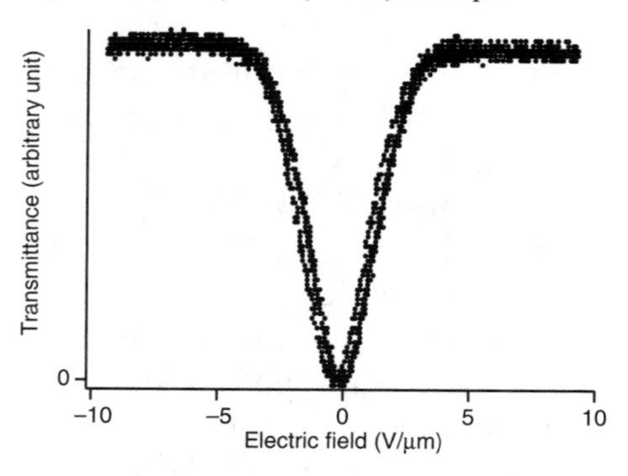

Figure 6.18 The V-shaped switching of compound I (a) and Mitsui mixture (b). The chemical structures and the phase sequences are shown above the plots. Sm*A*, smectic *A*; SmI*, chiral smectic I; AF*, chiral antiferroelectric phase; FI*, chiral ferroelectric phase; Sm*C**, smectic *C*.

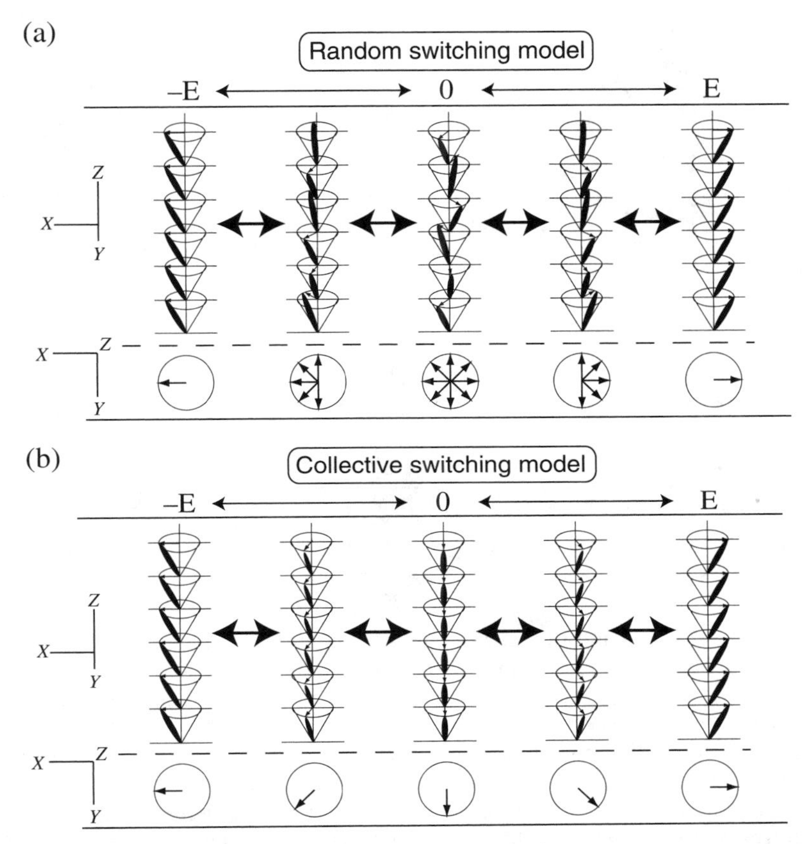

Figure 6.19 The models for the V-shaped switching. Random switching model (a) and collective switching model (b). The directors in each smectic layer are randomly distributed at the tip of V in the random model, but the directors are oriented in one direction within the *yz*-plane in the collective model.

The orientational structure at the tip of V can be analyzed by polarized Raman measurement. The V-shaped switching processes of compound I and Mitsui mixture (Figure 6.18) were investigated [29]. Figure 6.20 illustrates the polarized Raman profile of compound I as a function of the sample rotation angle, $I_{||}(\omega)$ and $I_{\perp}(\omega)$. The measurements were done at the initial state without any external field, at the ferroelectric SmC^* state that is induced by applying direct current fields, and at the tip of the V by applying a triangular wave electric field. The apparent orientational order parameters, $\langle P_2 \rangle_{app}$ and $\langle P_4 \rangle_{app}$, are summarized in Table 6.1.

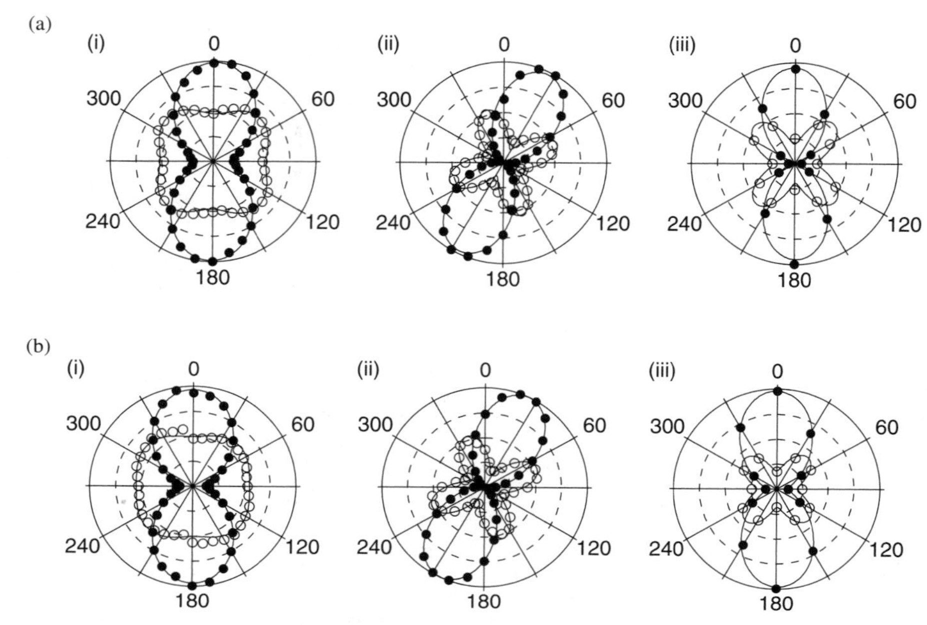

Figure 6.20 Polar plots of polarized Raman scattering intensity (in arbitrary units) versus incident laser polarization (in degree) for the phenyl line at 1600 cm^{-1}. (a) Compound I in antiferroelectric phase at 26 °C under statically applied electric fields of (i) $E = 0$ V, (ii) $E = 4.6$ V/μm, and (iii) at the tip of the V under dynamically applied electric field. (b) Mitsui mixture in antiferroelectric phase at 40 °C under statically applied electric fields of (i) $E = 0$ V, (ii) $E = 5.6$ V/μm, and (iii) at the tip of the V under dynamically applied electric field. Closed and open circles represent polarized and depolarized Raman intensities. The relative depolarized intensity is enlarged by twice as compared with polarized intensity. Solid lines show the best-fitting results with $\langle P_2 \rangle_{app}$ and $\langle P_4 \rangle_{app}$ given in Table 6.1.

Table 6.1 Apparent orientational order parameters of compound (A) at 26 °C and Mitsui mixture at 40 °C

	Electric Field (V/μm)	$\langle P_2 \rangle_{app}$	$\langle P_4 \rangle_{app}$
Compound (A)	0 (dc)	0.40 ± 0.01	-0.02 ± 0.01
	4.6 (dc)	0.78 ± 0.02	0.47 ± 0.04
	0 (1 Hz)	0.70 ± 0.03	0.35 ± 0.05
Mitsui mixture	0 (dc)	0.45 ± 0.01	-0.11 ± 0.01
	4.6 (dc)	0.78 ± 0.01	0.48 ± 0.02
	0 (1 Hz)	0.58 ± 0.03	0.20 ± 0.04

The distribution of the director can be extracted by simulating the order parameters using Equation 6.51. The typical distributions of local director in a smectic layer are assumed with a standard deviation σ:

Table 6.2 Simulated orientational order parameters

	$\langle P_2 \rangle_{app}$	$\langle P_4 \rangle_{app}$
Compound (A)		
Collective model	0.70	0.38
Random model	0.50	0.04
Mitsui mixture		
Collective model	0.74	0.41
Random model	0.55	0.09

The molecular tilt angles used for simulations were 26.3° at 40°C and 22.8° at 60°C, which were determined experimentally by applying DC electric field above the saturation value. See Hayashi et al. [29] for detail.

$$f_{\mathrm{dir}}(\varphi) = \frac{1}{\sqrt{2\pi}\sigma} \exp\left[-\frac{(\varphi + \pi/2)^2}{2\sigma^2} \right], \tag{6.53}$$

$$f_{\mathrm{dir}}(\varphi) = 1/(2\pi). \tag{6.54}$$

Equations 6.53 and 6.54 simulate the collective model with $\sigma = 0$ and the random model, respectively. It is further assumed that the orientational distribution of the molecules (hence, $\langle P_2 \rangle_{app}$ and $\langle P_4 \rangle_{app}$) is due to thermal fluctuations under the DC electric field. The simulated values according to Equation 6.50 are summarized in Table 6.2. Comparing the simulated and experimentally obtained values, it was found that the distribution of the directors of compound I agrees with the collective model rather than the random model. On the other hand, the distribution of the director of the Mitsui mixture was found to agree with the random model. These results showed that both V-shaped switching mechanisms explained by the random model and collective model are possible.

The microscopic analysis for the orientational structure at the tip of V showed that the difference of the molecular distribution results from the competition between the stability of the ferroelectric phase and the molecular interaction to the surface of the alignment layer [32].

6.6.5 Orientation of Molecular Local Structure

The orientational order parameters evaluated by polarized Raman spectroscopy are obtained for the local molecular structure. Hence, it

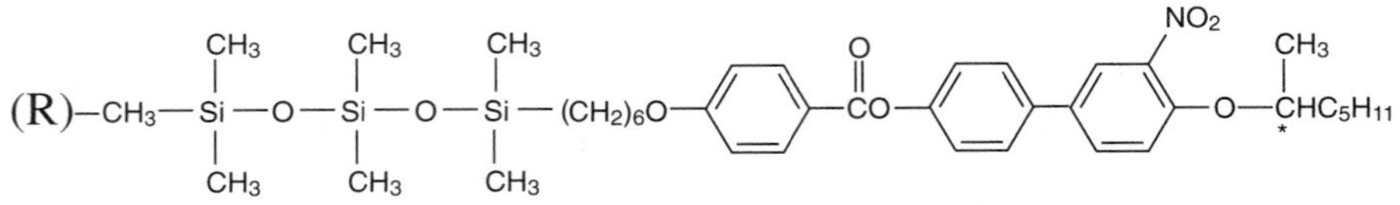

isotropic–56°C–Sm*A**–25°C–Sm*C**

Figure 6.21 Chemical structure, molecular shape, and phase sequence of a liquid crystal 4-[3′-nitro-4′-((R)-1-methylhexyloxy) phenyl] phenyl 4-(6-heptylmethyltrisiloxy-hexyloxy) benzoate (TSiKN65). Sm*A**, chiral smectic A; Sm*C**, chiral smectic C.

is possible to discuss the orientation of the local structure for the molecular long axis.

The molecular structure of the ferroelectric liquid crystal was investigated (Figure 6.21) [42]. The orientational order of phenyl rings at the core part ($1604\,\mathrm{cm}^{-1}$) was measured in the ferroelectric state induced by applying external electric field. The obtained apparent order parameters are 0.48 for $\langle P_2\rangle_{\mathrm{app}}$ and 0.12 for $\langle P_4\rangle_{\mathrm{app}}$. These values are extremely small compared with the typical values for Sm*A* or Sm*C* liquid crystals (cf. Table 6.1). Here, we should be aware that the longest principal axis of the Raman tensor possibly tilts somehow with respect to the molecular long axis; $\langle P_2\rangle_{\mathrm{app}}$ and $\langle P_4\rangle_{\mathrm{app}}$ may generally differ from $\langle P_2\rangle$ and $\langle P_4\rangle$ even in an orthogonal Sm*A* phase. Therefore, $\langle P_2\rangle_{\mathrm{app}}$ and $\langle P_4\rangle_{\mathrm{app}}$ depend not only on the molecular thermal fluctuations but also on the angle β_0 between the longest principal axis of the Raman tensor and the molecular long axis. The tilting angle of the core part was estimated as 28.5° from Equation 6.47, assuming that the typical values of $\langle P_2\rangle$ and $\langle P_4\rangle$ are 0.8 and 0.6, respectively. The large tilting angle of the core part can be deduced from the polarized Raman analysis (Figure 6.22).

6.7 RECENT DEVELOPMENTS OF THE RAMAN TECHNIQUE FOR LIQUID CRYSTAL STUDY

Spectroscopic investigations not only support the development of LCD but also elucidate the physics of liquid crystals, e.g. the molecular interaction between liquid crystal molecules and between the polymer and liquid crystal, and the mechanism of the reorientation of liquid crystal molecules induced by phase transitions or external field.

Polymer-dispersed liquid crystals are mainly developed as materials for display applications. The mechanism of the molecular orientation

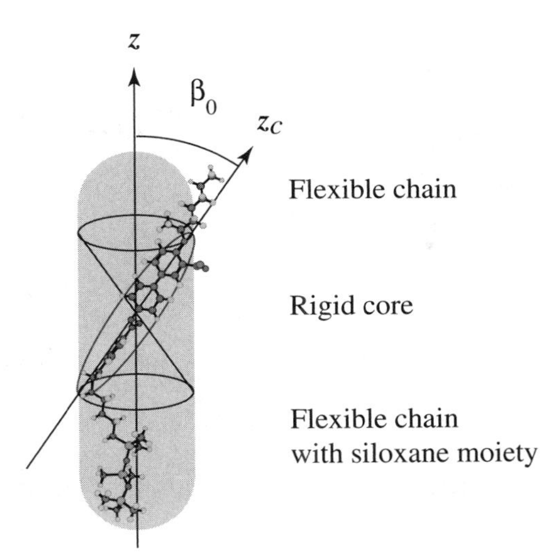

Figure 6.22 Schematic illustrations of the molecular structure of 4-[3′-nitro-4′-((R)-1-methylhexyloxy) phenyl] phenyl 4-(6-heptylmethyltrisiloxyhexyloxy) benzoate (TSiKN65). The molecular long axis is defined such that it gives the lowest moment of inertia (the z-axis). The longest principal axis of the Raman tensor is parallel to the long axis of the core part (the z_C-axis) when the core part is assumed to be a rigid rod with a uniaxial symmetry. β_0 is the angle between the z- and z_c-axes. The molecular orientational distribution has uniaxial symmetry around the long axes. The conformation of the chemical structure illustrated may not be real but is drawn just for the ease of understanding.

at the interface between the polymer skeleton and liquid crystal molecules was investigated by polarized Raman measurements [43].

Coherent anti-Stokes Raman scattering (CARS) microscopy was utilized for fast 3-D imaging of liquid crystal orientation with high spatial resolution and high sensitivity [44]. This directly visualized the orientation change of liquid crystal molecules induced by electric field. 3-D imaging enables a better understanding of the corporative phenomena of liquid crystals confined in a small volume.

Attempts to develop scanning near-field Raman spectroscopy (SNRS) with ultrahigh spatial resolution based on scanning near-field optical microscopy (SNOM) technology are being very actively pursued (e.g. Webster et al. [45]). The SNOM technique is used to investigate the reorientational behavior of liquid crystal at the surface of the alignment layer under applying external electric field [46]. SNRS will yield knowledge about the reorientational behavior in the spatial resolution of nanometer scale directly.

REFERENCES

1. Reinitzer, F. (1888). Beiträge zur Kenntniss des Cholesterins. *Monatsh. Chem.*, 9, 421.
2. Lehmann, O. (1889). Über fließende Krystalle. *Z. Physikal. Chem.*, 4, 462.
3. Friedel, G. (1922). Les états mésomorphes de la matière. *Ann. Phys.*, 18, 273.
4. Meyer, R.B., Liebert, L., Strzelecki, L., Keller, P. (1975). Ferroelectric liquid crystals. *J. Phys. Lett. (France)*, 36, L69.
5. Chandani, A.D.L., Hagiwara, T., Suzuki, Y., Ouchi, Y., Takezoe, H., Fukuda, A. (1988). Tristable switching in surface stabilized ferroelectric liquid crystals with a large spontaneous polarization. *Jpn. J. Appl. Phys.*, 27, L729.
6. Chandani, A.D.L., Ouchi, Y., Takezoe, H., Fukuda, A., Terashima, K., Furukawa, K., Kishi, A. (1989). Novel phases exhibiting tristable switching. *Jpn. J. Appl. Phys.*, 28, L1261.
7. Chandrasekhar, S., Sadashiva, B.K., Suresh, K.A. (1977). Liquid crystals of disc-like molecule. *Pramana*, 7, 471.
8. Niori, T., Sekine, T., Watanabe, J., Furukawa, T., Takezoe, H. (1996). Distinct ferroelectric smectic liquid crystals consisting of banana shaped achiral molecules. *J. Mater. Chem.*, 6, 1231.
9. Demus, D., Goodby, J., Gray, G.W., Spiess, H.-W., Vill, V., eds. (1998). *Handbook of Liquid Crystals*. Weinheim, Germany: Wiley-VCH.
10. Maier, V.W., Saupe, A. (1959). A simple molecular-statistics theory of the nematic liquid-crystalline phase, part I. *Z. Naturforsch.*, 14a, 882.
11. Maier, V.W., Saupe, A. (1960). A simple molecular-statistics theory of the nematic liquid-crystalline phase, part II. *Z. Naturforsch.*, 15a, 287.
12. Luckhurst, G.R. (1971). *Symp. Faraday Soc.*, 5, 46.
13. Humphries, R.L., James, P.G., Luckhurst, G.R. (1972). Molecular field treatment of nematic liquid crystal. *J. Chem. Soc. Faraday Trans. II*, 68, 1031.
14. Maier, V.W., Saupe, A. (1958). A simple molecular-statistics theory of the nematic liquid-crystalline state. *Z. Naturforsch.*, 13a, 564.
15. McMillan, W.L. (1972). Simple molecular model for the smectic A phase of liquid crystals. *Phys. Rev. A*, 4, 1238.
16. Galbiati, E., Zerbi, G. (1986). Molecular mobility and phase transitions in thermotropic liquid crystals: A spectroscopic study of dodecylcyano biphenyl. *J. Chem. Phys.*, 84, 3509.
17. Kirov, N., Dozov, I., Fontana, M.P. (1985). Determination of orientational correlation functions in ordered fluids: Raman scattering. *J. Chem. Phys.*, 83, 5267.
18. Fontana, M.P., Rosi, B., Kirov, N., Dozov, I. (1986). Molecular orientational motions in liquid crystals: A study by Raman and infrared band-shape analysis. *Phys. Rev. A*, 33, 4132.

19. Luckhurst, G.R., Veracini, C.A., eds. (1989). *The Molecular Dynamics of Liquid Crystals*. Dordrecht, the Netherlands: Kluwer Academic.
20. Jen, S., Clark, N.A., Pershan, P.S., Priestley, E.B. (1977). Polarized Raman scattering studies of orientational order in uniaxial liquid crystalline phases. *J. Chem. Phys.*, 66, 4635.
21. Constant, M., Decoster, D. (1982). Raman scattering: Investigation of nematic and smectic ordering. *J. Chem. Phys.*, 76, 1708.
22. Luckhurst, G.R., Yeates, R.N. (1976). Orientational order of a spin probe dissolved in nematic liquid crystals. An electron resonance investigation. *J. Chem. Soc. Faraday Trans. II*, 72, 996.
23. Miyajima, S., Hosokawa, T. (1995). 1H and 2H NMR studies of dynamic orientational, translational, and dipolar orders in the doubly reentrant liquid crystal 4-cyanobenzoyloxy-[4-octylbenzoyloxy]-p-phenylene and its deuterated analog. *Phys. Rev. B*, 52, 4060.
24. Nakai, T., Fujimori, H., Kuwahara, D., Miyajima, S. (1999). Complete assignment of 13C NMR spectra and determination of orientational order parameter for antiferroelectric liquid-crystalline MHPOBC. *J. Phys. Chem. B*, 103, 417.
25. Jang, W.G., Park, C.S., Maclennan, J.E., Kim, K.H., Clark, N.A. (1996). Orientational bias of carbonyl groups in the chiral smectic C phase. *Ferroelectrics*, 180, 213.
26. Zare, R.N. (1987). *Angular Momentum: Understanding Spatial Aspects in Chemistry and Physics*. New York: Wiley.
27. Lax, M., Nelson, D.F. (1971). Linear and nonlinear electrodynamics in elastic anisotropic dielectrics. *Phys. Rev. B*, 4, 3694.
28. Hayashi, N., Kato, T. (2001). Investigations of orientational order for an antiferroelectric liquid crystal by polarized Raman scattering measurements. *Phys. Rev. E*, 63, 021706.
29. Hayashi, N., Kato, T., Aoki, T., Ando, T., Fukuda, A., Seomun, S.S. (2002). Orientational distributions in smectic liquid crystals showing V-shaped switching investigated by polarized Raman scattering. *Phys. Rev. E*, 65, 041714.
30. Inui, S., Iimura, N., Suzuki, T., Iwane, H., Miyachi, K., Takanishi, Y., Fukuda, A. (1996). Thresholdless antiferroelectricity in liquid crystals and its application to displays. *J. Mater. Chem.*, 6, 671.
31. Hayashi, N., Kato, T., Aoki, T., Ando, T., Fukuda, A., Seomun, S.S. (2001). Probable langevin-like director reorientation in an interface-induced disordered SmC*-like state of liquid crystals characterized by frustration between ferro- and antiferroelectricity. *Phys. Rev. Lett.*, 87, 015701.
32. Hayashi, N., Kato, T., Ando, T., Fukuda, A., Kawada, S., Kondoh, S. (2003). Intrinsic aspect of V-shaped switching in ferroelectric liquid crystals: Biaxial anchoring arising from peculiar short axis biasing in the molecular rotation around the long axis. *Phys. Rev. E*, 68, 011702.

33. Yoshida, T., Tanaka, T., Ogura, J., Wakai, H., Aoki, H. (1997). *SID 97 Digest*. Boston, MA: Society for Information Display.
34. Okumura, H., Akiyama, M., Takatoh, K., Uematsu, Y. (1998). XGA TFT-AFLCD with quasi-dc driving scheme for monitor applications. *SID Digest*, 29(46), 1.
35. Fukuda, A. (1995). *Proc. Asia Display '95 (Society for Information Display, Santa Ana, CA)*. Hamamatsu, Japan.
36. Matsumoto, T., Fukuda, A., Johno, M., Motoyama, Y., Yui, T., Seomun, S.S., Yamashita, M. (1999). A novel property caused by frustration between ferroelectricity and antiferroelectricity and its application to liquid crystal displays-frustoelectricity and V-shaped switching. *J. Mater. Chem.*, 9, 2051.
37. Seomun, S.S., Takanishi, Y., Ishikawa, K., Takezoe, H., Fukuda, A. (1997). Evolution of switching characteristics from tristable to V-shaped in an apparently antiferroelectric liquid crystal. *Jpn. J. Appl. Phys.*, 36, 3586.
38. Takezoe, H., Chandani, A.D.L., Seomun, S.S., Park, B., Hermann, D.S., Takanishi, Y., Ishikawa, K. (1998). Proc. Asia Display '98 (Society for Information Display, Santa Ana, CA). Seoul, Korea.
39. Park, B., Nakata, M., Seomun, S.S., Takanishi, Y., Ishikawa, K., Takezoe, H. (1999). Molecular motion in a smectic liquid crystal showing V-shaped switching as studied by optical second-harmonic generation. *Phys. Rev. E*, 59, R3815.
40. Rudquist, P., Lagerwall, J.P.F., Buivydas, M. Gouda, F., Lagerwall, S.T., Clark, N.A., Maclennan, J.E., Shao, R., Coleman, D.A., Bardon, S., Bellini, T., Link, D.R., Natale, G., Glaser, M.A., Walba, D.M., Wand, M.D., Chen, X.-H. (1999). The case of thresholdless antiferroelectricity: Polarization-stabilized twisted SmC* liquid crystals give V-shaped electro-optic response. *J. Mater. Chem.*, 9, 1257.
41. Clark, N.A., Coleman, D., Maclennan, J.E. (2000). Electrostatics and the electro-optic behavior of chiral smectics C: "Block" polarization screening of applied voltage and "V-shaped" switching. *Liq. Cryst.*, 27, 985.
42. Hayashi, N., Kato, T., Fukuda, A., Vij, J.K., Panarin, Y.P., Naciri, J., Shashidhar, R., Kawada, S., Kondoh, S. (2005). Evidence of deVries structure in a smectic A liquid crystal observed by polarized Raman scattering. *Phys. Rev. E*, 71, 041705.
43. Blach, J.-F., Daoudi, A., Buisine, J.-M., Bormann, D. (2005). Raman mapping of polymer dispersed liquid crystal. *Vib. Spectrosc.*, 39, 31.
44. Saar, B.G., Park, H.-S., Xie, X.S., Lavrentovich, O.D. (2007). Three-dimensional imaging of chemical bond orientation in liquid crystals by coherent anti-Stokes Raman scattering microscopy. *Opt. Express*, 15, 13585.
45. Webster, S., Smith, D.A., Batchelder, D.N. (1998). *Vib. Spectrosc.*, 18, 51.

46. Tadokoro, T., Saiki, T., Toriumi, H. (2003). Two-dimensional analysis of liquid crystal orientation at in-plane switching substrate surface using a near-field scanning optical microscope. *Jpn. J. Appl. Phys.*, 42, L57.
47. Hirst, L.S., Watson, S.J., Gleeson, H.F., Cluzeau, P., Barois, P., Pindak, R., Pitney, J., Cady, A., Johnson, P.M., Huang, C.C., Levelut, A-M., Srajer, G., Pollmann, J., Caliebe, W., Seed, A., Herbert, M.R., Goodby, J.W., Hird, M. (2002). Interlayer structures of the chiral smectic liquid crystal phases revealed by resonant x-ray scattering. *Phys. Rev. E*, 65, 041705.

7

RAMAN APPLICATIONS IN FOAMS

Maher S. Amer, Ph.D.
Professor of Materials Science and Engineering, Wright State University

7.1 INTRODUCTION

Foams have been of scientific interest for at least a century [1]. They display unique properties such that they exhibit solid- or liquid-like behavior depending on the magnitude of the applied stress [2]. Under typical low-stress conditions, wet foam is a fragile structure that exhibits a shear modulus in the order of 10 Pa. The precise value of the shear modulus depends on a number of parameters including the bubble size, water content, and shear rate. The foam starts to flow as a viscous liquid under higher stresses. While foams are of significant technological importance in several applications such as fire fighting, ore segregation through flotation, enhanced oil recovery, and fractionation [3], they are also of increasing academic interest. For example, at one extreme of the length scale, it has been noted that the distribution of galaxy clusters in the universe has a foam-like structure [4], while at the other extreme, the metric of space–time itself has been described as foam-like at a scale near the Planck length ($\lambda_p = 10^{-35}$ m) [5]. At intermediate scales, more common in condensed matter physics, analogies with foam-like structures have again proven valuable in the description of a wide range of diverse systems, including grain growth in metals [6],

Raman Spectroscopy for Soft Matter Applications, Edited by Maher S. Amer

emulsions [7], magnetic fluids froth [8,9], Langmuir monolayer systems [10], and noncoalescing systems under microgravity conditions [11]. More interestingly, realizing the fact that for liquid films in foams, the correlation length, which is also a measure of a liquid–gas interface and a measure of the range of intermolecular forces, is usually in the range of 1–5 nm, and that the ratio of the line tension to the surface tension is usually in the range of 20 nm [12], raises another important aspect of investigating foam systems. Thin foam film systems provide an exceptional opportunity for the investigation of truly thermodynamic small systems, the core of what is known as "nanotechnology." In this chapter, we will discuss the history, physics, and general applications of foams as one of the important members of the soft matter family. We will also discuss the efforts to utilize Raman spectroscopy in analyzing and investigating foam systems and the potential of such characterization technique in investigating foams. In other words, in this chapter, we will try to discuss an important soft matter system that has its roots in the Renaissance era and is still active, important, and not fully explored in the nanotechnology era.

7.2 FOAM HISTORY

Foam is a two-phase material system in which the gas phase is enclosed in a liquid phase. This is different from, but greatly analogous to, emulsions and binary liquid foams in which two liquid phases are dispersed together. Liquid–gas foams are the focus of this chapter. The theory of liquid–gas foams is not quite elementary. However, a single material property, the surface tension, γ, of the liquid–gas interface, is all that matters for a solid start toward understanding the subject. The science of liquid foams emerged from the scientific study of liquid surfaces that dates as early as the fifteenth century, the time of Leonardo da Vinci, who pioneered the study of what is known today as the capillarity phenomenon [13]. The important concept of *surface tension* was introduced by J.A. Segner in 1751, and the concept of *contact angle* was introduced by Thomas Young in 1805, and independently by the famous mathematician the Marquis de Laplace. Since the fifteenth century, researchers have tackled foams, in the form of soap bubbles, from two different aspects. Researchers interested in physical, chemical, and biological sciences investigated soap bubbles for their macroscopic and molecular surface properties and their similarities to other systems of interest. Mathematicians, however, investigate foams due to their interest in problems that involve minimization of a surface area contained by a fixed boundary [14]. The two aspects were tackled separately and were uncorrelated. It was not until the nineteenth century that the two aspects of foam investigation started to unite, based upon the experimental results of the Belgian physicist Joseph Plateau, who showed that analogue solutions to the minimization problem could be obtained experimentally by dipping wire frames in soap solutions [15]. These experimental techniques and results inspired mathematicians to investigate new analytic solutions for the minimum area problem and its associated geometric properties. Around the turn of the twentieth century, Charles Boys was known as a great popularizer of soap foam. He gave numerous lecture demonstrations which intrigued everyone in the audience and published a fascinating book entitled *Soap Bubbles and the Forces Which Mould Them*, which has been popular among young people since its publication in 1890.

Only in the 1930s of the twentieth century important steps have been taken in the direction of the minimization problem by Douglas [16] and his contemporaries [17]. More recently, mathematical advances in understanding foam structure have been achieved utilizing computer simulations as we will discuss later in this chapter [18]. Since then, a large and very profitable chemical industry has grown up based upon foams and their surface properties. Consequently, large research programs in both industry and academia are still investigating the fascinating subject of foams (Figure 7.1).

Leonardo da Vinci
1452–1519

Joseph Plateau
1801–1883

Charles Boys
1855–1944

Figure 7.1 Da Vinci, Plateau, and Boys, three pioneers who started, established, and popularized the field of foam.

7.3 LIQUID–GAS INTERFACE AND FOAM STRUCTURE

In liquids, molecules experience forces from different directions due to interaction with neighboring molecules. The resultant force of all such forces, averaged over a time much longer than the collision time, will be zero. Molecules residing near the liquid–gas interface (liquid surface), however, do not experience the same forces from all different directions due to the presence of the interface. The result is a net force (also if averaged over time much longer than the collision time) acting on the surface molecules. This net force acts as if pulling the liquid molecules at the surface together, causing the surface to reduce its area providing the surface has such ability. Such net force at the liquid–gas interface creates what is known as *surface tension* (or surface energy per unit area), γ, and is also known to affect the local density of the liquid near the liquid–gas interface. Surface tension is perpendicular to any line drawn on the liquid surface. It is important to note that the surface tension of a liquid varies with temperature and behaves differently for associated and unassociated liquids. An unassociated liquid is a liquid consisting of unassociated molecules such as benzene and carbon tetrachloride. An associated liquid is a liquid consisting of associated molecules such as water, methanol, and formic acid. For an unassociated liquid under thermodynamic equilibrium with its vapor, which is usually not the case in foams, surface tension was found to follow the empirical equation

$$\gamma = \gamma_0 \left(1 - \frac{T}{T_c}\right)^n. \tag{7.1}$$

where γ_0 and n are constants that depend on the liquid nature. A typical value for n is 1.2. It is clear from the equation that the liquid surface tension decreases with increasing temperature and vanishes as the temperature reaches a critical value [19].

In the case of a liquid film (soap film, for example), surfactant molecules disperse their heads in the liquid while their tails are avoiding interaction with the liquid molecules (by sticking out of the liquid). The result is that a monolayer of the surfactant molecules is formed at the liquid–gas interface, and a thin layer of the liquid is trapped between the surfactant molecules in the form of a thin film forming a bubble boundary. In typical foam films, the film thickness ranges between 2×10^5 and 50 Å (20 μm to 5 nm). A schematic of a typical film structure is shown in Figure 7.2.

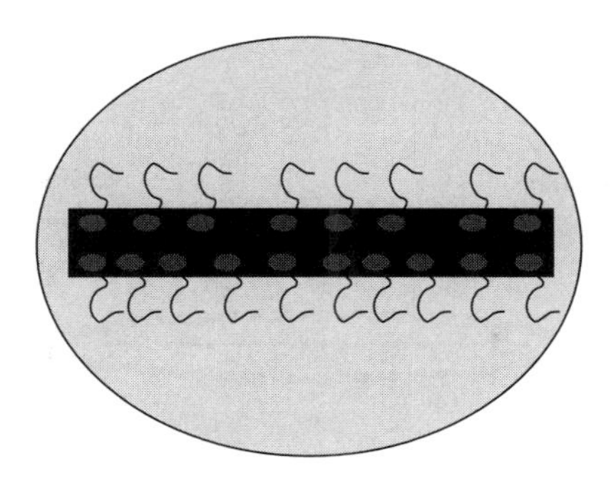

Figure 7.2 Schematic molecular structure of a liquid film in foams.

Since surfactant heads are typically charged, their mutual repulsion creates a pressure inside the liquid film known as the *disjoining pressure*. Such pressure maintains equilibrium in the film and prevents surface tension from thinning the film to zero thickness [20,21].

7.4 LAPLACE–YOUNG EQUATION

In 1805, Thomas Young [22] and, independently, Laplace [23] developed an equation correlating the excess pressure (Δp) across a curved liquid surface at a point with principal radii of curvature R_1 and R_2, with the liquid surface tension (γ) as follows:

$$\Delta p = \gamma\left(\frac{1}{R_1} + \frac{1}{R_2}\right). \tag{7.2}$$

In the case of liquid films, and due to the presence of two surfaces, it is convenient to introduce the concept of *film tension* (σ) defined as the force per unit length of film and equal to twice the surface tension (γ). For a single spherical bubble of radius (r), the forces acting are the film tension (σ) and the pressure difference between the inside and the outside of the bubble (Δp). Balancing the forces acting on a hemispherical section (as shown in Figure 7.3) of the bubble yields

$$\pi r^2 \Delta p = 2\pi r \sigma. \tag{7.3}$$

Hence,

$$\Delta p = \frac{2\sigma}{r} \quad \text{or} \quad \Delta p = \frac{4\gamma}{r}. \tag{7.4}$$

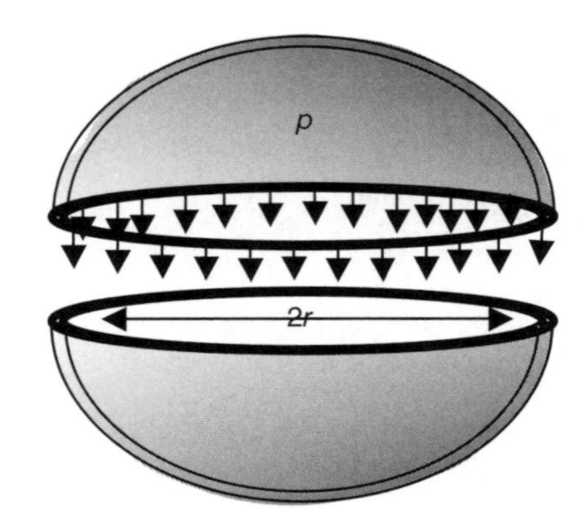

Figure 7.3 Spherical hemisphere equilibrium.

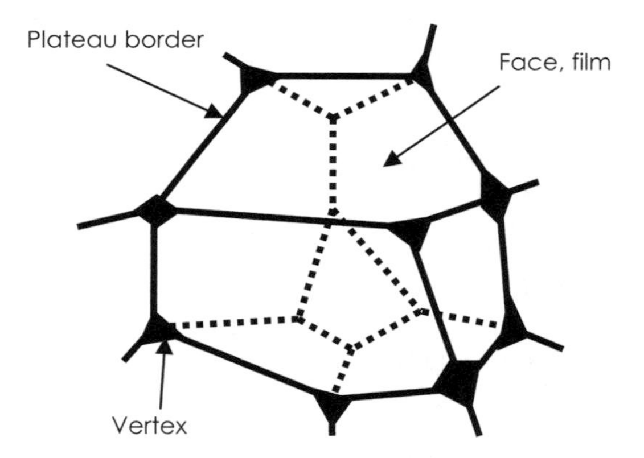

Figure 7.4 The structure of a foam cell.

It is important to note that the same result can be reached using the Laplace–Young equation (Equation 7.2) by substituting r (the spherical bubble radius) for both R_1 and R_2, and by taking into consideration the fact that the film has two surfaces.

7.5 FOAM STRUCTURE

Foam is a 3-D network of liquid films entrapping a number of gas cells in the form of a cellular structure as shown in Figure 7.4. The foam may contain more or less liquid depending on the circumstances. Very little liquid (usually less than a percent in weight) will result in what is known as *dry foam*. The entrapped gas bubbles are in a polyhedral cell form, with the liquid films as their faces, which are not flat. The films (faces) meet in lines (polyhedral edges) known as Plateau borders, and the borders meet at vertices. If the liquid content increases beyond a percent, the liquid is concentrated first in the Plateau borders, which become channels of finite width; polyhedral cells start to assume spherical shapes, and the foam turns into what we usually refer to as *wet foam*. Further increase in the liquid content results in bubble separation, and the foam loses its rigidity and turns into a bubbly liquid [24]. For academic purposes, 2-D foams have been utilized in a number of experimental and simulation investigations. Foam is considered 2-D if the thickness of the foam layer is smaller than the average size of the cell. A classical model used in experimental investigations is the thin foam layer sandwiched between two glass slides.

It is important to note that the pressure difference between adjacent cells can be calculated using Equation 7.4, which is based on the Laplace–Young equation. In 2-D foam, however, the equation has to be modified as follows:

$$\Delta p = \frac{2\gamma}{r}. \tag{7.5}$$

The disjoining pressure inside the liquid film has to be taken into consideration when calculating the pressure difference between adjacent cells using the Laplace–Young equation. However, in case of foams with negligible film thickness and fairly constant surface tension, disjoining pressures may be neglected when applying the theory [25]. Experimentally measured joining pressures have been found to depend on the film thickness and to range between a very modest 500 and 5000 Pa [26,27] (Figure 7.5).

For the study of foam structure and properties, theory and experiments can be focused on three different length scales depending on the nature of the study. For chemistry and stability of the liquid film investigation, molecular effects can be investigated on the nanometer length scale. For studies investigating foam physics and the stability of the structure, its equilibrium, coarsening, etc., foam can be investigated on the millimeter length scale (several cells limit). Finally, to investigate the mechanical behavior of foams and its engineering properties and to develop continuum models of foams, the investigation can be focused on the meter length scale.

As we mentioned before, understanding the physics of foam requires investigating its properties on the micro level (the cell level in Figure 7.6). It is necessary to understand the correlation between structure

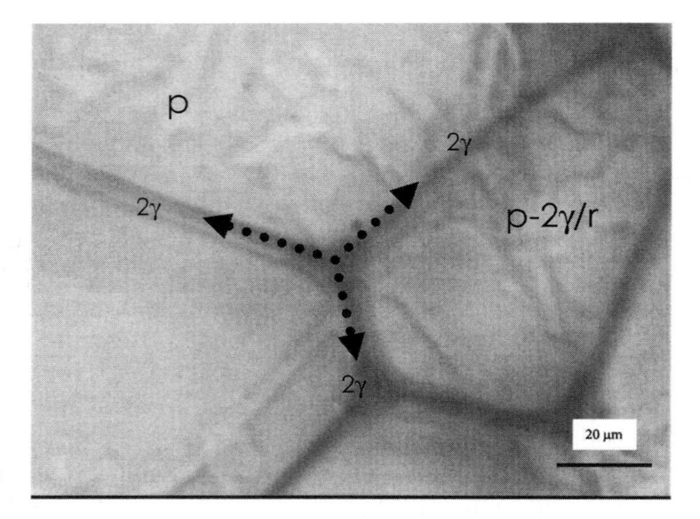

Figure 7.5 A photograph of a 2-D dry foam cell.

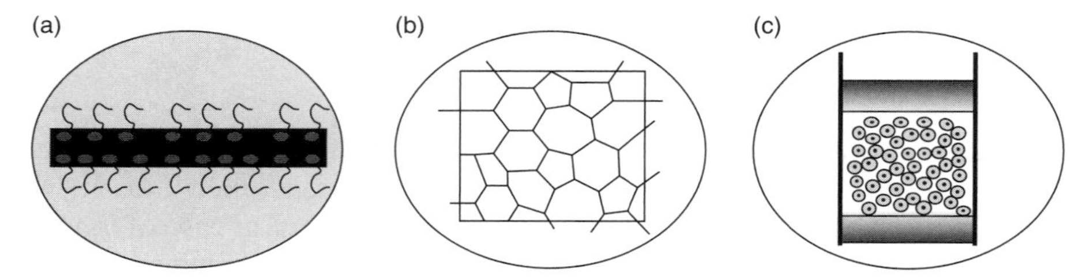

Figure 7.6 Different length scales to investigate different aspects of foams. (a) Nano-, (b) micro-, and (c) macro-length scales.

elements (such as cell shape, number of sides, size, cell strut angle) and physical properties of the foam, such as stability and mechanical behavior. According to the first equilibrium rule of the laws of Plateau, in dry foam, films can intersect only three at a time and the intersection angle must be 120°. It can be simply inferred that here Plateau was considering a perfect foam structure under thermodynamic equilibrium, which is seldom the case. In such a system, cells will be of the same size, six sided, and the strut angle should always be 120° as a consequence of a uniform, stable surface tension in all the films and balanced forces at all Plateau borders. Unfortunately, foam reality is different. Regardless of the preparation method and any subsequent treatment, foam is not ever in a metastable state due to continuous gas diffusion between cells, a process known as *coarsening*. Fortunately, the coarsening process is slow and occurs over tens of minutes. Hence, except for sudden instantaneous local topological changes, considering a foam as being "in equilibrium" is a useful but not very accurate assumption.

Knobler and his coworkers [28], using computer simulation and optical microscopy, have shown that in 2-D foams, the number of sides for foam cells has a distribution and ranges between 3 and 12, with a most probable value of 6 as shown in Figure 7.7a. Other investigators [29], using soap foams, have reported a very similar cell-side distribution but with five-sided cells as the most probable. Assuming equilateral cells, the number of cell sides (n) can be expressed as grain angle (φ) according to the equation

$$\varphi = \frac{\pi(n-2)}{n}. \tag{7.6}$$

The grain angle distribution yields a range between 60° and 150° (Figure 7.7b). This distribution of grain angles is very close to the 75°–175° range observed experimentally by Amer et al. for 2-D soap foam [30].

On the continuum level, foam is known to exhibit a mechanical behavior characteristic of an elastic solid at low stresses, then behaving as a plastic solid at moderate stresses, and finally exhibiting the flow of a liquid at higher stresses. The elastic solid behavior is mainly due to cell (gas bubble) deformation under low stresses. Observed plastic behavior is mainly associated with gas bubble rearrangement, and the

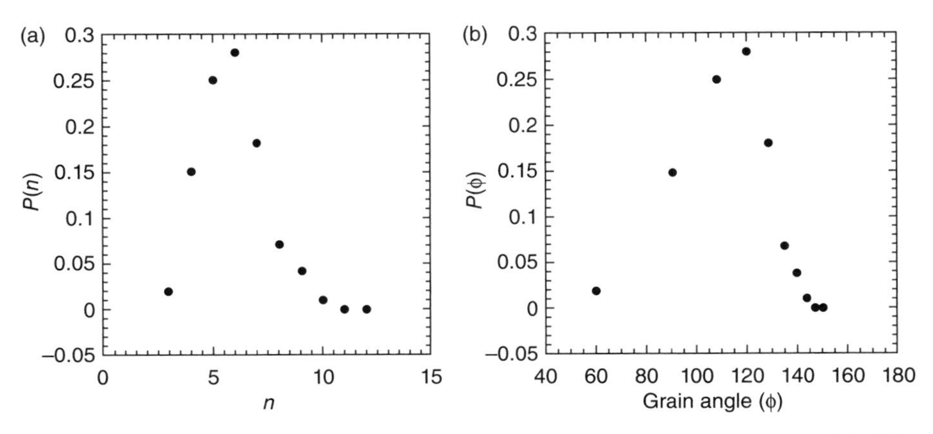

Figure 7.7 (a) Number of cell sides (n) distribution in 2-D dry foam as modeled by computer simulation (reproduced based on data published in [28]. (b) Corresponding grain (cell strut) angle distribution. P plotted in the vertical axis represents the probability of having a cell with (n) or (ϕ), respectively.

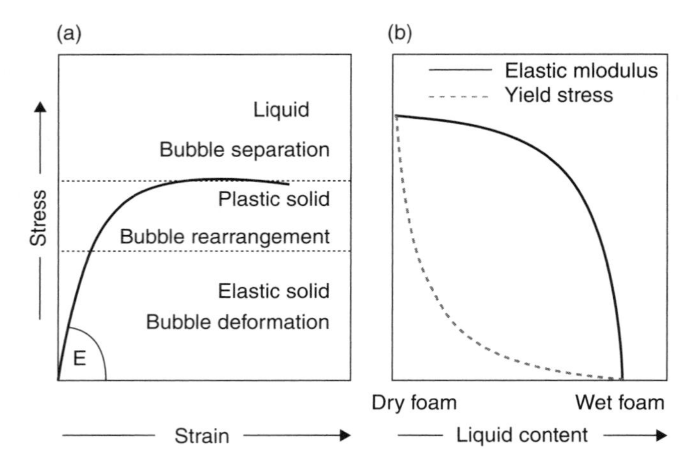

Figure 7.8 Schematic presentation of (a) the typical stress–strain relationship in foam and (b) the typical effect of foam liquid content on both elastic modulus (E) and yield stress.

liquid flow is associated with bubble separation. Figure 7.8a shows a schematic stress–strain relationship for a typical dry foam illustrating the behavior change among the three distinct states of the mechanical behavior of foam. Mechanical constants of foam have been found to be very sensitive to the liquid content of the foam. As the liquid content of the foam increases and the foam transforms from a dry foam to a wet foam, both elastic constant and yield stress have been found to drop dramatically as shown in Figure 7.8b.

Thus far, it is clear that the foam chemistry and interaction at the molecular level both dictate the film properties which in turn control the cell structure and affect the continuum-level mechanical behavior of the foam. For example, Overaker et al. [31] modeled the effect of cell strut angle on the mechanical behavior of 2-D foams. Their simulation showed that the elastic modulus of the foam is highest when the strut angle is 100 °.

7.6 EXPERIMENTAL TECHNIQUES FOR FOAM INVESTIGATION

In spite of the fundamental importance of understanding foam structure on all length scales mentioned above, very few experimental techniques have been utilized in its investigation. Recent simulation studies [32–36] aiming at elucidating the mechanics of foams under shear stresses have pointed out the need for more precise experimental techniques to gather crucial information on the structure of foams. Knobler [37] and Durian and coworkers [38–40] indicated that experimental measurements are critical for verification of simulation results and are essential for understanding the structure/behavior relationship in foams. Since the very famous Matzke's experiment [41] in 1945, most of the structure investigation of foam cells has been limited to direct observation using simple optical microscopy. This technique is more effective for 2-D foam samples but cannot be as successfully applied to 3-D foams, unless the foam is very dry, due to multiple scattering of light, which makes the foam opaquely white and its structure impossible to observe. To give the technique credits, it has been used effectively in coarsening investigations in 2-D foams.

Only a decade ago, the direct optical observation technique was advanced to include computerized optical tomography [42]. In this technique, an objective lens, with very small depth of field, is used to produce 2-D optical slices of the foam imaged on a CCD camera. A computer is used to generate a 3-D image of the foam.

Another modification of the direct optical observation method has been the optical glass fiber probe method [43]. In this technique, an optical glass fiber probe that is emitting light is inserted into the foam and reflected light intensity is recorded as a function of probe position. If the probe tip is into a gas bubble, light is reflected and intensity is recorded. If the probe tip, however, is in a liquid (film or Plateau border), there will be hardly any reflection and no intensity will be recorded. Statistical methods are then employed to transform the recorded intensity into bubble-size distribution through the foam. Concerns have been raised regarding the reliability of the technique, especially in light of the damaging effect of the probe tip on the foam structure and the ability of the tip to detect small bubbles as accurately as it does for larger bubbles.

Light scattering techniques based upon diffusing wave spectroscopy have also been employed to investigate foam structure [44]. Useful information regarding the liquid fraction in foam could be extracted using statistical methods from light transmission measurements through foam samples as a function of time. The addition of UV fluorescent dye and monitoring the fluorescence intensity has been shown to be an efficient technique for liquid fraction measurements, and is especially useful in making reliable measurements at the very low liquid fraction end of the measurements [45]. Magnetic resonance imaging (MRI) has also been used as a technique to map the liquid density through foam. The technique capability of producing detailed images has also been utilized in investigating coarsening phenomenon in 3-D foam [46].

In addition to optical techniques, a number of electrical-based techniques were utilized to investigate foam structure as well. Electrical-based techniques were all based on the change in electrical properties, namely, capacitance and resistance of the foam due to change in liquid fraction, and were all employed to measure the liquid fraction of the foam. The capabilities of such techniques ranged from a single point measurement to multipoint imaging using segmented capacitance plates [47]. To this end, it is clear that most of the experimental techniques utilized in characterizing foams targeted the cellular and continuum levels of foam investigation. More experimental techniques are required to investigate foam on the nanoscale, or in other words the film level of observation.

7.7 RAMAN SPECTROSCOPY IN FOAM CHARACTERIZATION

In spite of its powerful characterization capabilities, Raman spectroscopy has not been fully utilized in investigating foams. Since surfactant molecules play a central role in the formation and stabilization of foam, Raman spectroscopy with its ability to monitor molecular vibration can provide vital information on their packing, mobility, and conformation [48]; hence, a clear and informative picture of molecular structure within the liquid film inside foam can be obtained. Understanding the structure and performance of the liquid film (foam on the nano- or molecular level) enables a broader and better understanding of foam properties and behavior on macro and continuum levels, as we have discussed before. It can be confidently said that with its ability to extract vital information about the chemical, thermal, electrical, and mechanical properties of the system, Raman spectroscopy is still an unexplored frontier in the field of foam investigation.

As a laser beam impinges on a 3-D foam, the passage of light through this scattering medium is essentially a random walk with a mean free path l^* termed diffusive propagation. Such diffusive propagation results in what can be termed *diffusive excitation*, creating a distribution of elementary Raman scattering centers in the bulk of the foam. The Raman signal will, in turn, undergo a diffusive propagation in all directions, eventually reaching the foam cell boundaries, enabling the Raman signal from the bulk foam to be detected. The specific dimension of the Raman intensity distribution at scattering focal plane (usually the surface of the foam) is proportional to the transport mean free path l^*, which has been connected with the size of the foam bubble (d) according to the relationship [49]

$$l^* = 3.5d. \tag{7.7}$$

When foam bubble size (d) increases, the dimension of the Raman intensity in foam increases or, in other words, broadens, and hence, its maximum decreases. This leads to a decrease in the brightness of the image projected on the entrance slit of the spectrometer and, hence, a reduction in the detected Raman signal. Thus, for an optimum Raman signal collection, the spectrometer slit width should be adjusted to match the mean bubble size of foam [50].

Raman spectroscopy has been used to investigate the structure of the thin liquid films in stable 3-D foam [50]. The investigation, focusing on the Raman vibration fundamentals of the surfactant molecules, yielded important information regarding the structure of the liquid film. Two phases existing in the foam films were identified: a lamellar phase and an isotropic phase. The shape of the Raman spectra in the C–H stretching region (2800–3000 cm^{-1}), C–H bending region (1400–1500 cm^{-1}), and their characteristics indicated that the lamellar phase is an ordered multilayer hexagonal structure of the surfactant molecules in all-trans conformation. The spectra in the C–C stretching region (1050–1150 cm^{-1}) supported this conclusion as they showed the absence of features characteristic of gauche conformations. On the basis of Raman data, a model of foam molecular structure was proposed in which small bilayer lamellae dispersed in the liquid films after foam formation gradually self–organize around the bubbles in large shell-like bilayer structures. Figure 7.9a shows the Raman spectra in the C–H stretch region recorded at different times as wet foam ages. As shown

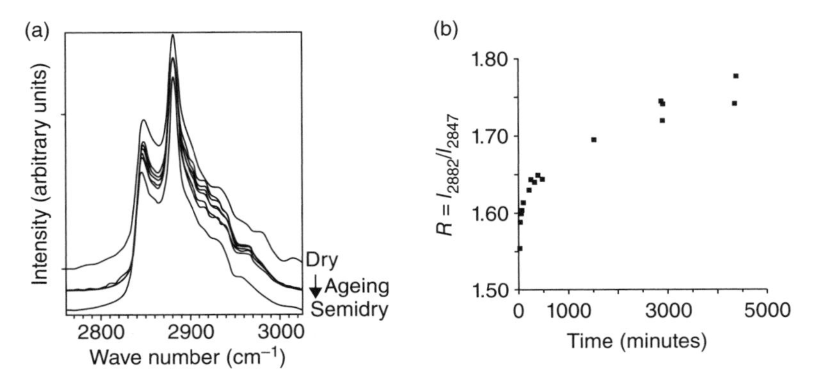

Figure 7.9 (a) Raman spectra in the C–H stretch region for foam as it dries and (b) the intensity ratio (R) as a function of time. (Reproduced from [50] by kind permission of the American Physical Society.)

in the spectra, the C–H stretch region contains two prominent peaks at 2882 and 2847 cm^{-1} that correspond to the C–H symmetrical and the antisymmetrical stretch modes [51]. The organization process of small lamellae into large bilayer structures could be followed by time-dependent changes in the intensity ratio ($R = I_{2882}/I_{2847}$) of the C–H band as shown in Figure 7.9b.

Recently, Raman spectroscopy was used to investigate 2-D foam utilizing fullerene nanospheres (C_{60}) as nanosensors [30]. The fullerene molecules were dispersed into the liquid film during foam formation and Raman spectroscopy was used to monitor the Raman active pentagon pinch mode (1469 cm^{-1}) throughout the foam junctions, and the peak position was plotted as a function of the cell strut angle (grain angle) as shown in Figure 7.10. Realizing that the shift in the fullerene Raman mode is mainly due to chemical interaction with surfactant molecules, and that the magnitude of this shift does depend on the concentration, conformation, and packing of such surfactants in the liquid film [52], it becomes clear from the demonstrated dependence of the Raman band position on the cell strut angle (Figure 7.10) that the film structure depends on the cell strut angle. It is interesting to realize that fullerene nanospheres were least disturbed at strut angle equal to 120°, the equilibrium strut angle as predicted by Plateau, and

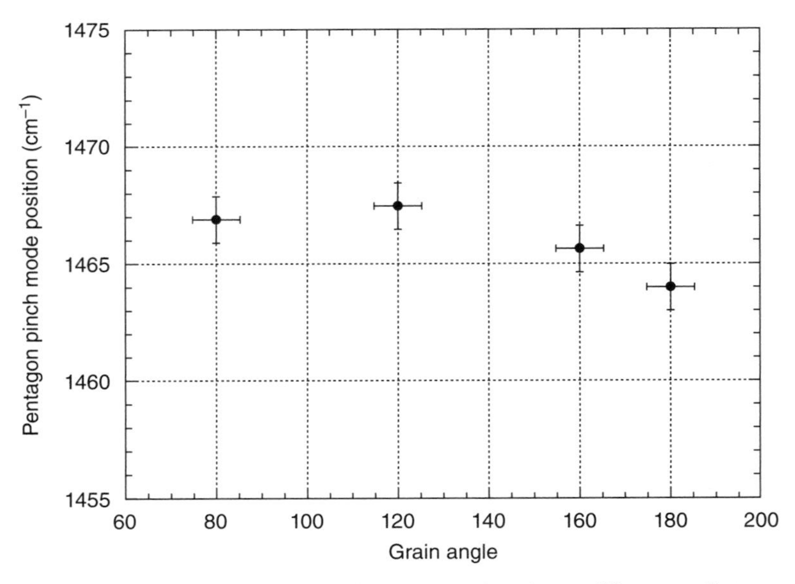

Figure 7.10 Raman band position of fullerene molecules at different cell strut angles in 2-D foam. (Reproduced from [30].)

the most probable strut angle as predicted by computer simulations (Figure 7.7b).

7.8 CONCLUDING REMARKS

As we have demonstrated in this chapter, soft foam is an important class of soft matter that resembles a large number of other important systems of interest in physics, biology, and cosmology, and is fundamentally related to nanotechnology. Understanding the structure–property relationship in foams on all three levels of length scale and correlating such length scales together is a challenging and fundamental problem in the physics of dispersed systems. Experimental techniques applicable to this field are much needed for its solution. Raman spectroscopy has been shown to be a method of great potential due to its noncontact nature and its ability to extract information about characteristic molecular vibrations, and to correlate such information to the liquid film structure and stability in the foam.

REFERENCES

1. Maxwell, J.C. (1874). Molecular motion. *Nature*, 10, 119.
2. Weaire, D., Hutzler, S. (1999). *The Physics of Foams*, Chapter 1. New York: Oxford University Press.
3. Aubert, J.H., Kraynik, A.M., Rand, P. B. (1986). Aqueous foams. *Sci. Am.*, 254, 74.
4. De Lapparent, V., Geller, M.J., Huchra, J.P. (1986). A slice of the universe. *Astrophys. J. Lett.*, 302, L1.
5. Weaire, D., Hutzler, S. (1999). *The Physics of Foams*, Chapter 15. New York: Oxford University Press.
6. Aboav, D.A. (1972). Foam and polycrystal. *Metallography* 5, 251.
7. Lacasse, M.D., Grest, G.S., Levine, D., Mason, T.G., Weitz, D.A. (1996). Model for the elasticity of compressed emulsions. *Phys. Rev. Lett.*, 76, 3448.
8. Babcock, K.L., Westervelt, R.M. (1989). Topological "melting" of cellular domain lattices in magnetic garnet films. *Phys. Rev. Lett.*, 63, 175.
9. Elias, F., Flament, C., Glazier, J.C., Graner, F., Jiang, Y. (1999). Foams out of stable equilibrium: Cell elongation and side swapping. *Philos. Mag. B*, 79, 729.
10. Akamatsu, S., Rondelez, F. (1991). Fluorescence microscopy evidence for two different LE-LC phase transitions in Langmuir monolayers of fatty acids. *J. Phys. II*, 1, 1309.

11. Monti, R., Dell'Aversana, P. (1994). Microgravity experimentation in non-coalescing systems. *Microgravity Q.*, 4, 123.
12. Rowlinson, J.S. (1983). The molecular theory of small systems, Faraday Lecture. *Chem. Soc. Rev*, 12, 251.
13. Porter, A.W. (1964). *Encyclopedia Britannica*, Vol. 21, 595.
14. Boys, C.V. (1890). In: London, E., Young, J.B., eds., *Soap Bubbles and the Forces which Mould Them*. New York: SPCK.
15. Taylor, R., ed. (1966). *Scientific Memoirs*, Vol. 5. New York: Johnson Reprint Corp., 584–712.
16. Douglas, J. (1931). Solution of the problem of plateau. *Trans. Am. Math. Soc.*, 33, 263–321.
17. Nitsche, J.C.C. (1974). Plateau's problems and their modern ramifications. *Am. Math. Monthly*, 81, 945–968.
18. Almgren, F.J., Jr., Taylor, J.E. (1976). Geometry of soap films and soap bubbles. *Sci. Am.*, 235, 82–93.
19. Harkins, W.D. (1952). *The Physical Chemistry of Surface Films*. New York: Reinhold.
20. Babak, V. (1998). Thermodynamics of plane-parallel liquid films. *Colloids Surf. A*, 142, 135.
21. Klitzing, R.V., Espert, A., Asnacios, A., Hellweg, T., Colin, A., Langevin, D. (1999). Forces in foam films containing polyelectrolyte and surfactant. *Colloids Surf. A*, 149, 131.
22. Young, T. (1805). An essay on the cohesion of fluids. *Phil. Trans. R. Soc. Lond.*, 95, 65.
23. Laplace, P.S. (1806). *Mécanique céleste*. Impr. Imperiale, Suppl. to 10th book. See also P.S. Laplace, *Celestial Mechanics*. Trans. N. Bowditch, Chelsea Pub. Co., 1966.
24. Weaire, D., Hutzler, S. (1999). *The Physics of Foams*. New York: Oxford University Press, 10.
25. Weaire, D., Hutzler, S. (1999). *The Physics of Foams*. New York: Oxford University Press, 22.
26. Bergeron, V. (1999). Measurement of forces and structure between fluid interfaces. *Curr. Opin. Colloid Interface Sci.*, 4, 249.
27. Karraker, K.A., Radke, C.J. (2002). Disjoining pressures, zeta potentials and surface tensions of aqueous non-ionic surfactant/electrolyte solutions: Theory and comparison to experiment. *Adv. Colloid Interface Sci.*, 96, 231.
28. Stine, K.J., Rauseo, S.A., Moore, B.G., Wise, J.A., Knobler, C.M. (1990). Evolution of foam structures in Langmuir monolayers of pentadecanoic acid. *Phys. Rev. A*, 41, 6884.
29. Kawasaki, K., Nagai, T., Nakashima, K. (1995). Intermittent flow behavior of random foams: A computer experiment on foam rheology. *Phys. Rev. E*, 51, 1246.

30. Maguire, J.F., Amer, M.S., Busbee, J. (2003). Exploring two-dimensional soap-foam films using fullerene (C60) nanosensors. *Appl. Phys. Lett.*, 82, 15, 2592.
31. Overaker, D.W., Cuitino, A.M., Langrana, N.A. (1998). Effects of morphology and orientation on the behavior of two-dimensional hexagonal foams and application in a re-entrant foam anchor model. *Mech. Mater.*, 29, 43.
32. Okuzono, T. Kawazaki, K. (1995). Intermittent flow behavior of random foams: A computer experiment on foam rheology. *Phys. Rev. E*, 51, 1246.
33. Durian, D.J. (1995). Foam mechanics at the bubble scale. *Phys. Rev. Lett.*, 75, 4780.
34. Durian, D.J. (1997). Bubble-scale model of foam mechanics: Melting, nonlinear behavior, and avalanches. *Phys. Rev. E*, 55, 1739.
35. Hutzler, S., Weaire, D., Bolton, F. (1995). The effects of Plateau borders in the two-dimensional soap froth III. Further results. *Philos. Mag. B*, 71, 277.
36. Tewari, S., Schiemann, D., Durian, D.J., Knobler, C.M., Langer, S.A., Liu, A.J. (1999). Statistics of shear-induced rearrangements in a two-dimensional model foam. *Phys. Rev. E*, 60, 4385.
37. Dennin, M. Knobler, C.M. (1997). Experimental studies of bubble dynamics in a slowly driven monolayer foam. *Phys. Rev. Lett.*, 78, 2485.
38. Gopal, A.D. Durian, D.J. (1999). Shear-induced "melting" of an aqueous foam. *J. Colloid Interface Sci.*, 213, 169.
39. Vera, M.U., Saint-Jalmes, A., Durian, D.J. (2000). Instabilities in a liquid-fluidized bed of gas bubbles. *Phys. Rev. Lett.*, 84, 3001.
40. Lemieu, P.A., Durian, D.J. (2000). From avalanches to fluid flow: A continuous picture of grain dynamics down a heap. *Phys. Rev. Lett.*, 85, 4273.
41. Matzke, E.B. (1945). The three-dimensional shapes of bubbles in foams. *PNAS*, 31, 281.
42. Thomas, P.D., Darton, R.C., Whalley, P.B. (1998). Resolving the structure of cellular foams by the use of optical tomography. *Ind. Eng. Chem. Res.*, 37, 710–717.
43. Ronteltap, A.D., Prins, A. (1989). Contribution of drainage, coalescence and disproportionation to the stability of aerated foodstuffs and the consequence for the bubble size distribution as measured by a newly developed optical glass-fibre technique. In: Bee, R.D., Richmond, P., Mingins, J., eds., *Food Colloids*. Cambridge: Royal Society of Chemistry.
44. Weitz, D.A., Pine, D.J. (1993). In: Brown, W., ed., *Dynamic Light Scattering*. Oxford, UK: Clarendon Press.
45. Koehler, S.A., Higenfeldt, S., Stone, H.A. (1999). Liquid flow through aqueous foams: The node-dominated foam drainage equation. *Phys. Rev. Lett.*, 82, 4232–4235.
46. Findlay, S. (1998). Experimental probes of liquid foams. PhD Dissertation, Trinity College.

47. Weaire, D., Hutzler, S. (1999). *The Physics of Foams*, Chapter 5. New York: Oxford University Press.
48. Abbate, S., Zerbi, G., Wunder, S.L. (1982). Fermi resonances and vibrational spectra of crystalline and amorphous polyethylene chains. *J. Phys. Chem.*, 86, 3140.
49. Durian, D.J., Weitz, D.A., Pine, D.J. (1991). Multiple light-scattering probes of foam structure and dynamics. *Science*, 252, 686.
50. Goutev, N., Nickolov, Zh.S. (1996). Raman studies of three-dimensional foam. *Phys. Rev. E*, 54, 1725.
51. Snyder, R.G., Scherer, J.R. (1979). Band structure in the C–H stretching region of the Raman spectrum of the extended polymethylene chain: Influence of Fermi resonance. *J. Chem. Phys.*, 71, 3221.
52. Amer, M.S. (2007). Raman spectroscopy and molecular simulation investigations of adsorption on the surface of single-walled carbon nanotubes and nanospheres. *J. Raman Spectrosc.*, 38 (Special Issue on Nanotechnology), 721.

8

RAMAN APPLICATIONS IN FOOD ANALYSIS

Nanna Viereck, Tina Salomonsen, M.Sc., Frans van den Berg, and Søren Balling Engelsen

Department of Food Science, Faculty of Life Sciences, University of Copenhagen

8.1 INTRODUCTION

Raman spectroscopy continues to be an attractive alternative instrumental technique for providing detailed information on chemical and physical properties of food products and food ingredients. The primary reasons are its extensive molecular information content, its signal being complementary to infrared (IR) analysis (Figure 8.1), its ease of sampling, its rapid nondestructive nature, the possible use of optical fibers, and its technological versatility. For these reasons, Raman spectroscopy holds great promise as a sensor for on-line or at-line applications in the food industry. Furthermore, direct measurements on foods enable and enhance scientific exploitation of the measurements, allowing exploratory studies of the more complex relationships within the sample. Suitable Raman active analytes in food include macrocomponents (fats, proteins, and carbohydrates) as well as minor components such as carotenoids, cyanogenic glucosides, inorganics and many others [1].

Raman Spectroscopy for Soft Matter Applications, Edited by Maher S. Amer

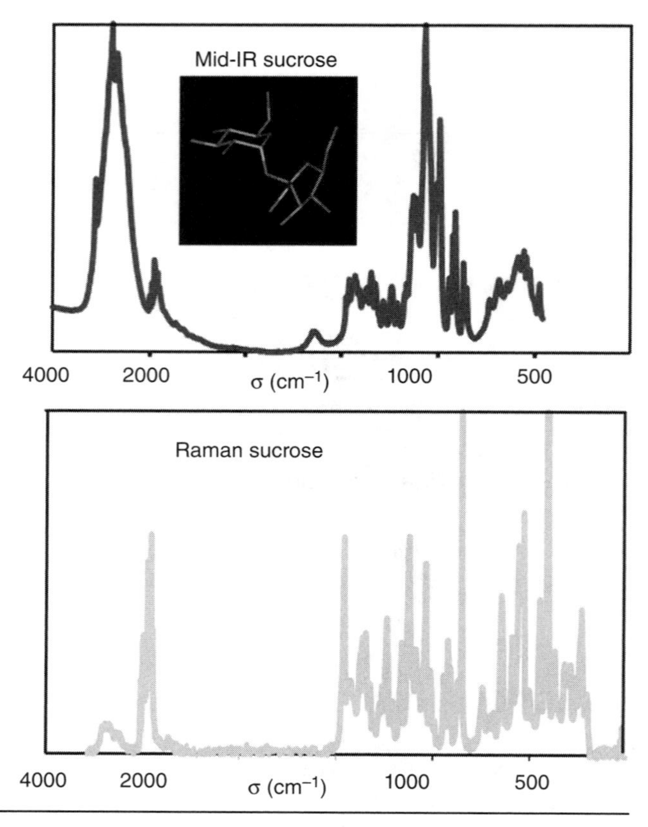

Figure 8.1 The complementary nature of Raman spectroscopy compared with infrared (IR) absorbance spectroscopy illustrated on a simple food molecule sucrose (β-D-fructofuranosyl-(2→1)-α-D-glucopyranoside). Sucrose contains 45 atoms that results in 3N-6 degrees of freedom corresponding to 129 normal vibrations. Obviously, there is a major difference in the selection rules.

This chapter provides an overview of the quantitative use of Raman spectroscopy on food systems with emphasis on the use of multivariate data analysis for the interpretation and quantification of the Raman signals. A comprehensive review of the possible quantitative applications of Raman spectroscopy within food research is given. The section will focus on the parameters that are important when measuring intact food samples, as these will illuminate the advantages and disadvantages of using Raman spectroscopy for food analysis. Finally, the current status and future potential of Raman spectroscopy as a process analytical technology (PAT) for process monitoring and control will be described.

8.2 UNIVARIATE APPLICATION OF RAMAN SPECTROSCOPY

Even to this date, most measurement series collected by Raman spectroscopy are evaluated and analyzed as univariate information using one wave number channel per sample. This method of analysis is reasonable for simple, chemically prepared samples with only one constituent, when the samples are not influenced by interferences or other difficulties. However, this approach is far from optimal when measuring intact food samples. Besides the obvious huge loss of information, there are a few problems that should be considered when using univariate methods for analyzing Raman spectra. A simplified example of some of the problems that can be encountered are displayed in Figure 8.2.

From the figure, it is obvious that if only the variable marked with a dot is recorded, spectrum A will be estimated to have a higher content of the compound in question than spectrum C, even though the two responses have equal amplitude relative to the signal baseline. If the whole spectrum is recorded, the baseline offset of spectrum A is easily detected, and the necessary action can be taken to correct or compensate for this offset. For this reason, univariate analysis of Raman spectra including interacting or coinciding signals can almost never be expected to produce useful and trustworthy results. Furthermore, when dealing with complex materials like foods or more general biological samples, the signals are of such complexity that the signal of a single chemical

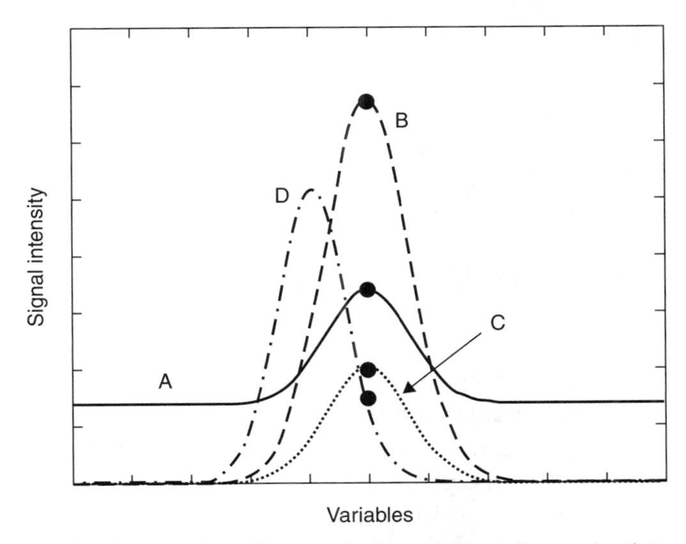

Figure 8.2 Artificial examples of issues that may arise when univariate rather than multivariate analysis is applied to spectral data.

component cannot be identified directly, and the property of interest can seldom be described by one single peak. For this type of question, it is important to look at the sample as a whole and to analyze the spectral data taking the matrix of the samples combined into consideration.

8.3 MULTIVARIATE APPLICATION OF RAMAN SPECTROSCOPY

Multivariate data analysis is based on the extraction of latent components or underlying *common structures* in the data and requires that a larger number of variables (typically a full Raman spectrum) be recorded for a set of similar objects (samples). For the analysis to make sense, the data must have common underlying structures! For the example in Figure 8.2, the structure, i.e. the peak, which is common for the first three samples (A, B, and C), is not present in the deviating sample (spectrum D). This could have one of two reasons: this sample *is* different and contains another chemical compound than the others, or it is due to some error in the spectral measurement, as the *spectral axis* is different for this sample, i.e. the different position of the signal is due to an instrumental shift. By making use of the full information, modest instrumental shifts can be corrected for or compensated in multivariate modeling, while univariate approaches in this situation might result in substantial errors.

Simultaneous analysis of complete Raman spectra, using multivariate statistics and mathematics, falls under the realm of chemometrics. The most common techniques, principal component analysis (PCA) and partial least squares (PLS) regression, are used for interpretation, outlier detection, and quantification [2]. A more detailed description of PCA and PLS is in Section 8.5.

8.4 RAMAN SPECTROSCOPY AND FOOD ANALYSIS

From a physical, chemical, or biological perspective, food products (and food processing) are complex multifactorial systems that are very difficult to sample and analyze representatively. From a chemical point of view, food is comprised of complex mixtures of heterogeneous classes of molecules (water, fats, proteins, and carbohydrates), while from a physical point of view, food consists of complex structures including amorphous solids, aqueous liquids, macromolecules, macro-organelles, cells, crystals, pores, etc. To analyze food for product quality, such as technological properties, nutritional value, or overall composition, represents a major challenge that is not always overcome by the investigators. Food matrices are rarely the transparent, nonfluorescent scatter matrices desired for the application of Raman spectroscopy.

The publically available research and literature related to the investigation of food using Raman spectroscopy are sparse. However, during the last decade, research in this area has been increasing. The two main difficulties with the Raman technique are that it is not easily applicable to materials that exhibit fluorescence, and its inherent poor signal-to-noise ratio. In addition, attempts to increase the signal-to-noise ratio by using high excitation monochromatic intensity often leads to small sampling areas and sample heating. Nevertheless, Raman spectroscopy has high selectivity (selection rules) for some important food quality traits (compounds) such as the C=C bond (Figure 8.3), and as many food systems primarily consist of water, it is a big advantage that water, in contrast to near infrared (NIR) and IR, gives rise to very weak signals in Raman due to its weakly polarizable bonds. For these reasons, Raman spectroscopy remains a promising tool for the quality control of food products and ingredients.

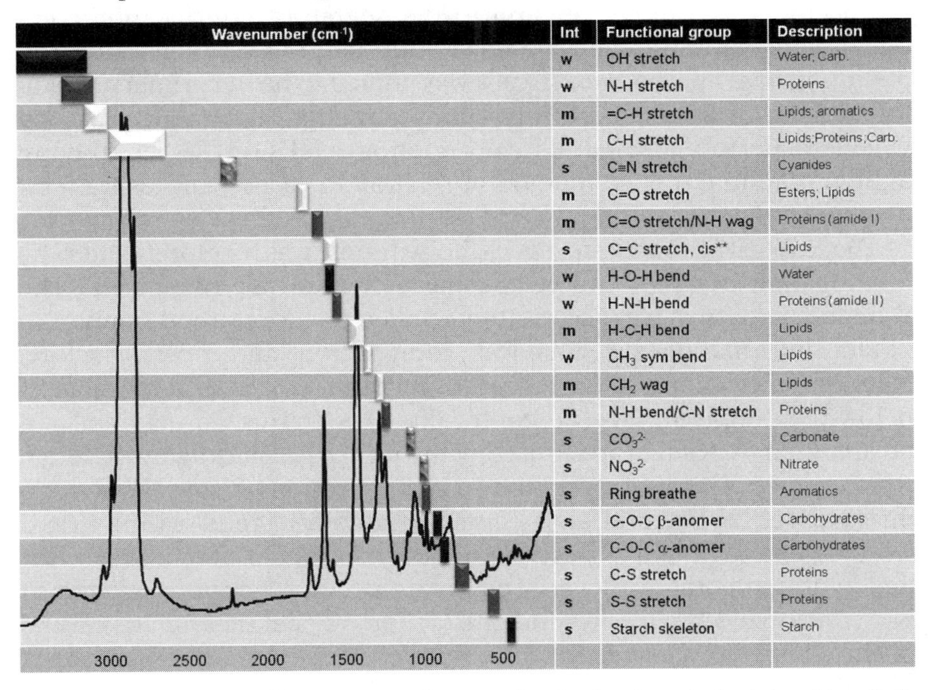

Figure 8.3 Important Raman group frequencies for food analysis. Intensities (Int): w = weak; m = medium; s = strong. Black bars = water and carbohydrates; gray bars = proteins; white bars = lipids; striped bars = others.

8.4.1 Carbohydrates

Next to the early applications of Raman spectroscopy to oils and margarines by Sadeghi-Jorabchi and coworkers [3,4] and to oils and carbohydrates by Góral and Zichy [5], one of the first attempts to apply multivariate Raman spectroscopy to a food relevant system was presented in the work by Engelsen and Nørgaard [6]. In the latter study, various quality parameters in industrial amidated pectin powders concerning the degree of esterification and amidation of the side bands were investigated by comparative vibrational spectroscopy. The inferior quantitative information in the fourier transform (FT)-Raman spectra compared with IR and NIR was attributed to the fact that Raman spectra are more sensitive to skeleton vibrations than to side-group vibrations (amide and methoxy), which were the primary target. In fact, the most prominent Raman information for the pectins was the 850-cm^{-1} band (Figure 8.4), which is the so-called α-anomeric band [7] of the α-(1,4)-linked polygalacturonic backbone that is partially methyl esterified. Indeed, the precise position of this band was found to be the strongest covariate spectral feature with the degree of esterification of the pectin. This indirect relation to the degree of esterification combined with the lower signal-to-noise ratio and the residual- and sampling-sensitive fluorescence signals was argued to be the primary reason for the inferior Raman performance. Even though Raman spectroscopy in general performed quantitatively inferior to NIR or IR spectroscopy, a great potential for the method in relation to carbohydrates was still documented.

The capacity of measuring carbohydrates was explored later by investigating living potato cells using a Raman microscope [8]. The Raman microscope was successfully applied to various food samples to obtain information on chemical composition and microstructure. Raman spectroscopy offers the possibility to focus on different planes below the sample surface, i.e. confocality. Due to this feature, food can be analyzed by Raman spectroscopy through, e.g. a packaging material, which can be very useful for industrial applications. Using the Raman microscope, Thygesen et al. [8] demonstrated for the first time that it was possible to obtain a spectrum of pectin *in situ*, located in the intact and hydrated potato cell wall. The potato pectin spectrum (Figure 8.5) is characterized by the prominent α-anomeric peak near 858 cm^{-1}, the position of which is indicative of a very low degree of esterification, by a peak at 1455 cm^{-1} due to ester $O{-}CH_3$ stretch, and by the unique galactoronic methyl ester peak in the area around 1745 cm^{-1} (Figure 8.3), where the surrounding and dominating starch granules in the cell

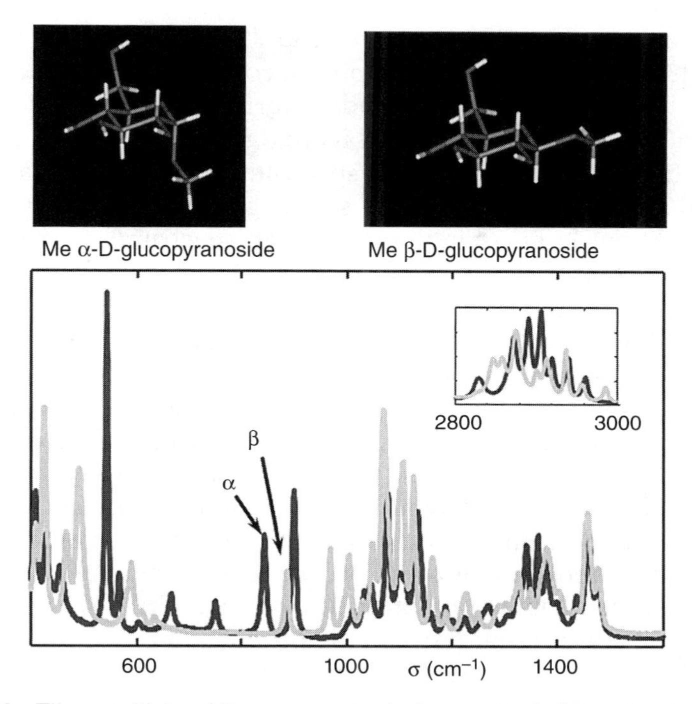

Figure 8.4 The sensitivity of Raman spectra to the anomeric form of carbohydrates. The spectra of methyl α-D-glucopyranoside and methyl β-D-glucopyranoside are compared. The molecules are identical except for the anomeric configuration at C1. The consequences in the Raman spectrum are dramatic. Practically all vibrational modes are changed including the bands in the C—H stretching region (insert). Note the anomeric band that is always between 800 and 860 cm^{-1} for the α-anomeric carbohydrates and between 860 and 900 cm^{-1} for the β-anomers.

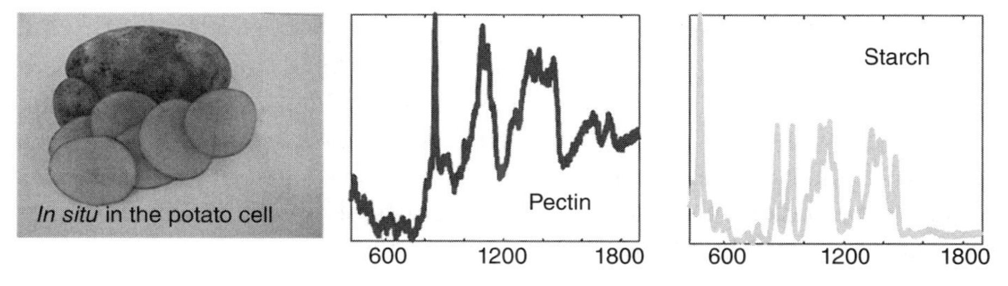

Figure 8.5 Raman spectra of pectin and starch obtained in the potato cell directly using a Raman microscope equipped with a 532-nm green laser.

do not have signals that interfere. The sharp signal near $1000\,cm^{-1}$ (mono-, 1,3-, and 1,3,5-substituted phenyl compounds at $1000 \pm 10\,cm^{-1}$) indicates that aromatic compounds interfere (and possible fluorescence) with the pectin spectrum (Figure 8.5).

In contrast to the *in situ* pectin spectrum, recording Raman spectra of intact and naturally hydrated starch granules is straightforward. Using a 532-nm laser excitation, a Raman spectrum of an individual starch granule, as shown in Figure 8.5, can be made in a few seconds. The starch spectrum is characterized by a strong band at $474\,cm^{-1}$ (Figure 8.3) and is of remarkable quality, equal to or even better than a typical Raman spectrum obtained from isolated and purified potato starch. This shows that the high crystallinity of the intact granule is well represented in the spectrum. Since it is composed of exclusive α-glucose units, the starch spectrum also shows the characteristic α-anomeric band at approximately $850\,cm^{-1}$ (medium intensity). Several reports have documented significant differences between Raman spectra (in particular the above-mentioned two bands) of starches of different botanical origin including potato and maize starches [9] and floridean starch of red algae, an impurity in the production of various hydrocolloids [10]. The fact that the two characteristic starch signals are sensitive to the crystallinity and conformation has led to the suggestion that Raman spectroscopy can be used for on-line monitoring of the gelatinization process of starch [11].

Raman spectroscopy has been used in other studies to elucidate detailed information about the properties of the amorphous or crystalline structure of carbohydrates [12]. This was for example demonstrated by Nørgaard et al. [13], who devised a multivariate Raman method for quantification of crystalline lactose in whey permeate powder. Lactose is used in many industrial products, especially in foods and pharmaceuticals, and lactose monohydrate is the main component in whey permeate powder with a typical content of 80–95% (w/w). Whey permeate powder is used as an ingredient in spreads, instant drinks, bakery products, and desserts. In whey permeate, lactose exists predominantly in the crystalline form, but whey permeate also contains some amorphous lactose (approximately 10%). Amorphous lactose in the glassy state is often undesirable in food and pharmaceutical products because it is metastable, which can lead to caking due to water adsorption followed by crystallization. While it was demonstrated that Raman spectroscopy is a useful method for quantifying the degree of crystallinity of lactose in whey permeate powder (prediction error, root mean square error of prediction [RMSEP] = 1.6% using six PLS factors; see Section 8.5 for definitions), the PLS models of the FT-Raman spectra predicting the crystallinity of lactose did not perform as well as the corresponding NIR models (RMSECV = 0.63% using six PLS factors; Nørgaard et al. [13]).

8.4.2 Edible Oils

Another important field for the application of Raman spectroscopy to foods is the edible oil area, in which the high sensitivity to the C=C bond (Figure 8.3) early on proved useful in determining the total level of unsaturation, the degree of conjugation, *cis/trans* isomer ratios, and the total number of double bonds in the hydrocarbon chains [4,14]. Later, we investigated the deterioration of frying oil by various spectroscopic and chemical/physical techniques including FT-Raman spectroscopy [15]. Daily oil samples were collected during a 4-week period of frying spring rolls, completing one round before the process was renewed with fresh oil. The Raman spectra displayed an increasing fluorescence toward low wave numbers as a function of increasing frying time. The Raman spectra included peaks at 1302 and 1440 cm^{-1} due to polymethylene twisting and scissoring, respectively. Carbon–carbon double-bond stretching was found as a sharp and relatively strong peak at 1656 cm^{-1}, characteristic of *cis* olefins (Figure 8.3; note that conjugation lowers σ(C=C) by 20–30 cm^{-1}), and only a broad and weak peak centered around 1748 cm^{-1} corresponding to the ester C=O stretch. In the aliphatic C–H stretching region, the two main peaks at 2852 and 2925 cm^{-1} correspond to polymethylene symmetric and asymmetric C–H stretchings, respectively, and a peak at 3009 cm^{-1} is due to olefinic =C–H stretchings in *cis* configurations. The quantitative models computed from the Raman spectra yielded good calibration models to triglycerides, free fatty acids, and iodine value but were inferior to the models obtained from IR and NIR, except for a model predicting the age of the frying oil expressed in days. The good performance of the latter Raman model was attributed to the increasing fluorescence in the signals at low wave numbers in the samples, originating from complex fluorescence phenomena in the frying oils correlated with time [15].

8.4.3 Cyanogenic Compounds

The content of triple bonds in food and food ingredients has also been the subject of numerous Raman studies. In one example, the content and distribution of the cyanogenic glucoside amygdalin in bitter almond cotylodons was investigated using the characteristic strong and narrow Raman active peak from the cyanide functional group (C≡N; Figure

8.3) in the molecule [8,16]. Many plant species contain cyanogenic glucosides that can cause cyanide toxification of humans and animals due to the release of HCN upon enzymatic degradation (hydrolysis). A noninvasive assay for measuring cyanogenic glucosides in plants by Raman spectroscopy was presented by Micklander et al. [16], which evolved from their study of bitter almonds. In bitter almonds, the nitrile group gave rise to a sharp Raman band at 2242 cm^{-1} (Figure 8.3). Fairly good prediction models for the endogenous amygdalin in bitter almond could be obtained from both the cyanide peak and bands related to the aromatic ring in the molecule. To demonstrate this, a multivariate interval-PLS (iPLS) calibration result of amygdalin content in bitter almond is presented (Figure 8.6; Nørgaard et al. [17]), in which the global Raman spectrum is subdivided into 37 intervals for each of which a new PLS model is calculated. The horizontal line represents the performance of a two-factor full-spectrum PLS model (RMSECV = 23.1 $nmol \cdot mg^{-1}$), and the vertical bars indicate the performance of each of the subinterval one-factor PLS models generated. The vertical bars corresponding to the areas around the bands at 3060 cm^{-1} (C–H

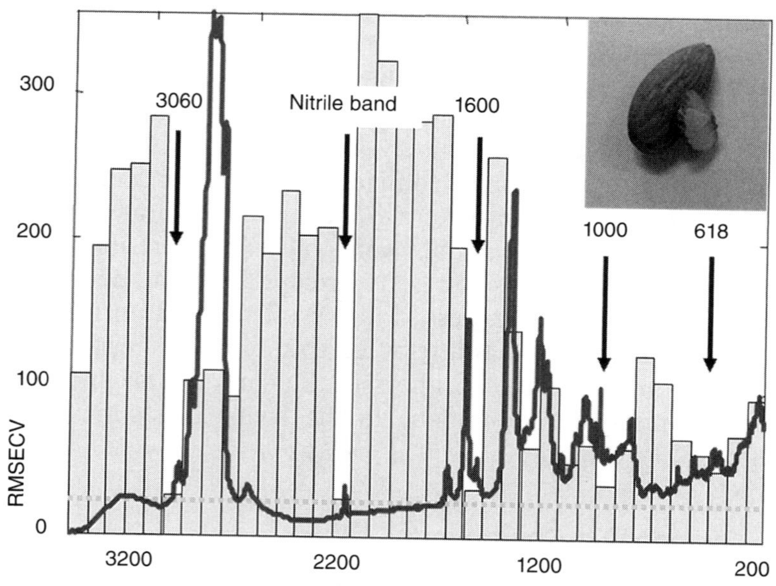

Figure 8.6 Interval-partial least squares regression (PLS) plot of the Raman calibration to the amygdalin content in almonds. The Raman spectrum is divided into 37 intervals, and the bars indicate the one-factor PLS performance of each interval. The horizontal line is the global/total spectrum two-factor (iPLS) performance. RMSECV = root mean squared error of cross validation.

stretching of the aromatic ring) and $2242\,cm^{-1}$ (C≡N–stretching of the nitrile group) are close to the horizontal line representing the RMSECV of the full-spectrum model. Thus, the simpler one-factor PLS models based on narrow spectral intervals yield a quantitative performance equal to the two-factor full-spectrum PLS model. We also observe from Figure 8.6 that the interval including the band at approximately $1600\,cm^{-1}$ produced by aromatic carbon–carbon stretching, the ring breathe mode at $1000\,cm^{-1}$ (Figure 8.3), and the out-of-plane mode at $618\,cm^{-1}$ of the benzene group yield quite good calibrations. The distribution of amygdalin in intact cross sections of the bitter almond has also been investigated using Raman microscopy; however, the variations between neighboring points in a sample and between samples were high [16].

The possibilities in using the distinctive cyanide peak in a Raman spectrum were further explored for the study of cyanogenic glucosides in plants in general [18]. In this study, using a green laser at 532 nm, it was found that the position of the signal from the C≡N triple bond of the cyanohydrins group was influenced by the nature of the side group and was above $2240\,cm^{-1}$ for the three cyanogenic glucosides that contain a neighboring aromatic ring and below or partially below $2240\,cm^{-1}$ for the nonaromatic cyanoglucosides. However, the low concentration of the cyanogenic glucoside in the plant tissue and the fluorescence from chlorophyll prevented detection of *in situ* Raman signals from cyanide. Instead, a flow injection method using surface-enhanced Raman spectroscopy (FI-SERS) was devised and demonstrated to be able to measure the cyanogenic potential of plant tissue extracts. As shown in Figure 8.7, the FI-SERS amplified the cyanide signal by a factor of 2000 that improves the detection limit by a factor of 10^5 (Figure 8.7). The figure reveals that if the concentration of cyanide is increased beyond a certain level ($\sim10^{-5}$ M), the intensity of the surface-enhanced signal starts to level off. Not until higher concentrations ($\sim10^{-2}$ M) does the unenhanced signal emerge. There is thus a range of concentrations that is too low to generate an unenhanced signal and too high to generate a surface-enhanced signal with a linear correlation between signal intensity and concentration. This situation is analogous to the problem of concentration quenching in fluorescence spectroscopy. The position of the CN band with and without surface enhancement is approximately 2135 and $2080\,cm^{-1}$, respectively. The signal for the CN bond in crystalline cyanogenic glucosides and for cyanogenic glucosides in solution (either *in situ* or *in vitro*) is positioned approximately $100\,cm^{-1}$ above the peak of the cyanide $C{\equiv}N^-$ ion in solution.

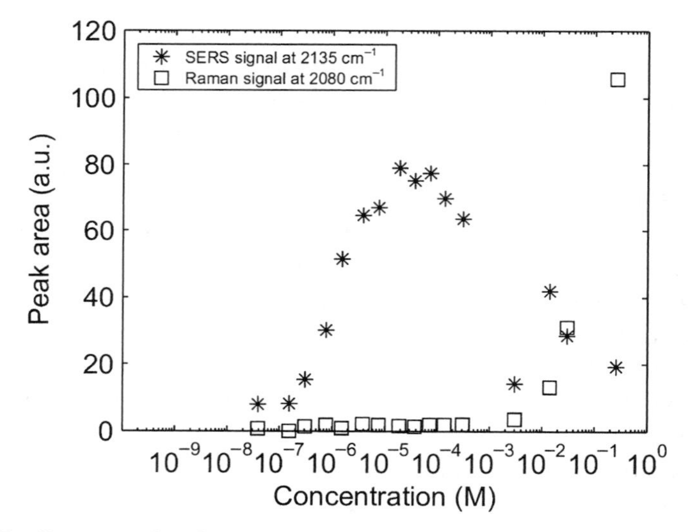

Figure 8.7 Concentration dependence of the Raman signal using Raman spectroscopy versus surface-enhanced Raman spectroscopy (SERS). At a cyanide concentration $>10^{-5}$ M, the intensity of the surface-enhanced signal at 2135 cm^{-1} starts to decrease. The unenhanced signal at 2080 cm^{-1} does not start to emerge until the concentration is $>10^{-2}$ M.

8.4.4 Proteins

Just like IR, Raman spectroscopy is very attractive in the study of proteins in general and for studying the peptide backbone structure in particular [19]. The sensitivity of the amide I band (Figure 8.3) toward the secondary structure moves the band from 1655 cm^{-1} for the α-helix to 1670 cm^{-1} for the antiparallel β-sheet structures. Raman spectroscopy can thus be used to examine protein secondary structure by observing minor shifts in Raman peak position. This principle was used in a Raman microscopic study of intact fibrillin-rich microfibrils by Haston et al. [20]. In this study, the filaments originating from the ovine eye tissue were in native, extended, or relaxed states when measured by Raman spectroscopy. The spectra of native and relaxed filaments were rather similar, indicating that change in the protein configuration due to tissue extension is a reversible process. However, Raman spectra of the extended tissue were significantly different, visualized in difference spectra of native minus extended spectra. The main spectral differences were seen in the amide I (carbonyl stretch) and III (C–N stretch) bands (Figure 8.3; reflecting protein backbone structure) and within the aromatic and aliphatic bands (reflecting the protein side-group structure). Overall, the extension of these filaments induces a reversible conformational change, expressed in decreased occurrence of random coil and β-turn secondary structure and an increase in α-helical content within the extended tissue. Furthermore, these changes also influence specific amino residues within the microfibrils, including changes in hydrogen bonding status and domain–domain interactions [20].

The use of Raman spectroscopy for the prediction of meat quality as expressed by the water-holding capacity (WHC) of fresh meat from pigs has also been explored [21]. WHC is an important quality trait for the meat industry, because meat is sold by weight including the contained water, and because WHC influences the appearance and sensory properties of the product. The reduction in WHC during slaughtering is influenced by pH and ongoing rigor development. The pH can be manipulated by preconditioning the animals with adrenaline and exercise; hence, the meat quality can vary as a consequence of the treatment of the animals before slaughtering. The preliminary result showed that Raman spectra had good potential for predicting the water drip loss in the meat samples with low prediction errors and correlation coefficients above 0.95. The most informative Raman regions contained N–H stretching of primary amides in proteins (3140 cm^{-1}) and the α-helical 940-cm^{-1} band, which both might indicate partial protein denaturation. The application of Raman spectroscopy to the fresh meat using a blue 632-nm HeNe laser resulted in good, well-resolved spectra, and it was found useful to employ sharp aromatic ring breathe vibration at 1000 cm^{-1} (Figure 8.3) as an "internal standard" for normalizing the Raman spectra. It was concluded that the Raman observations suggest coherence between WHC and pH, glycogen level, and protein conformation.

8.4.5 Inorganics

Another interesting and useful feature of Raman spectroscopy is the extraordinarily sharp spectra of inorganic compounds such as carbonate minerals, which give rise to sharp, characteristic, and intense Raman bands. The shape and precise wave number of this peak are characteristic for the type of carbonate and can thus be used to identify the polymorphic form. This was employed in the studies of "white spots," which develop on frozen shrimp shell during Atlantic storage [22,23]. The most intense peak in the carbonate Raman spectrum of the white spots was due to the totally symmetric carbonate CO stretch at 1086 cm^{-1} (Figure 8.3). In the study, the presence of the white spots was fully explained by the presence of α-chitin and calcium carbonate. It was

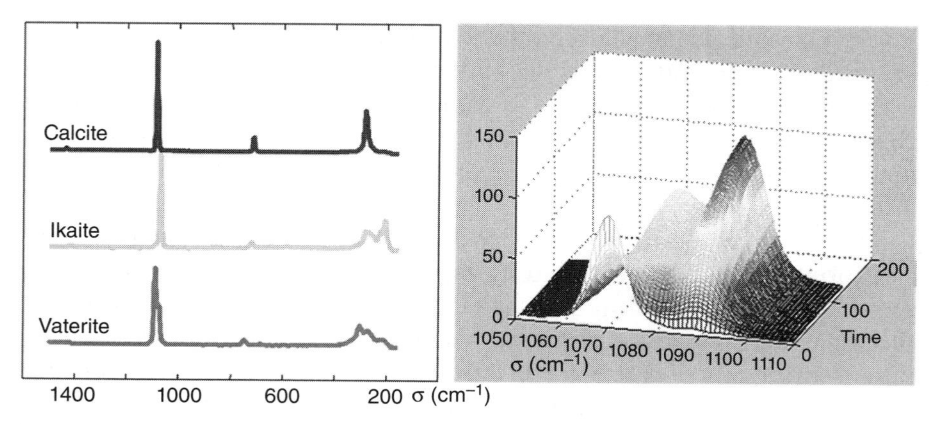

Figure 8.8 Time series of Raman spectra in the range 1110–1050 cm^{-1} during the dehydration process of ikaite at ambient temperature. The spectrum in the front with a single peak at 1070 cm^{-1} is the ikaite spectrum, and the spectrum in the rear with peaks at 1089 and 1075 cm^{-1} is identified as vaterite.

found that the weight ratio of α-chitin and calcium carbonate was constant, but the ratio of the two crystalline forms of calcium carbonate, vaterite and calcite, depended on prefreezing treatments of the shrimp. The dehydration of calcium carbonate hexahydrate (ikaite; 1070 cm^{-1}) to form anhydrous calcium carbonate in the crystal form of vaterite (1089 and 1075 cm^{-1}) was monitored by Raman spectroscopy (Figure 8.8), and it was found that dehydration of ikaite to vaterite and calcite occurred without the formation of an intermediate calcium carbonate. It was concluded that the white spots were due to the precipitation of ikaite.

8.5 MULTIVARIATE RAMAN STUDY OF ALGINATE POWDER—A CASE STUDY

Next to laboratory identification and qualification tasks on food and feed systems, Raman spectroscopy is used more and more for fast quantitative analysis. This is possible due to the introduction of chemometrics or multivariate data analysis. In this section, we will illustrate these concepts using a data set presented by Salomonsen et al. [24]. The potential of using Raman spectroscopy combined with chemometrics for reliable and rapid determination of the ratio of mannuronic-to-guluronic acid (M/G) ratio in commercial sodium alginate powders will be evaluated. The reference method for the quantification of the M/G ratio is the time-consuming, solution-state ^{1}H nuclear magnetic resonance (NMR) spectroscopy. A set of 100 commercial alginate powders (particle size of $<10^{-6}$ m) with an M/G ratio range of 0.5–2.1 was available for quantitative calibrations using PLS regression (standard deviation of the reference determined at 0.01–0.08; Salomonsen et al. [24]). Commercial alginates are extracted from brown seaweed (Phaeophyceae) and used as thickeners, stabilizers, and gelling agents in the food and pharmaceutical industries. Chemically, they are a family of binary copolymers of (1→4)-linked β-D-mannuronic acid and its C-5 epimer α-L-guluronic acid. Alginates are typically described by their M/G ratio, distribution of M- and G-units along the chain, and average molecular weight, since these parameters are closely related to functionality, i.e. solubility, interaction with metals, gel properties, and viscosity. The composition and sequential structures, and thereby also the functionality of alginates, vary according to season, age of population, species, and geographic location of harvesting. Because of these variations within the different alginate types, it is of great importance for the industry to obtain rapid, detailed knowledge about the composition of its products in order to design the functionality in the final use.

Solution-state NMR spectroscopy is a highly effective tool in compositional and structural analysis of alginates. However, at the concentrations required for a good signal-to-noise ratio, alginate solutions are too viscous to give well-resolved NMR spectra, even at elevated temperatures. To reduce the viscosity, it is necessary to partially degrade the alginate chain by a mild acid hydrolysis prior to analysis. This hydrolysis is relatively time-consuming and labor-intensive. Hence, solution-state NMR of alginates is rather elaborate and certainly not suitable for at- or on-line monitoring. Thus, the development of a faster and simpler method for the characterization of the alginate composition would be beneficial. Raman spectroscopy, in combination with multivariate data analysis/chemometrics, is a useful analytical tool that can reveal detailed information concerning the composition and properties of material at a molecular level. The main advantages of this technique compared with solution-state NMR are that it is rapid, nondestructive, and easy to operate, and, in most cases, requires no sample preparation. Due to these qualities, spectroscopic techniques are applicable for at-line analysis or, using optical quartz fibers, for on-line analysis.

FT-Raman spectra were collected on a Perkin Elmer System NIR FT-Raman interferometer (Perkin Elmer Instruments, Waltham, MA) equipped with an Nd:YAG laser emitting at 1064 nm with a laser power of 200 mW. Data were collected using an InGaAs detector and stored as Raman shifts in the range of 200–3600 cm^{-1}; here, we will work with the range of 700–1800 cm^{-1}. A 180° back-scattering arrangement was used, and no correction for the spectral response was applied. A total of 32 scans was averaged for each sample, and the resolution was 32 cm^{-1}. Each sample was measured in two repacked replicates, and the average values were used in the data analysis. Figure 8.9a shows the raw data for the 100 alginate powders where three samples with low, high, and medium M/G ratios are highlighted. These same three samples will be monitored throughout the section. As can be seen from the figure, the measurements have a high scatter background, obscuring the chemical information. This has to be handled before quantitative analysis. Also shown in the figure are the nonsquared correlation coefficient between signals for all samples and the M/G ratio reference values at one wave number. This shows that no single variable in the spectra is suitable for quantification.

Multiplicative scatter (or signal) correction (MSC) is one of the most widely used preprocessing techniques in spectroscopy. MSC in its basic form was first introduced by Martens and coworkers in 1983 [25] and further elaborated on by Geladi et al. [26]. The concept behind MSC is that by using a simple, linear preprocessing technique aimed at correcting spectroscopic artifacts/imperfections, undesirable scatter effect due

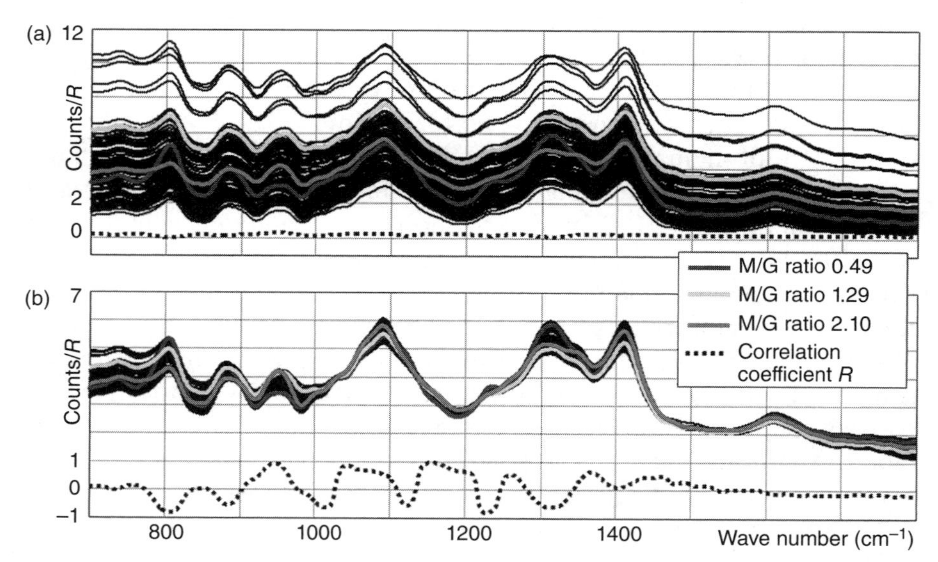

Figure 8.9 (a) Raw Raman spectra of 100 alginate powders and correlation coefficients for individual wave numbers with mannuronic-to-guluronic acid (M/G) ratio reference value; (b) multiplicative scatter (signal) correction (MSC) Raman spectra of the same sample set.

to particles and/or fluorescence signals can be compensated, removing them from the data matrix prior to further analysis. The preprocessing consists of two steps, estimation of the correction additive and multiplicative contributions:

$$\mathbf{x}_{\text{original}} = b_0 + b_1 \cdot \mathbf{x}_{\text{reference}} + \mathbf{e}, \tag{8.1}$$

correcting the recorded spectrum

$$\mathbf{x}_{\text{corrected}} = (\mathbf{x}_{\text{original}} - b_0)/b_1 = \mathbf{x}_{\text{reference}} + \mathbf{e}/b_1. \tag{8.2}$$

Here, $\mathbf{x}_{\text{original}}$ is one raw spectrum (sample) measured by Raman, $\mathbf{x}_{\text{reference}}$ is a reference spectrum used for preprocessing the data set (in this section, we use the average spectrum of the data set), $\mathbf{e}$ is the unmodeled part of $\mathbf{x}_{\text{original}}$, $\mathbf{x}_{\text{corrected}}$ is the corrected spectra, and b_0 and b_1 are scalar estimates, different for each sample. Figure 8.9b shows the data after MSC correction. It is observed that a large part of the unwanted variation is removed. The figure also shows that several parts of the spectral profiles correlate well (either negative or positive) with the M/G ratio reference values, indicating the potential of quantitative modeling.

PCA is a popular bilinear factor model used for exploring data tables in chemometrics and many other fields of research [27]. A PCA model finds the best least-squares low rank approximation of a data matrix **X** (samples × wave numbers) containing the Raman spectra:

$$\mathbf{x} = \mathbf{t}_1\mathbf{p}_1^{\mathrm{T}} + \mathbf{t}_2\mathbf{p}_2^{\mathrm{T}} + \mathbf{E} = \mathbf{T}\mathbf{P}^{\mathrm{T}} + \mathbf{E}$$
$$\text{minimize} \left\|\mathbf{x} - \mathbf{T}\mathbf{P}^{\mathrm{T}}\right\|^2. \tag{8.3}$$

In these equations, object-score vectors fulfill conditions $\mathbf{t}_i^{\mathrm{T}}\mathbf{t}_i = s_i$ and $\mathbf{t}_i^{\mathrm{T}}\mathbf{t}_j = 0$ (orthogonal score vectors; size s_i is equal to the square of the singular value), variable loadings satisfy the criteria $\mathbf{p}_i^{\mathrm{T}}\mathbf{p}_i = 1$ and $\mathbf{p}_i^{\mathrm{T}}\mathbf{p}_j = 0$ (orthogonal, normalized loading vectors), and **E** is the unmodeled part of **X** where we stop after two factors for ease of notation only. Hence, the first set of scores and loadings (i.e. the first principal component) is the best approximation of the original data, and the percentage explained variance captured from the original data matrix by this first pair expresses how well this approximation succeeded. Similarly, the second score-and-loading pair is the next best approximation of the original data table, etc. The scores can be seen as new pseudoconcentrations for the objects (Raman measurements). The loadings show the role of the original variables (on the wave number axis). Figure 8.10 shows the score plot of the first two principal components on the MSC preprocessed data set; M/G ratio reference numbers are indicated for all samples. A clear gradient is observed in the sample cloud in the PC2 direction, but we also see that the main direction (PC1) is not directly correlated with our prediction objective M/G ratio. Hence, other sources of (undesired) variation are still present in the matrix.

In PLS regression [2], the aim is not only to explain the variance in **X** (the spectral data), but also to do this in such a way that the explained variance is correlated as well as possible with variance in a second matrix/vector $\mathbf{y} = \mathbf{X}\,\mathbf{b}$ (M/G ratio reference values in this section). This can be achieved by solving the following equations:

$$\mathbf{x} = \mathbf{t}_1\mathbf{p}_1^{\mathrm{T}} + \mathbf{t}_2\mathbf{t}_2^{\mathrm{T}} + \mathbf{E}_{\mathrm{X}} = \mathbf{T}\mathbf{P}^{\mathrm{T}} + \mathbf{E}_{\mathrm{X}}$$
$$\mathbf{y} = \mathbf{u}_1 \mathrm{q}_1^{\mathrm{T}} + \mathbf{u}_2 \mathrm{q}_2^{\mathrm{T}} + \mathbf{E}_{\mathrm{Y}} = \mathbf{U}\mathbf{q}^{\mathrm{T}} + \mathbf{E}_{\mathrm{Y}}$$
$$\text{maximize}\left(\text{covariance}\left(\mathbf{t}_{\mathrm{i}}, \mathbf{u}_{\mathrm{i}}\right) \middle| \mathbf{X}\mathbf{w}_{\mathrm{i}} = \mathbf{t}_{\mathrm{i}}, \|\mathbf{w}_{\mathrm{i}}\| = 1\right). \qquad (8.4)$$

Here, two sets of object-score vectors (note that $\mathbf{t}_{\mathrm{i}}$ in PLS is not the same as for the PCA solution, and some of the orthogonality criteria

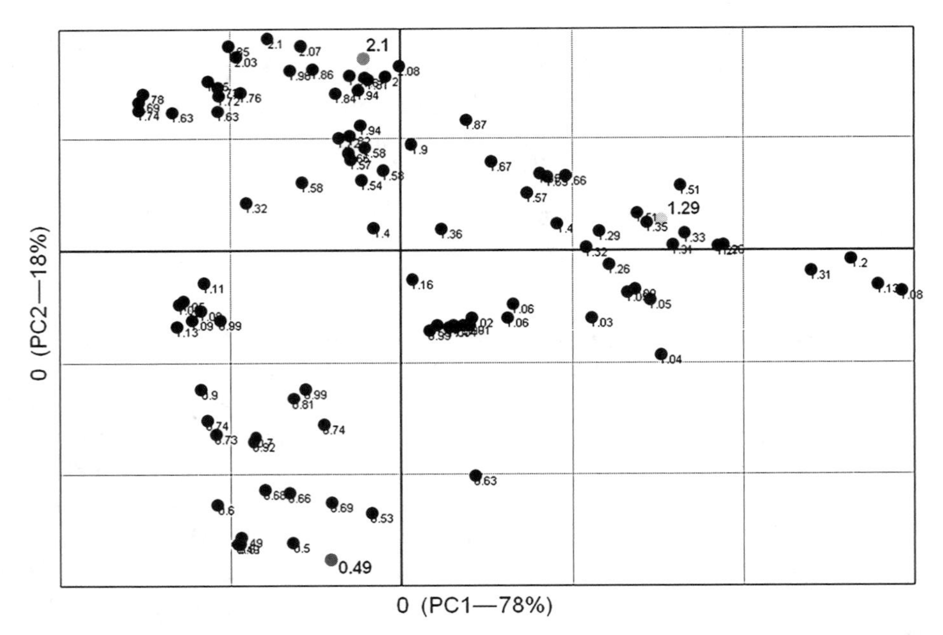

Figure 8.10 Principal component score plot of factor two (PC2, explaining 18% of the variation in the data set) versus factor one (PC1, 78%); numbers are reference mannuronic-to-guluronic acid (M/G) ratios.

from PCA are relaxed) and variable loadings $\mathbf{p}_i$ (for the Raman block) and q_i (for the reference vector) are obtained. Hence, the first set of scores and loadings is the best approximation of the original data matrices under the condition or constraint that the covariance between the two object-score vectors is maximized. The auxiliary vectors, $\mathbf{w}_i$, are used to compute the final regression model $\mathbf{y} = \mathbf{X}\,\mathbf{b} = \mathbf{X}\,\mathbf{W}\,(\mathbf{P}^T\,\mathbf{W})^{-1}$ [2]. The regression models can now be used to predict M/G ratios from Raman spectra. The correct number of factors to include in the final model will be determined from a leave-one-sample-out cross validation. Hence, each sample is removed once from the set, and its reference value is predicted from a model based on all other samples using 1, 2, ... factors in the PLS model. The best overall model can be found from a local minimum in the root-mean-squared error of prediction (RMSEP), which mimics a standard deviation and has the same units as the reference:

$$\mathrm{RMSEP} = \left(\sum \left(Y_i^{\mathrm{reference}} - Y_i^{\mathrm{predicted}}\right)^2 / I\right)^{0.5}. \tag{8.5}$$

Here, i indicates the test sample reference and prediction, and I is the number of samples used in the evaluation (equals 100 for our full data set). When RMSEP is found by cross-validation it is called root mean squared error of cross-validation (RMSECV). RMSECV profiles are shown in Figure 8.11 for the raw and MSC preprocessed spectra.

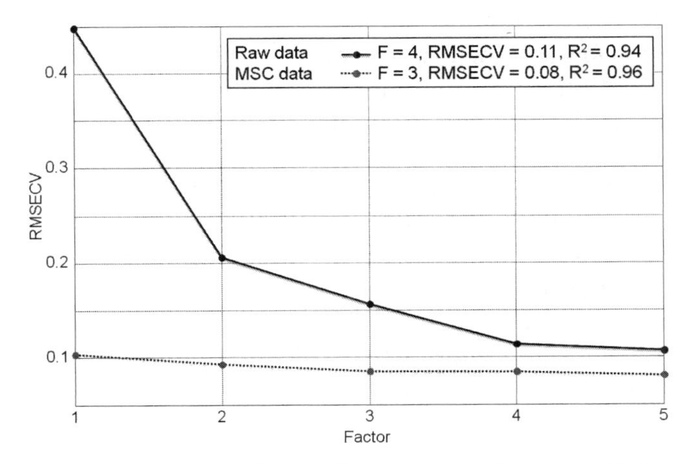

Figure 8.11 Root mean-squared error of leave-one-out cross-validation predictions versus number of factors in the partial least squares model for two data sets: raw and MSC data preprocessed spectra. RMSECV, root mean squared error of cross-validation.

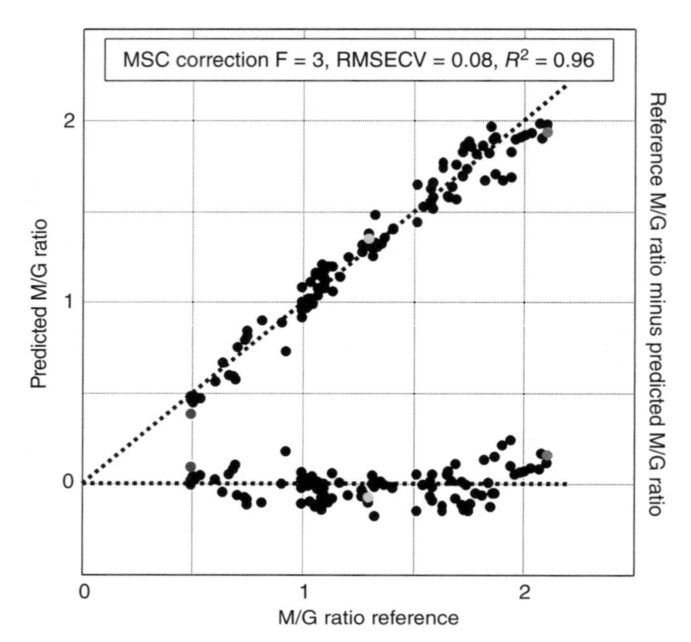

Figure 8.12 Predicted versus reference mannuronic-to-guluronic acid (M/G) ratio and reference minus predicted ratio based on multiplicative scatter correction (MSC) preprocessed spectra using a three-factor partial least squares model. RMSECV, root mean squared error of cross-validation.

As expected, MSC correction improves the predictive performance considerably. Figure 8.12 shows the cross validation (CV) predicted versus the reference plot. A good correlation is observed, and the prediction error is in the same order of magnitude as our reference method standard deviation maximum, though there is an (weak) indication of nonlinearity in the prediction errors, which can be solved by more advanced preprocessing methods [24].

The reader must realize that the solution of the PLS model in Equation 8.4 is a statistical estimator of the data set, and leave-one-out cross validation might seem too optimistic. To further test the capabilities of Raman spectroscopy in rapid, nondestructive quantitative prediction of M/G ratio in alginate powders, we tested another stratified resampling method. Since the samples are collected in random order, we split the set in four, where one quarter was used as test set in a calibration PLS model based on the other three splits. This gave us both an impression of the uncertainty in the RMSEP prediction error and the stability of the PLS model. Note that this is not an independent test set, because we used the same samples to fix out the PLS model complexity at three factors. It does, however, indicate what we can expect to be the

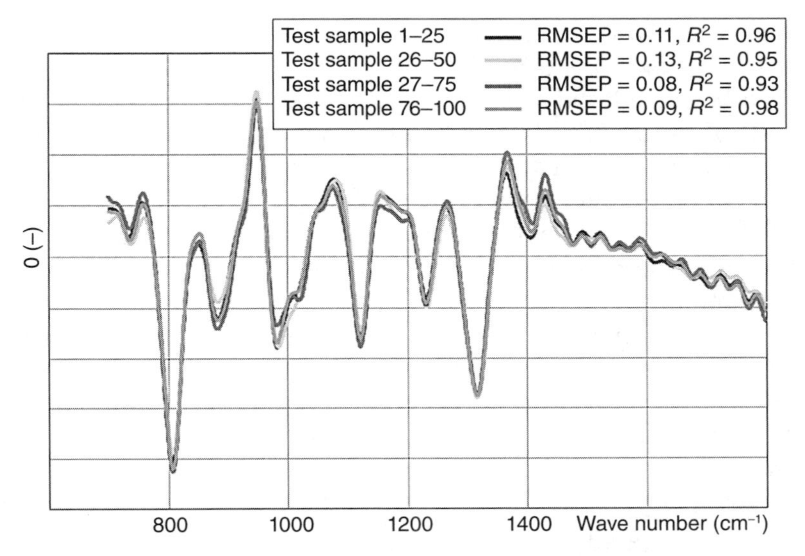

Figure 8.13 Partial least squares regression vectors (three factors) using four resamplings from the multiplicative scatter correction (MSC) preprocessed data set. RMSEP, root mean squared error of prediction.

future performance of our at- or on-line measurement system. Figure 8.13 shows the regression vectors **b** and prediction errors for the four test sets. The RMSEP average value is slightly higher, as can be expected from this more conservative testing regime. The regression vectors are very similar for the different models, but there is potential for (modest) improvement by wave number range selection and/or more advanced preprocessing.

8.6 RAMAN SPECTROSCOPY IN PROCESS ANALYTICAL TECHNOLOGY (PAT)

PAT is a concept recently introduced within the pharmaceutical industry. PAT refers to measuring the relevant information in *real time* during the production for (statistical) process monitoring, control, and optimization. This can involve the incoming raw material, process streams or parameters, and/or the finished product. This principle has been in focus in the food industry already for a number of years. The objective of PAT is to ensure final product quality and to improve production efficiency through total internal quality control by remote spectroscopic monitoring. In the PAT guidance to the pharmaceutical industry, it is stated that "for certain applications, sensor-based measurements can provide a useful *process signature* that may be related to the underlying process steps or transformations." In many cases, Raman spectra of food can be considered as a fingerprint of the food items that contains a wealth of information related to the raw materials and the process transformations. For this reason, Raman spectroscopy has a tremendous unexploited potential in process analytical chemistry and technology (PACT), since it is possible to observe or predict compounds related to the process conditions quickly and nondestructively [28]. What is more, wavelengths of visual illumination can be transmitted over long distances with almost no loss, and the Raman shift can thus facilitate the possibility for industrial on-line applications, since measurements can be taken at several points along the process line with only one spectrophotometer.

Raman spectra of intact food systems contain valuable information on various quality parameters and thus have a considerable potential both within research and as industrial on-line or at-line applications. The most important problems to be solved with regard to using Raman in relation to PACT are (1) the inherent poor signal-to-noise ratio, (2) the serious fluorescence disturbances, and (3) the challenge to make applicable standardization techniques, since the Raman signal is not measured relative to an incident light beam. Of these problems, the high sensitivity to fluorescence is *the* major obstacle for the introduction of Raman sensors, since small changes in raw materials or changes taking place during production may influence the fluorescence intensity significantly. Novel concepts, such as shift excitation difference spectroscopy, show promising results in eliminating these disturbing fluorescence signals [29,30].

Raman scattering was a phenomenon [31] that was discovered long before the necessary technology was available for routine analysis of soft matters such as food. Consequently, Raman spectroscopy has been regarded as an exotic technique with no broad utility. However, with the current developments in laser and spectrometer technology, simple and powerful spectrometers are being developed that are on the edge of taking their rightful place as important tools for food research and quality control.

REFERENCES

1. Li-Chan, E.C.Y. (1996). The applications of Raman spectroscopy in food science. *Trends in Food Science and Technology*, 7, 361–370.

2. Martens, H., Næs, T. (1993). *Multivariate Calibration*. New York: Wiley.
3. Sadeghi-Jorabchi, H., Hendra, P.J., Wilson, R.H., Belton, P.S. (1990). Determination of the total unsaturation in oils and margarines by Fourier transform Raman spectroscopy. *Journal of the American Oil Chemists' Society*, 67(8), 483–486.
4. Sadeghi-Jorabchi, H., Wilson, R.H., Belton, P.S., Edwards-Webb, J.D., Coxon, D.T. (1991). Quantitative analysis of oils and fats by Fourier transform Raman spectroscopy. *Spectrochimica Acta*, 47A, 1449–1458.
5. Góral, J., Zichy, V. (1990). Fourier transform Raman studies of materials and compounds of biological importance. *Spectrochimica Acta*, 46A(2), 253–275.
6. Engelsen, S.B., Nørgaard, L. (1996). Comparative vibrational spectroscopy for determination of quality parameters in amidated pectins as evaluated by chemometrics. *Carbohydrate Polymers*, 30(1), 9–24.
7. Cael, J.J., Koenig, J.L., Blackwell, J. (1974). Infrared and Raman spectroscopy of carbohydrates. Part IV. Identification of configuration- and conformation-sensitive modes for D-glucose by normal coordinate analysis. *Carbohydrate Research*, 32, 79–91.
8. Thygesen, L.G., Løkke, M.M., Micklander, E., Engelsen, S.B. (2003). Vibrational microspectroscopy of food. Raman versus FT-IR. *Trends in Food Science and Technology*, 14(1–2), 50–57.
9. Bulkin, B.J., Kwak, Y., Dea, I.C.M. (1987). Retrogradation kinetics of waxy-corn and potato starches: A rapid, Raman-spectroscopic study. *Carbohydrate Research*, 160, 95–112.
10. Yu, S., Blennow, A., Madsen, F., Olsen, C.E., Engelsen, S.B. (2002). Physicochemical characterization of floridean starch of red algae. *Starch/Stärke*, 54(2), 66–74.
11. Schuster, K.C., Ehmoser, H., Gapes, J.R., Lendl, B. (2000). On-line FT-Raman spectroscopic monitoring of starch gelatinisation and enzyme catalysed starch hydrolysis. *Vibrational Spectroscopy*, 22(1-2), 181–190.
12. Soderholm, S., Roos, Y.H., Meinander, N., Hotokka, M. (1999). Raman spectra of fructose and glucose in the amorphous and crystalline states. *Journal of Raman Spectroscopy*, 30(11), 1009–1018.
13. Nørgaard, L., Hahn, M., Knudsen, L.B., Farhat, I.A., Engelsen, S.B. (2005). Multivariate near-infrared and Raman spectroscopic quantifications of the crystallinity of lactose in whey permeate powder. *International Dairy Journal*, 15(12), 1261–1270.
14. Ozaki, Y., Cho, R., Ikegaya, K., Muraishi, S., Kawauchi, K. (1992). Potential of near-infrared Fourier transform Raman spectroscopy in food analysis. *Applied Spectroscopy*, 46, 1503–1507.
15. Engelsen, S.B. (1997). Explorative spectrometric evaluations of frying oil deterioration. *Journal of the American Oil Chemists' Society*, 74(12), 1495–1508.

16. Micklander, E., Brimer, L., Engelsen, S.B. (2002). Non-invasive assay for cyanogenic constituents in plants by Raman spectroscopy: Content and distribution of amygdalin in bitter almond (Prunus amygdalus). *Applied Spectroscopy*, 56(9), 1139–1146.
17. Nørgaard, L., Saudland, A., Wagner, J., Nielsen, J.P., Munck, L., Engelsen, S.B. (2000). Interval partial least squares regression (iPLS): A comparative chemometric study with an example from the near infrared spectroscopy. *Applied Spectroscopy*, 54(3), 413–419.
18. Thygesen, L.G., Jørgensen, K., Møller, B.L., Engelsen, S.B. (2004). Raman spectroscopic analysis of cyanogenic glucosides in plants: Development of a flow injection surface-enhanced Raman scatter (FI-SERS) method for determination of cyanide. *Applied Spectroscopy*, 58(2), 212–217.
19. Lippert, J.L., Tyminski, D., Desmeules, P.J. (1976). Determination of secondary structure of proteins by laser Raman spectroscopy. *Journal of the American Chemical Society*, 98(22), 7075–7080.
20. Haston, J.L., Engelsen, S.B., Roessle, M., Clarkson, J., Blanch, E.W., Baldock, C., Kielty, C.M., Wess, T.J. (2003). Raman microscopy and X-ray diffraction, a combined study of fibrillin-rich microfibrillar elasticity. *Journal of Biological Chemistry*, 278(42), 41189–41197.
21. Pedersen, D.K., Morel, S., Andersen, H.J., Engelsen, S.B. (2003). Early prediction of water-holding capacity in meat by multivariate vibrational spectroscopy. *Meat Science*, 65(1), 581–592.
22. Mikkelsen, A., Engelsen, S.B., Hansen, H.C.B., Larsen, O., Skibsted, L.H. (1997). Calcium carbonate crystallization in the a-chitin matrix of the shell of pink shrimp, Pandalus borealis, during frozen storage. *Journal of Crystal Growth*, 177(1–2), 125–134.
23. Mikkelsen, A., Rønn, B., Skibsted, L.H. (1997). Formation of white spots in the shell of raw shrimps during frozen storage. Seasonal variation and effects of some production factor. *Journal of the Science of Food and Agriculture*, 75, 433–441.
24. Salomonsen, T., Jensen, H.M., Stenbæk, D., Engelsen, S.B. (2008). Chemometric prediction of alginate monomer composition. A comparative spectroscopic study using IR, Raman, NIR and NMR. *Carbohydrate Polymers*, 72(4), 730–739.
25. Martens, H., Jensen, S.A., Geladi, P. (1983). Multivariate linearity transformation for near-infrared reflectance spectrometry. In: Christie, O.H.J., ed., *Conference Proceedings of the Nordic Symposium on Applied Statistics*. Stavanger, Norway: Stokkand Forlag Publishers, 205–234.
26. Geladi, P., McDougall, D., Martens, H. (1985). Linearization and scatter-correction for near-infrared reflectance spectra of meat. *Applied Spectroscopy*, 39(3), 491–500.
27. Jolliffe, I.T. (2002). *Principal Component Analysis*, 2nd Edition. New York: Springer-Verlag.

28. Dyrby, M., Engelsen, S.B., Nørgaard, L., Bruhn, M., Lundsberg-Nielsen, L. (2002). Chemometric quantitation of the active substance (containing C N) in a pharmaceutical tablet using near-infrared (NIR) transmittance and NIR FT-Raman spectra. *Applied Spectroscopy*, 56(5), 579–585.
29. Bell, S.E.J., Bourguignon, E.S.O., Dennis, A.C., Fields, J.A., McGarvey, J.J., Seddon, K.R. (2000). Identification of dyes on ancient Chinese paper samples using the subtracted shifted Raman spectroscopy method. *Analytical Chemistry*, 72(1), 234–239.
30. Osticioli, I., Zoppi, A., Castellucci, E.M. (2006). Fluorescence and Raman spectra on painting materials: Reconstruction of spectra with mathematical methods. *Journal of Raman Spectroscopy*, 37(10), 974–980.
31. Raman, C.V., Krishnan, K.S. (1928). A new type of secondary radiation. *Nature*, 121, 501–502.

MEDICAL APPLICATIONS

9

RAMAN APPLICATION IN BONE IMAGING

Sonja Gamsjäger, Ph.D.
Ludwig Boltzmann Institute of Osteology at the Hanusch Hospital of WGKK and AUVA Trauma Centre Meidling

Murat Kazanci, Ph.D.
Max Planck Institute of Colloids and Interfaces and St Francis Xavier University

Eleftherios P. Paschalis, Ph.D.
Senior Scientist, Ludwig Boltzmann Institute of Osteology at the Hanusch Hospital of WGKK and AUVA Trauma Centre Meidling

Peter Fratzl
Professor, Max Planck Institute of Colloids and Interfaces

9.1 INTRODUCTION

Bone is a composite material with a highly complex hierarchical structure (Figure 9.1) based on a collagen-rich organic matrix and embedded mineral particles [1]. Many bones such as the femoral head (Figure 9.1a) have a dense external shell (cortical bone, Figure 9.1b) and a spongy interior (cancellous bone, Figure 9.1c), where only about 20% of the volume is filled with bone material and the rest with bone marrow. The basic building block is the mineralized collagen fibril which is usually assembled in lamellae (Figure 9.1d) in a rotated plywood-like fashion. All hierarchical levels are important for mechanical performance, and

Raman Spectroscopy for Soft Matter Applications, Edited by Maher S. Amer

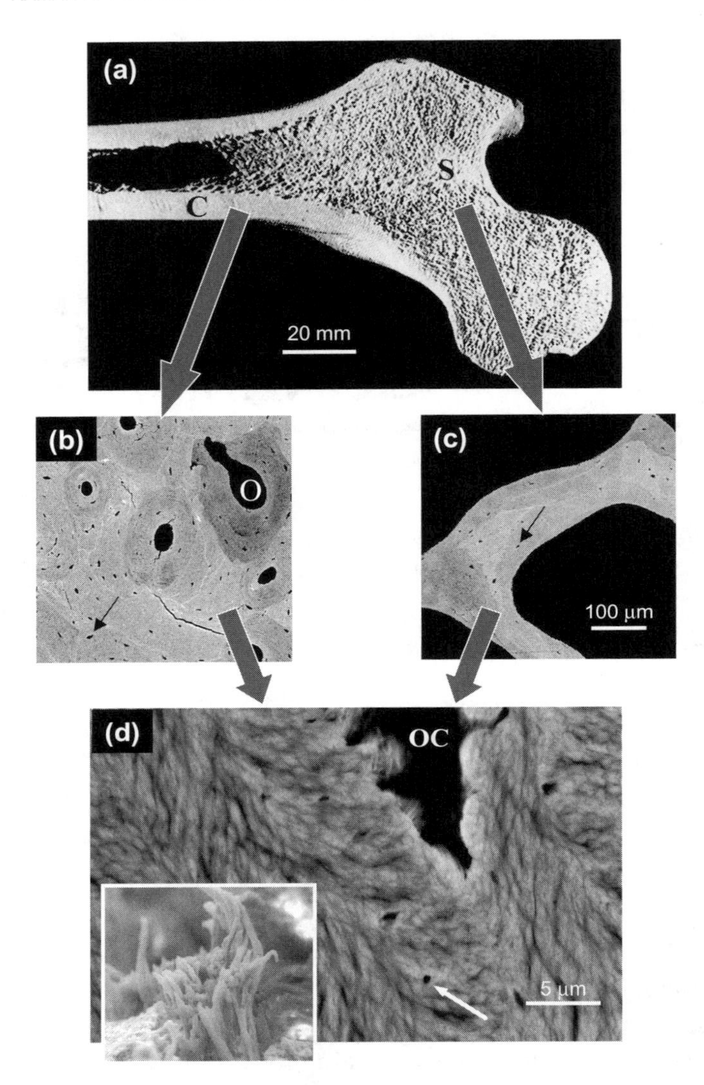

Figure 9.1 (a) Section through a femoral head showing the shell of cortical bone (C) and the spongious bone (S) inside. (b) Enlargement of the cortical bone region visualized by backscattered electron imaging (BEI), revealing several osteons (O) corresponding to blood vessels surrounded by concentric layers of bone material. (c) BEI picture of a single trabeculum from the spongious bone region. The arrows in both (b) and (c) indicate osteocyte lacunae (OCs) where bone cells have previously been living. (d) Further enlargement showing the lamellar and fibrillar material texture around an OC as visible in scanning electron microscopy (SEM) (see white arrow). The lamellae are formed by bundles of mineralized collagen fibrils (insert).

a great variety of structures are obtained due to the adaptation of bone structure to the actual needs. Nevertheless, the basic material is always a collagen–mineral nanocomposite, which has already quite remarkable properties well suited for its mechanical function. The composite contains mineral platelets (essentially nanocrystalline carbonated hydroxyapatite [HA]), protein (predominantly collagen type I), and water. This material combines two components with extremely different properties, namely, the mineral, which is stiff but brittle, and the (wet) protein, which is tough but much less stiff than the mineral, resulting in outstanding fracture resistance to the bone material despite the inherent brittleness of the mineral [1].

As a consequence, the overall structure of bone (from the material level [Figure 9.1d] to shape and architecture [Figure 9.1a]) needs to be assessed in diseases which affect bone fragility, such as osteoporosis. Osteoporosis is a disease which affects roughly 75 million people in Japan, the United States, and Europe [2]. Loss of bone mass, measured clinically as change in bone mineral density (BMD), is considered an important risk factor for bone fragility. However, BMD is not the sole predictor of whether an individual will experience a fracture [3,4], and there is considerable overlap in BMD between populations that do and do not develop fractures [5–7]. It has been demonstrated that for a given bone mass, an individual's risk to fracture increases with age [8]. Consistent with these findings, numerous investigators have shown that mechanical variables directly related to fracture risk are either independent [9] or not totally accounted for bone mass itself [10–14]. Epidemiological evidence also shows considerable overlap of bone density values between fracture and nonfracture groups, suggesting that low bone quantity alone is an insufficient cause of fragility fractures [15–17].

It is becoming evident then, that in addition to BMD, bone quality should also be considered when assessing bone strength and fracture risk. Bone quality is a broad term encompassing a plethora of factors such as geometry and bone mass distribution, trabecular bone microarchitecture, microdamage, increased remodeling activity, along with genetics, body size, environmental factors, and changes in bone mineral and matrix tissue properties [6,7]. In addition, new fracture risk factors are emerging and it is still unclear how these are manifested in BMD changes (if at all). For example, recent clinical/epidemiological data [18–21] show a definite correlation between homocysteine in the blood and fracture risk. Homocysteine is known to interfere with lysyl oxidase action [22], thus altering collagen posttranslational modifications and thus collagen cross-link profiles. Moreover, BMD measurements may not be able to distinguish between bone volume and matrix mineralization changes, both of which play an important role in determining fracture risk [23].

One of the obstacles to be circumvented when assessing mineral and matrix tissue properties is tissue heterogeneity at the microscopic level. In normal humans, cortical bone constitutes approximately 80% of the human skeletal mass and trabecular bone approximately 20% [24]. Bone surfaces may be undergoing formation or resorption, or they may be inactive. These processes, which can be visualized microscopically, occur throughout life in both cortical and trabecular bone [24]. Bone remodeling is a surface phenomenon, and in humans occurs on periosteal, endosteal, Haversian canal, and trabecular surfaces [24–27]. The rate of cortical bone remodeling, which may be as high as 50% per year in the mid-shaft of the femur during the first 2 years of life, eventually declines to a rate of 2–5% per year in the elderly. Rates of remodeling in trabecular bone are proportionately higher throughout life and may normally be 5–10 times higher than cortical bone remodeling rates in the adult [28–55]. This information is critical when evaluating bone at the microscopic level.

Utilizing techniques such as small-angle X-ray scattering (SAXS), quantitative backscattered electron imaging (qBEI), and Fourier transform infrared microspectroscopy (FTIRM) and Fourier transform infrared imaging (FTIRI), the analysis of bone mineral (poorly crystalline HA) at the microscopic level and the contribution of mineral crystallinity (crystallite size) and maturity (chemical composition) to bone strength are being actively pursued. Based on such studies, models for the importance of mineral crystallite shape and size in determining bone strength have been put forth. Moreover, an important role for the organic matrix in the determination of biomechanical properties is predicted. Specifically, the matrix is proposed to play an important role in alleviating impact damage to mineral crystallites and to matrix/mineral interfaces, behaving like a soft wrap around mineral crystallites, thus protecting them from the peak stresses caused by impact and homogenizing stress distribution within the bone composite [32–40].

Considerably less attention has been directed at collagen, although there are several publications in the literature reporting altered collagen properties associated with fragile bone, in both animals and humans [56–64]. A limitation of such studies is the fact that biochemical methods requiring tissue homogenates were utilized to determine collagen properties such as cross-links; thus, it was not feasible to study the spatial variation of such properties at the microscopic level as a function of surface metabolic activity (forming, resorbing, quiescent) and thus tissue age.

Another technique that has reemerged in the analysis of bone and, more specifically, bone quality is Raman spectroscopy.

9.2 A TOOL FOR CHARACTERIZING BONE MATERIAL

One of the difficulties in assessing the bone material quality is its hierarchical and complicated structure. Lamellar cortical bone, for example, comprises mineralized collagen fibers with alternating fiber orientation in successive lamellae. One approach to address this issue is Raman microspectroscopy and imaging. This offers several technical advantages. It has excellent spatial resolution (0.5–1.0 μm) and, in general, the vibrational bands are narrow so that small frequency shifts and band shape changes often are observed easily. Raman microspectroscopy evolved rapidly once it was pointed out that the intensity of Raman light should be independent of sample volume and should remain essentially constant with decreasing sample size down to the dimension determined by the diffraction limit, and hence the wavelength, of the laser excitation. With routine limits of detection in the nanogram range and high molecular selectivity, micro-Raman spectroscopy has now become a major analytical technique. It may be applied to characterize bone material properties at the microscopic level as a function of anatomical location and bone surface metabolic activity, the analysis of biologically important details such as individual cement lines, individual lamellae, and boundaries around microcracks.

Since Raman spectroscopy is nondestructive, the same specimen can be examined by a multitude of different techniques. Information from the mineral component and the organic matrix is obtained simultaneously, providing a complete picture of the major bone constituents in the area surveyed. Because of the heterogeneous nature of bone, single-point Raman microspectroscopy cannot adequately describe the chemical microstructure of bone. Spatial information is needed. For this reason, Raman spectroscopic imaging is increasingly popular for the analysis of complex organized systems such as bone and teeth [65]. Raman spectroscopy and imaging are valuable tools for the analysis of bone, the determination of protein secondary structure, and the study of the composition of crystalline materials. The information from the mineral component and the organic matrix is obtained simultaneously, creating a complete picture of bone composition.

9.2.1 Typical Raman Lines from Bone

Figure 9.2 shows a typical Raman spectrum of bone with the phosphate (PO_4) ν_1 band at 960 cm^{-1}, 431, 440, 449 (ν_2PO_4), 581, 592, 608 (ν_4PO_4), 1031, 1048, 1076 (ν_3PO_4), and the bands associated with collagen (amide III at 1250 cm^{-1}, CH_2 wag at 1450 cm^{-1}, and amide I at 1665 cm^{-1}) are of particular interest for compositional studies [66–69]. The fine spatial resolution possible in Raman analysis allows the observation of phenomena occurring at the microscopic level. The narrow vibrational bands allow one to distinguish between various chemical species. There is little hindrance from water, and it is possible to obtain Raman spectra since mechanical, chemical, and photochemical degradation as well as fluorescence of the samples can be effectively eliminated [70,71]. A short time dependence of fluorescence background obtained with the 532-nm laser can be seen in Figure 9.3. In a study by Golcuk et al. [72], a 532-nm laser (widely used in Raman spectroscopy) irradiation has been used efficiently to remove the fluorescence background from Raman spectra of cortical bone, and they have shown that photochemical bleaching reduces over 80% of the fluorescence background after 2 hours and was found to be nondestructive within 40 minutes. The use of a near-infrared (NIR) laser, such as a 785-nm diode laser which excites little bone fluorescence compared to the green (532 nm) lasers, results in longer acquisition times (Figure 9.3). The long acquisition time is a result of the inverse fourth power dependence of Raman intensity with excitation wavelength.

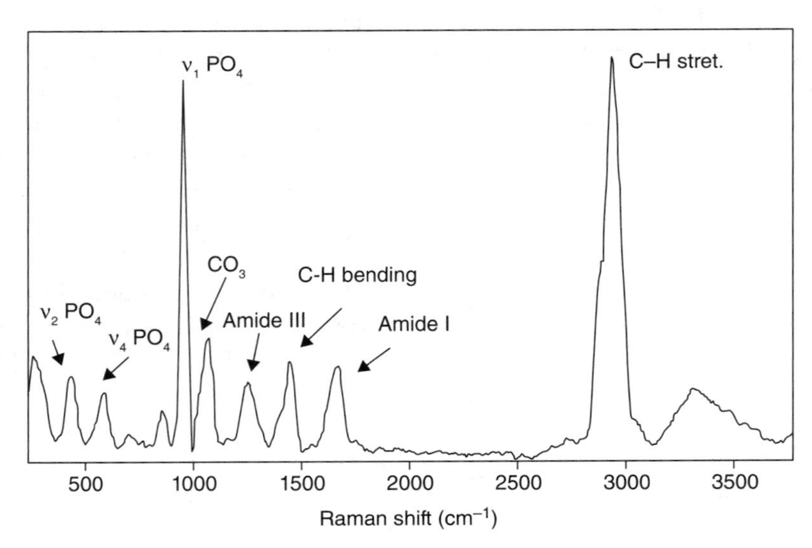

Figure 9.2 A typical Raman spectrum of bone.

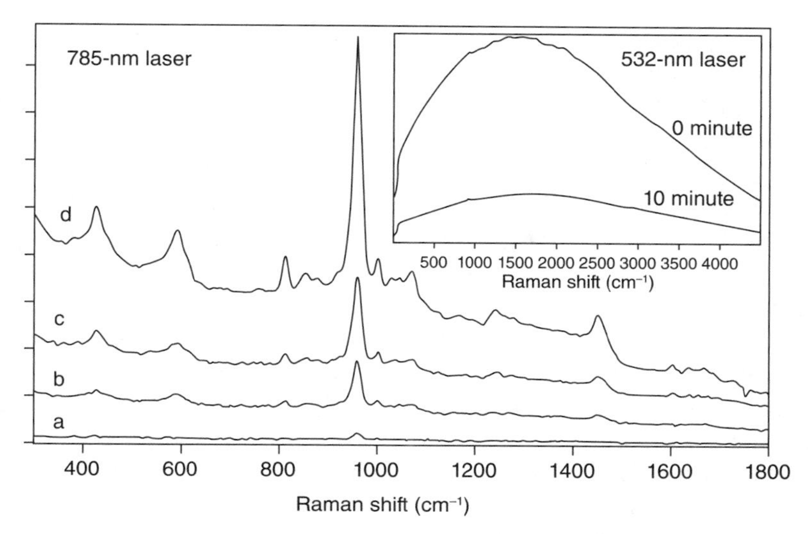

Figure 9.3 Raman spectra of bone at different photobleaching times (0 and 10 min) obtained with the 532-nm laser and Raman spectra with different integration times obtained with the 785-nm laser: (a) Integration time 1, Co-addition 1; (b) Integration time 10, Co-addition 1; (c) Integration time 20, Co-Addition 1; and d) Integration time 60, Co-Addition 3; laser power 10 mW.

Raman spectroscopy can been used to assess the degree of collagen mineralization, mineral maturity (and thus cystallinity in an indirect manner), and the presence of nonstoichiometric substitutions in the mineral crystals of bone [68,73–75].

9.2.2 Specimen Preparation

The advantages of Raman spectroscopy are its high specificity and its versatility. It is a nondestructive technique and requires, in general, only minimal or no sample preparation. Frozen or embedded specimens with thickness ranging from less than 1 μm can be analyzed. Yeni et al. [76] studied the effect of ethanol and glycerol as fixatives and a variety of embedding media (Araldite, Eponate, Technovit, glycol methacrylate, polymethylmethacrylate [PMMA], and LR White) on Raman spectral properties, and by extension bone mineral characteristics such as mineralization, crystallinity, and carbonation, measured in the cortical bone of mouse humeri. Raman spectra of fixative solutions and embedding media were also recorded separately in order to examine the specifics of potential spectral overlap. The study found significant effects of fixation, embedding, and anatomical location on Raman spectral properties. These alterations may be attributable to leaching of the mineral component into solutions utilized during fixation and embedding, the modifications in the vibrations of collagen bonds through alteration of their chemical environment and the overlap of the embedding medium-related peaks with bone-related peaks, which may confound the calculation of intensity ratios. The replacement of tightly bound water molecules in the immediate vicinity of mineral and collagen molecules by fixatives may be excluded as a source of spectral error [77]. On the other hand, removal of loosely bound and unbound water molecules interferes with the conformation of tightly bound water and affects the supramolecular packing pattern of collagen molecules. Removal of water by bench or oven drying is known to result in closer spacing of the collagen matrix [78,79]. Changes in the lateral spacing of collagen molecules and the bound water structure perturb the secondary structure of collagen and the peptide bond conformation. Therefore, the primary effect of fixation would be on the amide bands of the Raman spectral pattern, which are mostly composed of peptide vibrations. Comparison of nonembedded control, ethanol-fixed, and glycerol-fixed bones revealed that fixation had a significant effect on mineralization and crystallinity while carbonation was not affected. Glycerol-fixed bones had a significantly greater PO_4/amide I ratio. None of the Raman parameters differed between freeze-dried bones and wet controls, with the only exception of an increase in PO_4/amide I in freeze-dried bone. The results show that ethanol fixation is relatively favorable compared to glycerol fixation. The effects of the embedding medium on the Raman spectrum of bone tissue may occur by band overlap and/or by interaction of chemical groups between bone matrix and embedding medium. The latter has been ruled out at least for PMMA; however, other embedding media may induce shifts in the wave numbers of mineral or collagen bands [76].

9.3 STRUCTURE VERSUS COMPOSITION

The Raman signal is dependent upon the dual influence of the composition and the structure of the sample [80]. For this reason, the polarization of Raman scattering relative to the input laser polarization has been studied extensively and can be of significant analytical value. The laser induces a polarization in the sample that is parallel to the incident electric field. For a totally symmetric vibration, the Raman scattered light retains the polarization of the incident light. As a consequence, the observed intensity is strongly dependent on the observation direction [81]. Several aspects of Raman polarization measurements are valuable for chemical analysis. First, polarization provides a means to determine the symmetry of the vibrations underlying observed spectral features. Second, any nonrandom molecular orientation will affect the observed polarizations, so Raman may provide information about such orientation. Third, polarization may be used to distinguish preferentially oriented molecules from randomly oriented molecules [81].

9.3.1 Polarization Dependence of Raman Scattering

The ν_1 phosphate stretching vibration at $961\,cm^{-1}$ is the strongest marker for bone mineral. The ν_2 and ν_4 phosphate bending vibrations are visible at 438 and $589\,cm^{-1}$, respectively. There is a strong band at $1075\,cm^{-1}$, indicating type-B carbonate substitution in the bone specimen (carbonate substituting for phosphate in the apatite lattice). The bands observed in the high-frequency region are amide III ($\sim1256\,cm^{-1}$), the C—H bending mode ($\sim1458\,cm^{-1}$), amide I ($\sim1677\,cm^{-1}$), and C—H stretching ($\sim2937\,cm^{-1}$). The amide I at $\sim1677\,cm^{-1}$ and the amide III $\sim1256\,cm^{-1}$ peaks are mainly due to the presence of collagen, while the C—H bending bands at ~1458 and $\sim2937\,cm^{-1}$ are present in both collagenous and noncollagenous organic moieties as well as PMMA [65,82,83]. However, all these bands have completely different polarization behaviors. The amide I band is mostly associated with the C=O stretching vibration, whereas the amide III mode has a major contribution from the C—N vibration [84]. As amide I bands' scattering is more intense in the direction perpendicular to the fiber axis, it is suggested that the Raman tensor for this vibration may be preferentially aligned perpendicular to the fiber axis [84], whereas amide III has two different C—N vibration modes, one representative of perpendicular conformation ($1270–1300\,cm^{-1}$) and the second one of parallel conformation ($1230–1260\,cm^{-1}$) [85]. Parallel conformation of the amide III shows a similar orientation pattern with ν_1PO_4, and perpendicular conformation of the amide III shows a similar pattern with amide I. For the mineral component of the bone, intensities of the Raman bands are solely dependent on the *c*-axis orientation [86]. A preliminary Raman study on crystallites near the intact outer surface of human enamel has also shown the strong orientational dependency similar to those of HA single crystals [80].

9.3.2 Quantifying Collagen Fibril Orientations

As shown in Figure 9.4, the Raman scattering effect is highly dependent on polarization. Since pump-laser diodes have a highly polarized

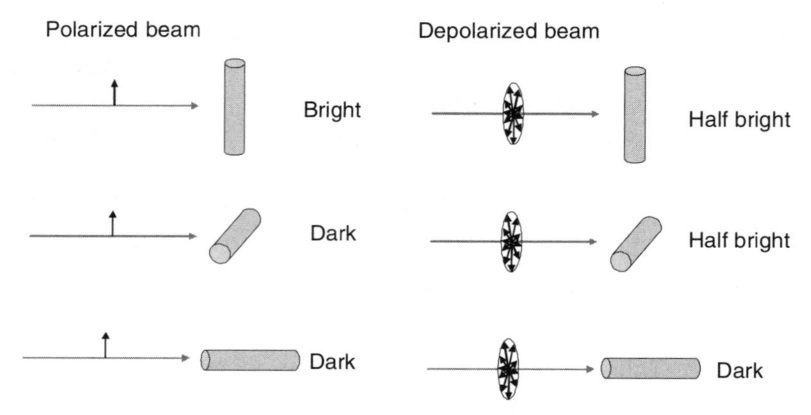

Figure 9.4 Schematic representation of the intensity of Raman signal (from dark to bright) for a vibration parallel to the collagen fibril orientation (indicated by the cylinders). Clearly, depolarization of the beam removes some but not all of the orientation dependence. However, a rotation of the specimen around the axis of the beam does reveal the anisotropy only when the incident laser light is polarized.

output, depolarizers have attracted a lot of attention in recent years in order to overcome the problem. Their deployment to date is restricted because of their technical limitations. The basic function of a depolarizer is to convert a polarized light input into a randomly polarized output. Actually, the light is depolarized on average. There is no such thing as depolarized light—at any given moment in time, light always has a state of polarization. The function of a depolarizer is to scramble the polarization so that on average over time, there appears to be no specific polarization state.

For the experimental section, fresh bovine bone was obtained from the femur of an ox, and all soft tissue was removed. The bone was sawed into sections with continuous water irrigation and was wrapped in phosphate-buffered saline (PBS)-soaked gauzes. The sample was fixed in 70% ethanol, dehydrated in a graded series of ethanol solutions, embedded in PMMA, and the surfaces made planar and parallel to each other by grinding and polishing. Osteonal tissue sections were chosen for structural analysis in longitudinal and transverse directions.

In order to investigate the effect of a depolarizer on the Raman analysis of bone, a $\lambda/4$ plate was used as a depolarizer and bovine bone sections for Raman spectral mapping. These sections were obtained as in Gupta et al. [87], from plexiform bovine bone where regions of highly parallel collagen fibrils are sandwiched between osteonal structures. Raman images of the parallel fibered regions were obtained by first placing the samples with the collagen direction parallel to the polarization direction of the incident light. This position was called $\alpha = 0°$. The sample was then turned in steps of 15° toward $\alpha = 90°$ within the plane perpendicular to the incident light beam, keeping the polarization fixed. Raman data were collected in each of the positions. The experiment was repeated without a $\lambda/4$ plate. As a second step, the longitudinal bone sections (Figure 9.6, top) were replaced by cross sections with the collagen orientation perpendicular to the surface and, thus, parallel to the beam direction. This gives the opportunity to examine each tissue section from two different orthogonal planes.

The results show that for a longitudinal section with the collagen orientation perpendicular to the incident laser beam (Figure 9.5), ν_1PO_4/amide I ratio decreases from ~8.0 (collagen fibril orientation parallel to the laser polarization) to ~3.5 (collagen fibril orientation perpendicular to the laser polarization) when no depolarizer is used (black circles in Figure 9.5). When the experiment was repeated with a

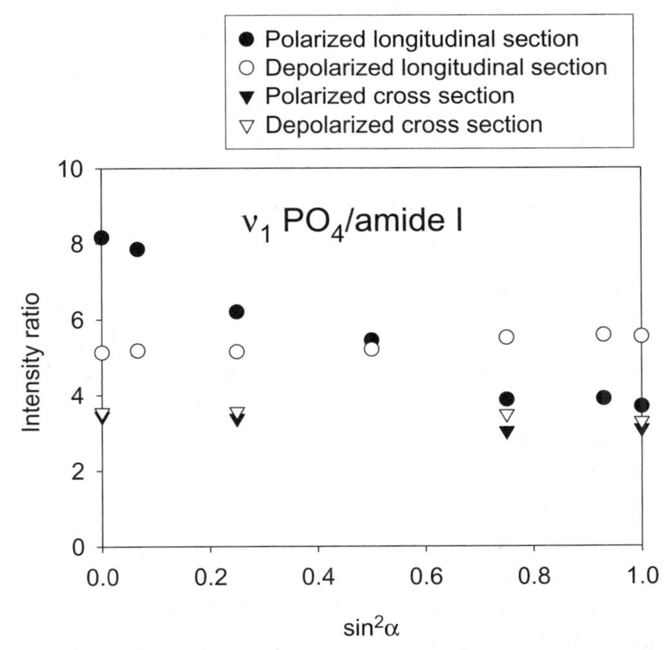

Figure 9.5 ν_1PO_4/amide I ratio obtained from longitudinal and cross sections under polarized and depolarized light. The angle α indicates the rotation of the specimen around the incident laser beam keeping its polarization fixed. In the longitudinal section, $\alpha = 0$ corresponds to the collagen direction parallel to the laser polarization. In the cross sections, the collagen direction is parallel to the laser beam and, therefore, always perpendicular to the polarization.

depolarizer, no significant difference was detected on the ν_1PO_4/amide I ratio at different angles in perpendicular plane to the beam direction. In addition, the (constant) intensity measured with the depolarizer precisely corresponds to the average value of the intensity as a function of angle. However, when a cross section was analyzed with the collagen direction parallel to the beam direction, there was no angular dependence and no difference between the measurements with and without the depolarizer (triangles). But even using a depolarizer in both cases, the intensity ratio in the cross section is considerably decreased compared to the intensity ratio in the longitudinal section (from ~5.50 to ~3.50). The reason for this is simple. In the cross section, the collagen fibrils are always perpendicular to the laser beam polarization, while in the longitudinal sections, one measures an average between fibrils parallel and perpendicular to the laser polarization, which gives a higher intensity. This is confirmed by the fact that the measurement at $\alpha = 90°$ ($\sin^2\alpha = 1$) in the longitudinal section (full circles) gives the same value close to 3.5 than the measurements at all angles in the cross section (triangles). Indeed, the collagen fibrils are perpendicular to the laser polarization in all these cases. Since collagen structures are 3-D, orientation effects cannot be generally removed by the use of a depolarizer. Figure 9.6 shows the angular variation of amide I intensity (correlation of $\sin^2\alpha$ versus amide I intensity) relevant to the polarization direction of the beam. The amide I band is the most polarization-susceptible band among all other Raman bone bands.

9.3.3 Matrix Orientation in Osteons

In an experimental work by the authors [88,89], human female femoral mid-shaft bone was investigated after it was fixed, embedded, and the surfaces made planar and parallel to each other by grinding and polishing. A continuous laser beam with diode pumped green laser excitation (532 nm) was microfocused on the sample using 100× (NA = 0.9, Nikon) microscope objectives. Detector integration time for each pixel was 2 seconds per sample and images were created using 1 μm step size.

The fiber orientation theory by Wagermaier et al. [90] in Figure 9.7a and the sketch of a partial protein structure in Figure 9.7b were used in this study in order to explain the differences in contrast images. Figure 9.7b shows the typical sequence in the collagen molecule with glycine at every third residue and a large number of prolines and hydroxyprolines at the intermediate positions. The C=O bonds (amide I) are perpendicular to the backbone, while C–N bonds (amide III) are

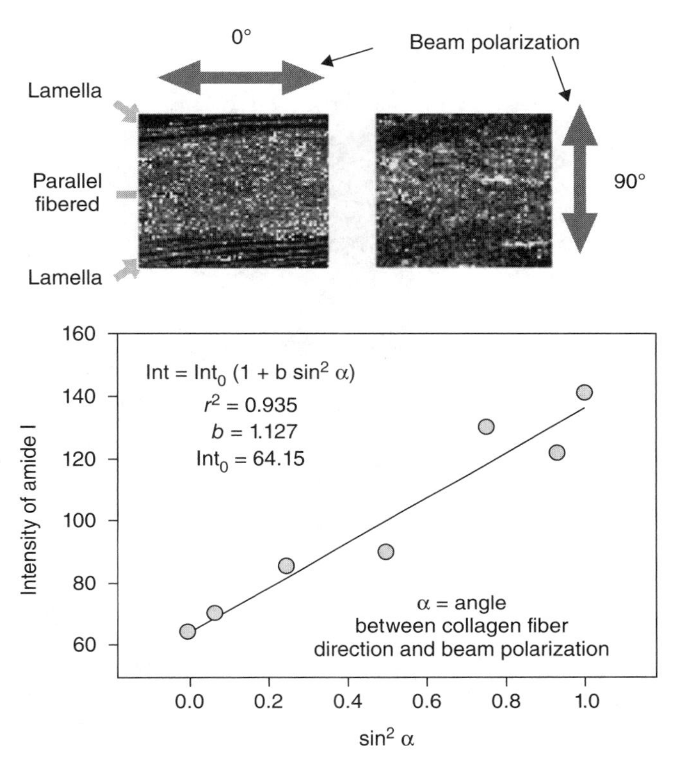

Figure 9.6 Orientation of bovine plexiform bone in relation to the polarization direction and angular variation of the amide I band.

both in the backbone and perpendicular to it in the prolines and hydroxyprolines.

In this study, it was demonstrated for the first time that each cylindrical sub-layer in the lamellae structure of osteons having similar orientation gives the same color contrast in Raman images. Moreover, for all the contrast images, the orientation contrast vanishes when the layer structure of lamellar bone is perpendicular to the polarization direction of the incident light. The reason is that the collagen orientation is always within the plane of the layers. When the structure becomes parallel to the polarization direction, different color contrasts for single vibration modes were observed. It is worth mentioning two Raman lines in particular. The first one is the ν_1PO_4 stretching vibration, which is strongest parallel to the axis of the fiber [80]. The other strongly orientation affected vibration type is amide I (C=O) stretching. Even though collagen fibers and mineral *c*-axis have parallel orientations [91,92], the C=O (amide I) vibration is perpendicular to the fiber axis.

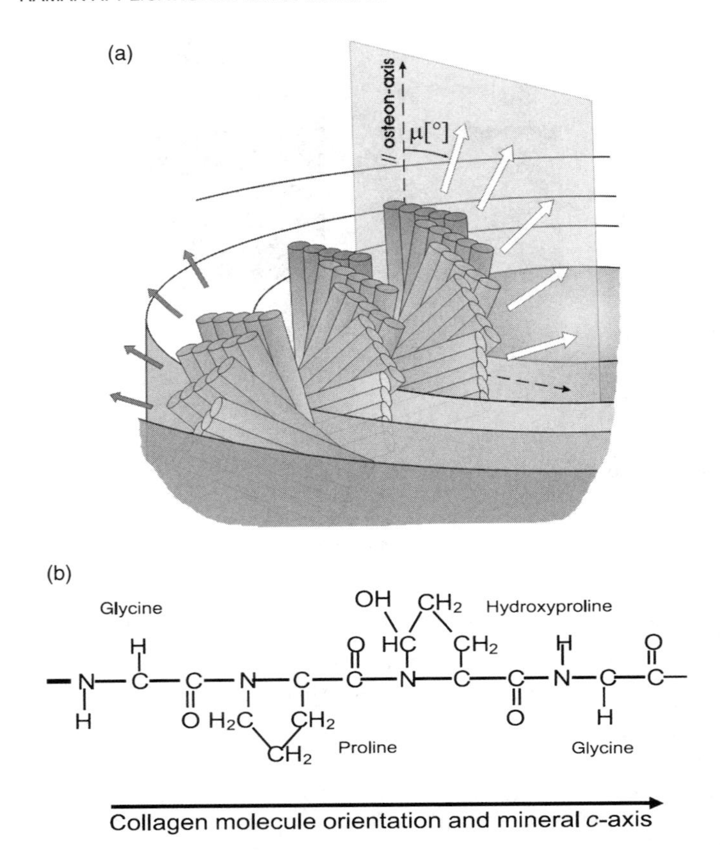

Figure 9.7 (a) Model of fibrillar orientation relative to the osteon axis by Wagermaier et al. and (b) a typical sequence in the collagen molecule with glycine at every third residue and a large number of prolines and hydroxyprolines at the intermediate positions.

Since the amide III band encompasses different vibration modes, its contrast image does not display any orientation effect, unlike ν_1PO_4 and amide I vibration modes. The same argument is also valid for ν_2PO_4 and ν_4PO_4 bands. It has previously been demonstrated by Tsuda and Arends [86] and by Leroy et al. [80] that those bands also consist of different Raman polarization tensors. Therefore, when ν_1PO_4, amide I, or their ratios with different substitutes were calculated and mapped, they mainly displayed orientation of the structure relative to the polarization direction of the incident light, even though it is common to use them for estimating the chemical contents of the bone in the literature [83]. However, ν_2/amide III and CO_3/ν_2PO_4 ratios displayed variation in bone composition.

9.4 BONE COMPOSITION

9.4.1 Mineral-to-Matrix Ratio (MTMR)

Raman spectroscopy is useful in the study of normal and diseased bone, as in the case of osteoporosis and osteogenesis imperfecta. In recent years, different animal and human models were studied by using Raman spectroscopy, especially Raman microspectroscopy, to investigate the relevant processes at the ultrastructural level.

One of the most direct measurements of Raman analysis of bone is the MTMR, obtained through the ratio of the integrated areas or peak heights of any of the phosphate and amide peaks. It is a measure of BMD. Raman microspectroscopy is used as a probe of ultrastructural (molecular) changes in both the mineral and matrix (protein and glycoprotein, predominantly type I collagen) components in real time of murine cortical bone as it responds to elastic deformation by Morris et al. [65]. They loaded murine femora in a custom-made mechanical tester that fits on the stage of a Raman microprobe. Average load and strain were measured using a load cell. These devices ensure that specimens were not loaded to or beyond the yield point. Changes occurred in the mineral component of bone as a response to loading in the elastic regime. They propose that the mineral apatitic crystal lattice was deformed by movement of calcium and other ions. With Raman microspectroscopy, they showed that bone mineral is not a passive contributor to tissue strength.

De Carmejane and coworkers [69] determined the relative response during loading and unloading of mineral versus matrix, and within the mineral, phosphate versus carbonate, as well as proteinated versus deproteinated bone. For all mineral species, shifts to higher wave numbers were observed as pressure increased. The change in vibrational frequency with pressure for the more rigid carbonate was less than for phosphate and was caused primarily by movement of ions within the unit cell. Deformation of phosphate, on the other hand, results from both ionic movement as well as distortion. Changes in vibrational frequencies of organic species with pressure were greater than for mineral species, and were consistent with changes in protein secondary structures such as alterations in interfibril cross-links and helix pitch [69]. In the study of Hofmann et al. [93], acoustic impedance distributions of single osteon sections cut at various directions relative to the femoral long axis were spatially mapped with hyperspectral (one spectral and two spatial dimensions) Raman images. The spatial distribution of phosphate ν_1 and amide I bands was directly linked to elastic anisotropy represented by the acoustic impedance. Dehring et al. [94] presented Raman spectra of femur condyles and observed mineral bands that arise from the subchondral bone. In two separate experiments, transgenic mouse models of early-onset osteoarthritis (OA) and lipoatrophy were compared to tissue from wild-type mice. Raman spectroscopy was used to identify chemical changes in the mineral of subchondral bone that may accompany or precede morphological changes that can be observed by histology. The transgenic mice were compared to age-matched wild-type mice. Subtle alterations in the mineral or collagen matrix were observed by Raman spectroscopy using established Raman markers such as the carbonate-to-phosphate ratio, MTMR, and amide I ratio. The Raman microscope configuration enabled rapid collection of Raman spectra from the mineralized layer that lies under an intact layer of nonmineralized articular cartilage. The effect of the cartilage layer on the collection of spectra was also discussed. The technique proposed is capable of providing insight into the chemical changes that occur in the subchondral bone on a molecular level [94].

Glucocorticoid-treated mice were studied by Lane et al. [95]. Changes in trabecular bone structure, elastic modulus, and MTMR of the fifth lumbar vertebrae were assessed in prednisolone-treated mice and placebo-treated controls for comparison with estrogen-deficient mice and sham-operated controls. There was a loss of mineral as shown by a halo of the amide I (lower value) and phosphate (higher value) peaks around the osteocyte lacunae (OCs) from the prednisolone-treated animals compared with the placebo group. The study of Kubisz et al. [96] presents investigations of Raman scattering from the bone irradiated with γ-radiation up to the dose of 1000 kGy. Results of Raman spectra studies of γ-irradiated bone allowed to show that the inorganic components of the animal bone are more sensitive to γ-radiation than the organic component. The changes in the irradiated bone were evaluated on the basis of the changes in the intensities of line, changes in their position, and changes in some important ratios. Independent determination of the protein content and the content of some important amino acids were compared to the data obtained in the Raman studies. The majority of the bands assigned to the organic component of bone were affected by doses higher than 100 kGy.

Ramasamy and Akkus [97] studied the mineralization and mineral quality and the overall collagen orientation with quantitative polarized imaging of standardized coupon-shaped specimens from femurs of Swiss Webster mice (9 weeks). Raman microspectroscopic analysis indicated that the MTMR, mineral crystallinity, and carbonation did not vary between the quadrants and that the orientation of collagen fibers with respect to the anatomical loading axis has a profound effect on the uniaxial mechanical function of murine bone.

Nevertheless, when considering such an outcome, one has to bear in mind two potential pitfalls. First, the polarization components (HA crystals and collagen) depend on the orientation of the axes with respect to the plane of polarization of the input laser light as well as on the relative polarization of the input and the observing polarizer [85,98]. In 1984, Gevorkian and coworkers [98] studied the bone tissue as well as the collagen fibers of the mouse tail tendon and oxyapatite crystals using polarized Raman spectra. They showed that polarized Raman spectra allow the collection of information on the spatial orientation of collagen fibers and oxyapatite microcrystals in various parts of the bone tissue and that in the diaphis osteon of human femur, the collagen fibers and crystallographic axis *c* of oxyapatite crystals are directed along the long axis of the bone in the direction of the main mechanical loads on it.

Kazanci and coworkers [88] studied 12 osteons in tissue sections from the femoral mid-shaft of a healthy human female (Figure 9.8). Spatial changes in the amount of mineral and organic matrix as well as the variation in the mineral content were determined, imaged, and plotted as a function of the polarization of incident light. It was shown that the ν_1PO_4/amide I ratio mainly displays lamellar bone orientation (Figure 9.8c,f), and ν_2PO_4/amide III and CO_3/ν_2PO_4 ratios display variation in bone composition. The ν_2PO_4/amide III ratio is higher in the interstitial bone region, whereas the CO_3/ν_2PO_4 ratio has lower values in the same region. Raman bands (ν_2PO_4, ν_4PO_4, and amide III) should also be taken into consideration for compositional analysis of bone structures, especially ones with unknown orientational features. The results demonstrated the versatility of the analytical technique and provided insights into the organization of bone tissue at the ultrastructural level [88]. Furthermore, lamellar bone orientation and variation in bone composition from normal human femoral mid-shaft bone cube-like specimen were investigated by Raman microspectroscopy and imaging (Figure 9.9). Identical bone lamellae in both longitudinal and transverse directions were analyzed, enabling the discrimination between orientation and composition changes, providing new insights into the organization of lamellar bone tissue from two orthogonal orientations [89].

Second, this outcome, although a measure of density, may or may not agree with the outcomes of more widely used techniques to estimate BMD such as dual energy X-ray absorptiometry (DXA), quantitative backscattered electron imaging (qBEI), and microradiography. The reason is the latter technique outcomes report the amount of mineral

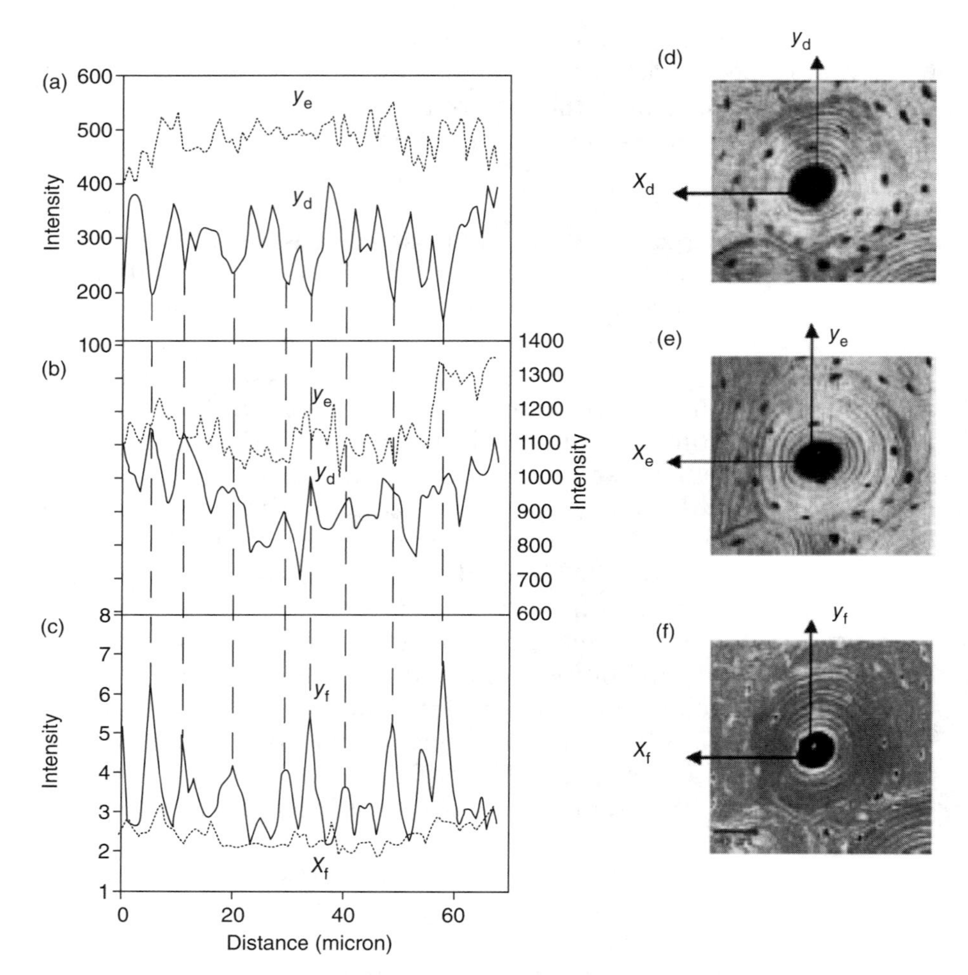

Figure 9.8 (a) Amide I, (b) ν_1PO_4 line intensities, and (c) ν_1PO_4/amide I ratio of the osteonal tissue. (d) Osteon's spectral image (amide I contrast) before rotation, (e) osteon's spectral image (amide I contrast) that was obtained by rotating the sample 90° relative to the scanning direction, and (f) ν_1PO_4/amide I ratio for the same osteon. Straight and dotted lines correspond to line intensity scores along the indicated directions. The polarization direction of the laser light was horizontal in (d) and (f) and vertical in (e).

per unit volume, while the Raman outcome reports the amount of mineral per unit volume per amount of collagen in this same volume.

9.4.2 Mineral Maturity

Bone mineral is a carbonated, ion-substituted, poorly crystalline apatite. Once it is laid down by the osteoblast, it undergoes a series of chemical

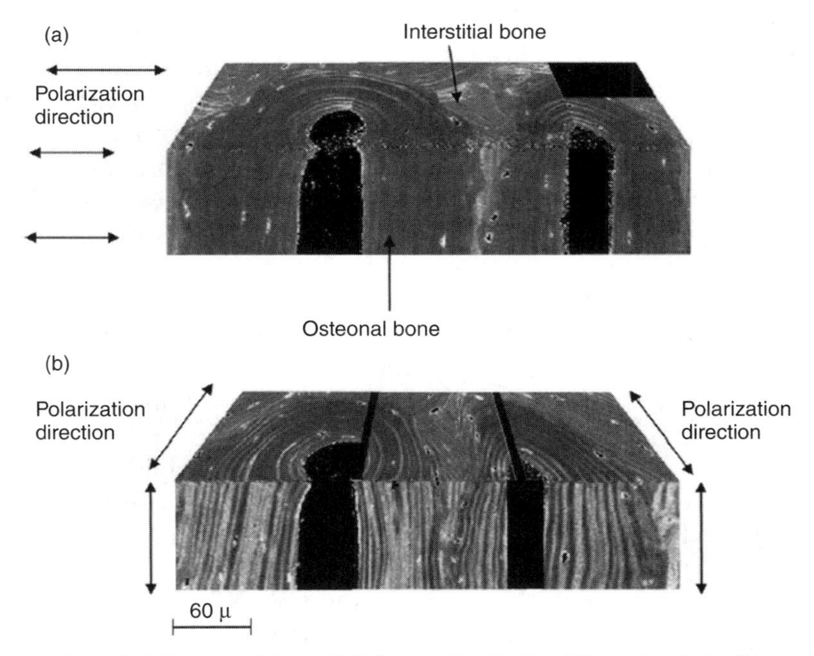

Figure 9.9 (a) 3-D view of the ν_1PO_4/amide I ratio for different polarizations of the incident laser beam as indicated in the figure by double arrows. Same lamellae show different contrast depending on polarization direction of the beam in panel (b).

modifications (mineral maturation); thus, its chemical characteristics are dependent on tissue age and disease.

About 40 apatitic compounds $(Ca10-_xPb_x(PO_4)_6(OH)_2$ were investigated using Raman spectroscopy [100], and this study has shown that the Raman analysis is sensitive to changes in chemical composition and lattice structure of apatitic compounds. A detailed understanding of changes in the mineral phase as HA $(Ca_{106}[OH]_2)$ matures is essential for understanding how normal bone achieves its remarkable mechanical performance and how it is altered in disease, as well as the effects of therapeutic interventions. A model system for investigation of the *in vivo* maturation of HA is available, namely, the *in vitro* conversion of amorphous calcium phosphate (ACP) to HA in a supersaturated solution of calcium and phosphate ions. In another study, Pasteris et al. [101] characterized the degree of hydroxylation and the state of atomic order in several natural and synthetic calcium phosphate (CAP) phases, including apatite of biological (human bone, heated human bone, mouse bone, human and boar dentin, and human and boar enamel), geological, and synthetic origin. They agreed with others, who used nuclear magnetic resonance (NMR), infrared (IR) spectroscopy, and inelastic neutron scattering, that—contrary to the general medical nomenclature—bone apatite is not hydroxylated and is therefore not hydroxyapatite (HAP), and they hypothesized that the body biochemically imposes a specific state of atomic order and crystallinity on its different apatite precipitates (bone, dentin, enamel) in order to enhance their ability to carry out tissue-specific functions. 2-D Raman spectroscopy was applied to analyze changes in the ν_1 region of the Raman spectrum during the maturation of HA following the solution-mediated conversion of ACP to HA [102]. Finally, they analyzed an HA crystallinity in a human iliac crest biopsy sample. Trabecular bone contains a fraction of HA that is more crystalline and mature than could be achieved *in vitro* during the room-temperature ACP-to-HA interconversion.

In the study by Kazanci et al. [103], the well-established chemical conversion of ACP to HAP, a chemical process that results in the *in vitro* production of mineral crystallites, which are similar in chemistry and size distribution to the ones in bone, was analyzed by Raman spectroscopy and wide-angle X-ray scattering (WAXS). The results show that as the crystallites mature in the 002 and 310 crystallographic axes, both the full width at half-height and the band position of the Raman peaks change as a function of reaction extent and crystallite maturation, size, and shape (Figure 9.10). In the experiment, the precipitation began with ACP (ν_1PO_4 around 950 cm^{-1}), and at the transition point around 90 minutes (at the nucleation stage), octacalcium phosphate (OCP) began to be observed based on the appearance of the charac-

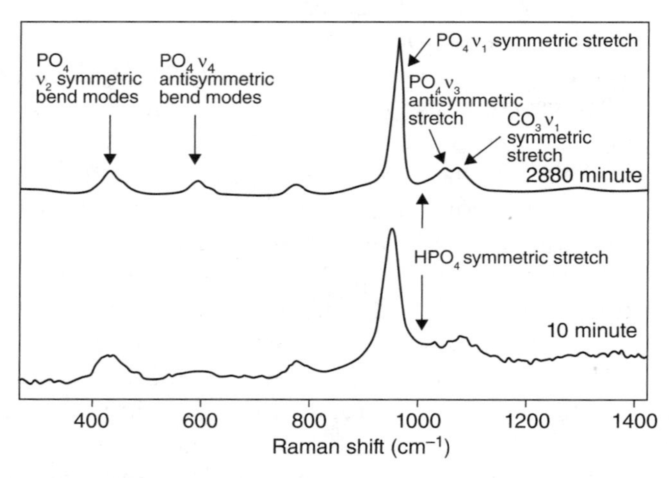

Figure 9.10 Raman spectra of the ν_1, ν_2, ν_3, and ν_4 phosphate, monohydrogen phosphate regions, and the ν_1 carbonate region of amorphous calcium phosphate (ACP). Times (minutes) shown are reaction times.

teristic peak of ν_1PO_4 at 955 cm^{-1}. After 90 minutes, the structure completed its maturation and the band position shifted to ν_1PO_4 at 960 cm^{-1} (crystalline stage) [103] as seen in Figure 9.11a. This finding of an OCP intermediate in this *in vitro* system is in accordance with observations by Crane et al. [104] and Weiner et al. [105], who observed the presence of minor quantities of OCP in living murine calvarial tissue.

Carbonate ions CO_3 substitute for two anionic sites in the apatite structure: monovalent OH− sites and trivalent phosphate PO_4 sites, known as type-A and type-B substitutions, respectively. The ratio of the amount of type-B carbonate and the amount of the phosphate allows estimating the age (maturation status) of mineral crystals in bone.

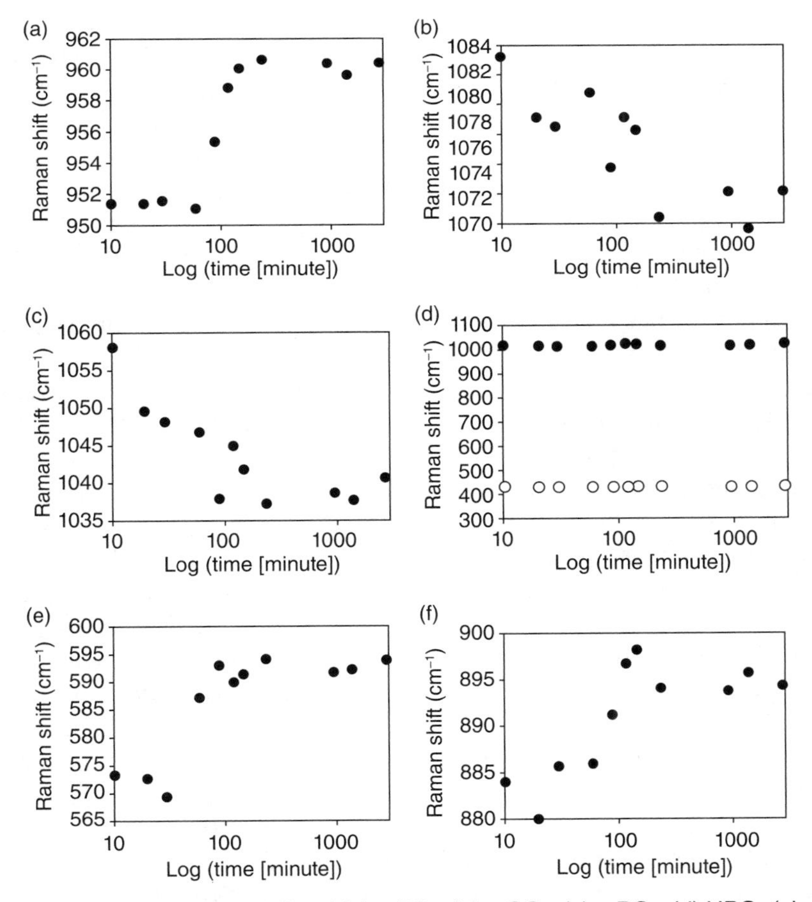

Figure 9.11 Raman band shifts of (a) ν_1PO_4, (b) ν_1CO_3, (c) ν_3PO_4, (d) HPO_4 (shaded symbols) and ν_2PO_4 (open symbols), (e) ν_4PO_4, and (f) shoulder of ν_1PO_4 by log of reaction time from the chemical conversion of amorphous calcium phosphate (ACP) to HAP.

A peak at $1071\,cm^{-1}$ was assigned to a combination of the carbonate ν_1 mode at $1070\,cm^{-1}$ with a phosphate ν_3 mode at $1076\,cm^{-1}$. The carbonate ν_4 bands at 715 and $689\,cm^{-1}$ identify the samples as B-type carbonated apatite. The carbonate content of apatite was calibrated to a Raman band, and the method was used to determine the carbonate content of a sample of bovine cortical bone [106]. Furthermore, Penel and coworkers [67] had shown that two different bands due to the carbonate ν_1 mode were identified depending on the carbonate substitution site A or B, at 1107 and $1070\,cm^{-1}$, respectively. Their results, compared with the IR data already reported, suggesting that even low levels of carbonate substitution induce modifications of the HA spectrum.

Raman studies on pressure-induced changes in phosphate minerals, including HA, monocalcium phosphate, and dicalcium phosphate, show that the OH stretching mode of the HA shifts to higher wave numbers with increasing pressure, while the associated OH librational mode shifts in the opposite direction [107]. Raman transects, microspectra taken at equal intervals along a line, are used to explore the microstructure of human cortical bone. Transects of 50 spectra taken at 2.5-μm intervals across an osteon showed spatial differences in local mineral and protein composition as different physiological structures are traversed. Differences in mineral composition are seen near the rim of an osteon and further out in the lamellae. The blood vessel wall, primarily composed of collagen and elastin, is detected inside the Haversian canal. Factor analysis was used to explore the data set and reveals differences in mineral composition, bone matrix component, an osteoidal tissue component, and one blood vessel protein component [83].

Raman spectroscopic markers have been established for fatigue-related microdamage in bovine bone by Timlin et al. [108]. Microdamage was induced using a cyclic fatigue loading regime. After loading, the specimens were stained en bloc with basic fuchsin to facilitate damage visualization and to differentiate fatigue-induced damage from cracks generated during subsequent histological sectioning. Three regions were defined: tissue with no visible damage, tissue with microcracks, and tissue with diffuse damage. Raman transects, lines of 150–200 Raman spectra, were used for initial tissue surveys. Raman imaging confirms the qualitative relationship between the Raman spectral signature of bone mineral and the type of microdamage in bovine bone [108].

Bohic et al. [109] investigated the characterization of the trabecular rat bone mineral after ovariectomy and bisphosphonate treatment. Chemical analyses of the crystals, combined with data from serum analysis, did indicate small changes in the mineral phase after ovariectomy, which was reversed after treatment with bisphosphonates. To study the effect of mouse age and *in vitro* fluoride treatment on the bone mineral (i.e. mineral phase and degree of crystallinity), Raman microprobe analysis was used to analyze microscopically small regions of bioapatite in mouse femora. The results indicate that the laser Raman microprobe readily detects the conversion of carbonated hydroxylapatite to carbonated fluorapatite, as well as changes in crystallinity of either mineral phase, in microscopically small regions of a bone sample [110].

Tarnowski et al. [111] used Raman microspectroscopy to determine the composition of the mineral environments present in mouse calvaria, the flat bones that comprise the top of the skull. They acquired Raman transects (lines of point spectra) from mouse calvaria during a developmental time course ranging from embryo to 6 months of age. Exploratory factor analysis revealed the presence of a variety of apatitic mineral environments throughout the tissue series. The presence of a heterogeneous mineralized tissue in the postnatal specimens suggests that ionic incorporation and crystal perfection in the lattice vary as the mouse develops. This variation is indicative of the presence of both recently deposited mineral and more matured remodeled mineral. Raman imaging was increasingly used to provide chemical structure information on tissue specimens. Morris and coworkers [112] reported the use of Raman microscopy to image mouse calvaria stained with hematoxylin, eosin and toluidine blue, thus allowing the direct correlation of histological and spectral information.

Akkus and coworkers [113] studied age-related alterations in the physicochemical properties of rat cortical bone that affect the organ and tissue-level mechanical function. They quantified age-related changes in the mineral crystallinity, the degree of mineralization, and the extent of carbonate substitution in phosphate locations using Raman microspectroscopy. Sahar et al. [74] review the use of laser scanning confocal microscopy (LSCM), scanning electron microscopy (SEM), transmission electron microscopy (TEM), and Raman spectroscopic imaging for visually observing microdamage in bone, and discuss the current understanding of damage mechanisms derived from these microscopic and spectroscopic techniques. Insight into failure mechanisms of bone, particularly at the ultrastructural level, is facilitated by the development of improved means of defining and measuring tissue quality. Included in these means are microscopic and spectroscopic techniques for the direct observation of crack initiation, crack propagation, and fracture behavior.

Raman spectroscopic evidence for OCP and other transient mineral species deposited during intramembranous mineralization was obtained by Crane and coworkers [104] on B6CBA F1/J wild-type mice. During the first 24 hours, the spectra contained bands of OCP or an OCP-like mineral. The main phosphorus–oxygen stretch was at 955 cm^{-1} instead of the 957–959 cm^{-1} seen in bone mineral, and there was an additional band at 1010–1014 cm^{-1}, as expected for OCP. A broad band was found at 945 cm^{-1}, characteristic of a highly disordered or even amorphous calcium phosphate. An increased amount of mineral was observed in sutures treated with the fibroblast growth factor FGF2, but no qualitative differences in Raman spectra were observed between experimental and control specimens [104]. In osteoporotic bone from mice van Apeldoorn et al. [75] studied the decrease of BMD and the structural deterioration of bone tissue. Raman microscopy was used to analyze bone composition in eight wild type and eight trichothiodystrophy (TTD) animals, and decreased levels of phosphate and carbonate were observed in the cortex of femora isolated from TTD mice. In contrast, the bands representing the bone protein matrix were not affected in these mice.

9.4.3 Collagen and Other Proteins

The organic matrix of bone is 85–90% type I collagen fibers. A specific amino acid sequence of collagen allows the formation of a triple helix, which is generally composed of the amino acid sequence repeat (X–Y–glycine)$_n$, with proline and hydroxyproline often present at positions X and Y. Other frequently occurring amino acids in collagen include alanine, lysine, arginine, leucine, valine, serine, phenylalanine, and threonine.

The amide I mode is primarily a C=O stretching band (at 1665 cm^{-1}. It does have some contributions from CN stretching and CCN deformation, and the amide III mode (at 1250 cm^{-1}) is the in-phase combination of NH and CN stretching modes, with small contributions from CC stretching and CO. Bandekar [114] has summarized the amide modes and protein conformations in a review. According to Frushour and Koenig, two lines at 1271 and 1248 cm^{-1} of the amide III band are related to nonpolar, prolinerich and polar, proline-poor, denaturated regions of collagen polypeptide chains [115]. The band at about 1263 cm–1 assigned to characteristic vibrations of amide III is used to estimate conformation characteristic of protein backbone. The band disappearance means that the hydrogen bonds between the triple helices of collagen break, and the disorder of the protein structure increases [116].

In general, amide bands are weak, except when UV resonance Raman is used. UV resonance Raman spectra excited within the amide transitions has a superior signal-to-noise ratio, thus providing more information. The UV Raman technique avoids the fluorescence background usually found with visible and NIR excitation and, due to resonance Raman effects, is particularly sensitive to the organic component of bone. Raman spectra show numerous bands from vibrational modes localized within the amide fragments of the peptide or protein. Each of these bands shows unique spectral dependencies on conformation. These differing dependencies result from the different atomic displacement coordinates within each vibration and from the unique responses of each of these displacements to changes in the amide backbone conformation and hydrogen bonding [114]. In the work of Asher and coworkers [117], the physical origin of the frequency and the resonance Raman enhancement dependence of the amide III and the CR_2 bending modes on the peptide secondary structure were elucidated. These vibrations are clearly the most structure-sensitive of the amide backbone vibrational modes.

Ager et al. [118] studied with deep-UV excitation (244 nm) human cortical bone obtained from donors over a wide age range (34–99 years). Spectral changes in the amide I band at $1640\,cm^{-1}$ were shown to correlate with both donor age and with previously reported fracture toughness data obtained from the same specimens. These results are discussed in the context of possible changes in collagen cross-linking chemistry as a function of age, and are deemed important to further our understanding of the changes in the organic component of the bone matrix with aging.

Kale et al. [119] analyzed MG63 cells by Raman microspectroscopy. This system of *ex vivo* bone formation provides important information on the physiological, biological, and molecular basis of osteogenesis. The role of 3-D cellular interactions is well understood in embryonic osteogenesis, but *in vitro* correlations are lacking. They reported that *in vitro* serum-free transforming growth factor (TGF)-b1 stimulation of osteogenic cells immediately after passage results in the formation of 3-D cellular condensations (bone cell spheroids) within 24–48 hours. Several Raman spectra were acquired at 2-mm intervals along a randomly selected line. The multiple randomly acquired spectra (known as Raman transects) were measured along a line across each microspicule and were analyzed using factor analysis, a multivariant statistical analysis in which the data are reduced to a specific number of (fewer) linear combinations, or factors, that describe the data. This statistical approach is a powerful means of "averaging" spectra [119].

Tarnowski et al. [120] studied bone deposition on bovine pericardia, fixed according to three different protocols and either implanted subcutaneously or not implanted (controls). A lightly carbonated apatitic phosphate mineral, similar to that found in bone tissue, was deposited on the surface of a glutaraldehyde-fixed, implanted pericardium. Collagen secondary structure changes were observed on glutaraldehyde fixation by monitoring the center of gravity of the amide I envelope. It was proposed that the decrease in the amide I center of gravity frequency for the glutaraldehyde-fixed tissue compared to the nonfixed tissue was due to an increase in nonreducible collagen cross-links (1660 cm^{-1}) and a decrease in reducible cross-links (1690 cm^{-1}). They found a secondary structure change to the pericardial collagen after implantation; an increase in the frequency of the center of gravity of amide I is indicative of an increase in cross-links.

Transcutaneous bone Raman spectroscopy has been optimized in rat and chicken tissues. For rat tibia, the carbonate/phosphate ratio can be measured at a depth of 1 mm below the skin with an error of 2.3% at an integration time of 120 seconds and within 10% at a 30-second integration time. For chicken tibia, 4 mm below the skin surface, the error is less than 8% with a 120-second integration time [121].

A mouse model (SAMP6) was analyzed by Silva and coworkers [122] to examine the link between bone material properties and skeletal fragility based on the mechanical, histological, biochemical, and spectroscopic properties of bones from a murine model of skeletal fragility. Intact bones from SAMP6 mice are weak and brittle compared to SAMR1 controls, a defect attributed to reduced strength of the bone matrix. The matrix weakness was attributed primarily to poorer organization of collagen fibers and reduced collagen content.

9.5 EMERGING CLINICAL APPLICATIONS

Taken from the viewpoint of clinicians and medical analysts, the potential of Raman spectroscopic techniques as new tools for biomedical applications is promising [123]. The great appeal of Raman spectroscopy lies in the potential for continuous patient monitoring, replacing "random" biopsy of tissues with guided biopsy and monitoring the effects of therapies. Raman spectroscopy may fulfill a role in all of these fields and has already become an accepted analytical method with many applications in different scientific fields [98,124].

9.5.1 Implants and Prostheses

The interaction between biomaterials used in surgical procedures and the host bone is not yet perfectly understood. Given Raman spectroscopy's spatial resolution and minimal specimen preparation requirements, this is a field that could benefit greatly by such spectroscopic analyses, both in terms of characterizing various coatings as well as revealing the properties/characteristics of the device–bone interface.

A quantitative determination of the mineralization of bone tissue and HA coatings of hip and knee prostheses was performed by Dippel et al. [125]. The distribution of the HA content in the coatings investigated was found to be similar all the time. This result was independent of the composition of the coatings and of the history of the whole prosthesis. In the immediate vicinity of the prosthesis, a large HA content that decreased to a minimum toward the periphery of the coating and increased at the site of the on grown bone could be observed. For the interface between bone and HA coating, a transitional zone was observed at a lateral distance of 30–40 microns to the implant [125]. In 1999, He et al. [126] obtained porous beta-tricalcium phosphate (TCP) bioceramics implanted into rabbit femurs and studied the boundary between the ceramic and the host bone by Raman spectroscopy (514.5- and 623.8-nm laser). They demonstrated that besides CAP, collagen protein and lipids also existed in the implants and the boundary forming components of organic bone tissues. The results indicated that when implanted into rabbit femur, the beta-TCP bioceramic was partly dissolved and degraded and the new bone tissue was formed on the surface and in the pores of implanted bioceramics.

The integration of HA-coated implants in dog femur was studied by NIR Fourier transform Raman microscopy by Dopner et al. [127]. Raman spectra were taken in lateral scans in step widths of 10–40 μm from the implant surface up to a distance of 320 μm into the bone tissue. The spectra were subjected to a component analysis for the quantitative determination of the protein and the inorganic components. This quantitative analysis showed for the first time the quantitative distinction between the HA form of mature bone tissue and synthetic HA introduced by the implant coating. It was demonstrated that full mineralization of growing bone is not achieved after 6 months. In contrast, after a residence time of 18 months in the body, the Raman spectra reveal a complete calcification of the new bone tissue as indicated by content of biological HA that was the same as in mature bone tissue throughout the whole implant/bone interface. On the other hand, the content of synthetic HA is strongly reduced in the sample prepared after 18 months implantation, whereas for the shorter implantation time, substantial contributions of synthetic HA are found even at positions beyond the thickness of the implant coating. These results indicate that the coating material is actively involved in the mineralization of on growing bone [127]. Bioactive surface coatings are becoming increasingly important for the development of more biocompatible implants and biosensors. Metal organic chemical vapor deposition (MOCVD) has been widely used to prepare coatings in the semiconductor industry and has shown several benefits compared to other coating methods. Darr and coworkers [128] studied CAP-based coatings such as HA [$(Ca_{10}(PO_4)_6(OH)_2)$]. The presence of carbonate is suggested by the peaks at 1417 and 860 cm^{-1}. These peaks are comparable to the reference spectrum of bone that shows carbonate bending (ν_3) and asymmetric stretching (ν_2) peaks at 1417 and 873 cm^{-1}. The aim of the study by Lopes et al. [129] was the incorporation of HA of calcium (CHA, approximately 960 cm^{-1}) on the healing bone around dental implants. The process of maturation of the bone is important for the success of dental implants, as it improves the fixation of the implant to the bone. Low-level laser therapy (LLLT) has been suggested as a means of improving bone healing because of its biomodulatory capabilities. The results showed significant differences in the concentration of CHA on irradiated and control specimens at both 30 and 45 days after surgery, and it was concluded that LLLT does improve bone healing, and that this can be safely assessed by Raman spectroscopy [129].

Penel et al. [130] have utilized intravital Raman spectroscopy to evaluate the composition of bone and biomaterials in living animals. A new model for the intravital study of the composition and structure of membranous bone by Raman microspectroscopy was described. Titanium bone chambers equipped with a fused-silica optical window were implanted transcutaneously in the calvaria of New Zealand rabbits. The implanted optical windows were well tolerated, and spectral acquisitions were performed without any additional invasive procedure. Bone and implanted apatitic biomaterials were analyzed at different times after surgery. All Raman bands were unambiguously identified in the bone and biomaterial spectra. The main PO_4 and CO_3 Raman bands in bone spectra were consistent with those found in the carbonated apatite spectrum. The changes observed in bone varied as a function of time and location. The composition and structure of all of the biomaterials studied seemed to remain stable over time and location. Penel and coworkers reported for the first time the complete intravital study of the Raman spectra of bone and CAP biomaterials over a period of 8 months [130].

9.5.2 A Clinical Diagnostic Tool

Craniosynostosis, the premature fusion of the skull bones at the sutures, is the second most common human birth defect in the skull. Raman microspectroscopy was used to examine the composition, relative amounts, and locations of the mineral and matrix produced in mouse skulls undergoing force-induced craniosynostosis. The calvaria that comprise the top of the skull are most often affected, and craniosynostosis is a feature of over 100 human syndromes and conditions. This study found that osteogenic fronts subjected to uniaxial compression had decreased relative mineral content compared to unloaded osteogenic fronts because of new and incomplete mineral deposition. Increased matrix production in osteogenic fronts undergoing craniosynostosis was observed. This was the first report in which Raman microspectroscopy was used to study musculoskeletal disease [131].

Draper et al. [132] demonstrated that Raman spectra of bone in wild type and oim/oim could be obtained through 1.1 mm of skin using picosecond time-resolved spectroscopy; the aim of this study was to test the spectral features of both the mineral and organic phases of bone specimens with known differences in material properties. Their study conclusively showed that major components of the Raman spectrum of bone can be detected without excision of bone or even cutting the skin by using time-resolved Raman spectroscopy. Bone tissue compositional differences in women with and without osteoporotic fracture were described by Raman spectroscopy. Fifteen iliac crest biopsies (per group) were obtained from women who had sustained a fracture and from normal controls. Raman spectroscopy was used to determine measures of chemical composition of trabecular and cortical bone. Femoral trabecular bone in fractured women had a higher carbonate/amide I area ratio than in unfractured women. Iliac crest biopsies revealed a higher carbonate/phosphate ratio in cortical bone from women who had sustained a fracture. Results suggest that the chemical composition of bone tissue may be an additional risk factor for osteoporotic fracture [133].

Preliminary work on the development of a novel detection method for osteoporosis was done by Pillay et al. [134], Towler et al. [135], and Moran et al. [136]. Patients can suffer osteoporotic fractures despite normal BMD, partly because of as-yet-undetected influences of both the protein and mineral phases of bone that are affected in osteoporosis. There is currently no clinically applicable method of evaluating the status of the protein phase. The proteins in human nail (keratin) and bone (collagen) require sulfation and disulfide bond (S–S) formation for structural integrity, and disorders of either sulfur metabolism or cystathione beta-synthase can lead to structural abnormalities in these tissues. Raman protein spectra provide a method of noninvasive measurement of the degree of sulfation of structurally related proteins that may be indicative of bone health. Raman spectroscopy was used to evaluate the disulfide (S–S) content of fingernails. Pillay et al. [134] examined the fingernails of two groups of patients, with ($n = 9$) and without ($n = 13$) osteoporosis at either the hip or lumbosacral spine. They performed nanoindentation to assess the degree of nail brittleness and Raman spectroscopy to assess the disulfide bond content of nail. The spectroscopy data showed differences between the two sets of nails. The disulfide bond content of the nails sourced from osteoporotic patients was lower than that from healthy patients ($P = 0.06$ between groups). Towler and coworkers [135] studied nail samples from 169 women (84 premenopausal and 85 postmenopausal), of which 39 had a history of osteoporotic fracture. This suggests that measurements of change in the protein phase of structural proteins such as keratin in the human nail may be correlated with clinically relevant changes in bone proteins that are important in fracture risk [135]. Moran and coworkers [136] used both mechanical (nanoindentation) and chemical (Raman spectroscopy) methods to evaluate differences between fingernails sourced from osteoporotic and nonosteoporotic patients. The disulfide bond content of fingernail samples from each group was measured by Raman spectroscopy, and the disulfide bond content of fingernail was found to be significantly lower in the osteoporotic group.

It can be concluded that a relationship between the mechanical and chemical properties of nail and bone may exist in a measurable way [134–136].

9.6 CONCLUSIONS

In conclusion, micro-Raman spectroscopic imaging has emerged as a potent and versatile technique in the study of bone and, in particular, bone quality. Its advantages lie in the spatial resolution, minimal sample preparation requirements, acquisition times, and lack of major spectral overlaps. It offers information both on bone density expressed as mineral to matrix, and qualitative information on the mineral and organic matrix components. Nevertheless, caution should be exercised when interpreting the obtained results. As mentioned previously, mineral to matrix is an index of bone density, but it is expressed as amount of mineral per amount of collagen per volume, as opposed to the amount of mineral per unit volume obtained through techniques such as microradiography and qBEI. Similarly, the direct outcome of any vibrational technique such as IR and Raman spectroscopy is mineral maturity, and careful calibration against a technique such as small-angle X-ray scattering (SAXS) and/or X-ray diffraction (XRD) is required before one can make arguments on mineral crystallite shape and size. Moreover, choosing the appropriate lines and taking care of polarization direction of the incident light, both fibril orientation and composition may be extracted from the same bone Raman spectrum.

REFERENCES

1. Fratzl, P., Gupta, H.S., Paschalis, E.P., Roschger, P. (2004). Structure and mechanical quality of the collagen-mineral composite in bone. *J. Mater. Chem.*, 14, 2115–2123.
2. WHO. (2003). *Prevention and Management of Osteoporosis*. Geneva, Switzerland: World Health Organization Study Group.
3. Boyce, T.M., Bloebaum. R.D. (1993). Cortical aging differences and fracture implications for the human femoral neck. *Bone*, 14(5), 769–78.
4. Marshall, D., Johnell, O., Wedel, H. (1996). Meta-analysis of how well measures of bone mineral density predict occurrence of osteoporotic fractures. *BMJ*, 312(7041), 1254–9.
5. Cummings, S.R. (1985). Are patients with hip fractures more osteoporotic? Review of the evidence. *Am. J. Med.*, 78(3), 487–94.
6. McCreade, R.B., Goldstein, A.S. (2000). Biomechanics of fracture: Is bone mineral density sufficient to assess risk? *J. Bone Miner. Res.*, 15(12), 2305–2308.
7. Manolagas, S.C. (2000). Corticosteroids and fractures: A close encounter of the third cell kind [editorial; comment]. *J. Bone Miner. Res.*, 15(6), 1001–5.
8. Hui, S., Slemenda, C.W., Johnston, C.C. (1988). Age and bone mass as predictors of fracture in a prospective study. *J. Clin. Invest.*, 81, 1804–9.
9. Jepsen, K.J., Schaffler, M.B. (2001). Bone mass does not adequately predict variations in bone fragility: A genetic approach. *Trans. Orthop. Res. Soc. 47th Annual Meeting*, San Francisco, 114.
10. Parfitt, A.M. (1987). Bone remodeling and bone loss: Understanding the pathophysiology of osteoporosis. *Clin. Obstet. Gynecol.*, 30, 789–811.
11. Mosekilde, L., Mosekilde, L., Danielsen, C.C. (1987). Biomechanical competence of vertebral trabecular bone in relation to ash density and age in normal individuals. *Bone*, 79–85.
12. McCabe, F., Zhou, L.J., Steele, C.R., Marcus, R. (1991). Noninvasive assessment of ulnar bending stiffness in women. *J. Bone Miner. Res.*, 6, 53–9.
13. Kanis, J.A., Melton, L.J.I., Christiansen, C., Johnston, C.J., Haltaev, N. (1994). Perspective: The diagnosis of osteoporosis. *J. Bone Miner. Res.*, 9, 1137–1142.

14. Kann, P., Graeben, S., Beyer, J. (1994). Age-dependence of bone material quality shown by the measurement of frequency of resonance in the ulna. *Calcif. Tissue Int.*, 54, 96–100.
15. Schnitzler, C.M. (1993). Bone quality: A determinant for certain risk factors for bone fragility. *Calcif. Tissue Int.*, 53, S27–31.
16. Ott, S.M. (1993). When bone mass fails to predict bone failure. *Calcif. Tissue Int.*, 53(Suppl. 1), S7-S13.
17. Cummings, S.R., Black, D.M., Nevitt, M.C., Browner, W.S., Cauley, J.A., Genant, H.K., Mascioli, S.R., Scott, J.C., Seeley, D.G., Steiger, P., Vogt, T.M. (1990). Appendicular bone density and age predict hip fracture in women: The study of osteoporotic fractures research group. *JAMA*, 263, 665–668.
18. Raisz, L.G. (2004). Homocysteine and osteoporotic fractures—culprit or bystander? *N. Engl. J. Med.*, 350(20), 2089–90.
19. McLean, R.R., Jacques, P.F., Selhub, J., Tucker, K.L., Samelson, E.J., Broe, K.E., Hannan, M.T., Cupples, L.A., Kiel, D.P. (2004). Homocysteine as a predictive factor for hip fracture in older persons. *N. Engl. J. Med.*, 350(20), 2042–9.
20. van Meurs, J.B., Dhonukshe-Rutten, R.A., Pluijm, S.M., van der Klift, M., de Jonge, R., Lindemans, J., de Groot, L.C., Hofman, A., Witteman, J.C., van Leeuwen, J.P., Breteler, M.M., Lips, P., Pols, H.A., Uitterlinden, A.G. (2004). Homocysteine levels and the risk of osteoporotic fracture. *N. Engl. J. Med.*, 350(20), 2033–41.
21. Gram Gjesdal, C., Vollset, S.E., Ueland, P.M., Refsum, H., Meyer, H.E., Tell, G.S. (2007). Plasma homocysteine, folate and vitamin B12 and the risk of hip fracture. The Hordaland Homocysteine Study. *J. Bone Miner. Res.*, 22(5), 747–56.
22. Liu, G., Nellaiappan, K., Kagan, H.M. (1997). Irreversible inhibition of lysyl oxidase by homocysteine thiolactone and its selenium and oxygen analogues. Implications for homocystinuria. *J. Biol. Chem.*, 272(51), 32370–7.
23. Fratzl, P., Roschger, P., Fratzl-Zelman, N., Paschalis, E.P., Phipps, R., Klaushofer, K. (2007). Evidence that treatment with risedronate in women with postmenopausal osteoporosis affects bone mineralization and bone volume. *Calcif. Tissue Int.*, 81(2), 73–80.
24. Einhorn, T.A. (1996). The bone organ system: Form and function. In: Marcus R, Feldman D, Kelsey J, eds. *Osteoporosis*. New York: Academic Press Inc., pp. 3–22.
25. Bullough, P. (1990). The Tissue Diagnosis of Metabolic Bone Disease. *Orthop. Clin. North Am.*, 21, 65–79.
26. Bullough, P. (1992). *Atlas of Orthopaedic Pathology*. New York: Gower Medical Publishing.
27. Raisz, L.G., Kream, B.E. (1983). Regulation of bone formation. *N. Engl. J. Med.*, 309, 29–35.

28. Camacho, N.P., Rinnerthaler, S., Paschalis, E.P., Mendelsohn, R., Boskey, A.L., Fratzl, P. (1999). Complementary information on bone ultrastructure from scanning small angle X-ray scattering and Fourier-transform infrared microspectroscopy. *Bone*, 25(3), 287–93.

29. Fratzl, P., Groschner, M., Vogl, G., Plenk, H., Jr., Eschberger, J., Fratzl-Zelman, N., Koller, K., Klaushofer, K. (1992). Mineral crystals in calcified tissues: A comparative study by SAXS. *J. Bone Miner. Res.*, 7(3), 329–34.

30. Fratzl, P., Schreiber, S., Klaushofer, K. (1996). Bone mineralization as studied by small-angle x-ray scattering. *Connect. Tissue Res.*, 34(4), 247–54.

31. Misof, B.M., Roschger, P., Tesch, W., Baldock, P.A., Valenta, A., Messmer, P., Eisman, J.A., Boskey, A.L., Gardiner, E.M., Fratzl, P., Klaushofer, K. (2003). Targeted overexpression of vitamin D receptor in osteoblasts increases calcium concentration without affecting structural properties of bone mineral crystals. *Calcif. Tissue Int.*, 6, 6.

32. Gao, H., Ji, B., Jager, I.L., Arzt, E., Fratzl, P. (2003). Materials become insensitive to flaws at nanoscale: Lessons from nature. *Proc. Natl. Acad. Sci. U.S.A.*, 100(10), 5597–600.

33. Roschger, P., Gupta, H.S., Berzlanovich, A., Ittner, G., Dempster, D.W., Fratzl, P., Cosman, F., Parisien, M., Lindsay, R., Nieves, J.W., Klaushofer, K. (2003). Constant mineralization density distribution in cancellous human bone. *Bone*, 32(3), 316–23.

34. Zizak, I., Roschger, P., Paris, O., Misof, B.M., Berzlanovich, A., Bernstorff, S., Amenitsch, H., Klaushofer, K., Fratzl, P. (2003). Characteristics of mineral particles in the human bone/cartilage interface. *J. Struct. Biol.*, 141(3), 208–17.

35. Misof, B.M., Roschger, P., Cosman, F., Kurland, E.S., Tesch, W., Messmer, P., Dempster, D.W., Nieves, J., Shane, E., Fratzl, P., Klaushofer, K., Bilezikian, J., Lindsay, R. (2003). Effects of intermittent parathyroid hormone administration on bone mineralization density in iliac crest biopsies from patients with osteoporosis: A paired study before and after treatment. *J. Clin. Endocrinol. Metab.*, 88(3), 1150–6.

36. Tesch, W., Vandenbos, T., Roschgr, P., Fratzl-Zelman, N., Klaushofer, K., Beertsen, W., Fratzl, P. (2003). Orientation of mineral crystallites and mineral density during skeletal development in mice deficient in tissue nonspecific alkaline phosphatase. *J. Bone Miner. Res.*, 18(1), 117–25.

37. Roschger, P., Grabner, B.M., Rinnerthaler, S., Tesch, W., Kneissel, M., Berzlanovich, A., Klaushofer, K., Fratzl, P. (2001). Structural development of the mineralized tissue in the human L4 vertebral body. *J. Struct. Biol.*, 136(2), 126–36.

38. Grabner, B., Landis, W.J., Roschger, P., Rinnerthaler, S., Peterlik, H., Klaushofer, K., Fratzl, P. (2001). Age- and genotype-dependence of bone

material properties in the osteogenesis imperfecta murine model (oim). *Bone*, 29(5), 453–7.

39. Roschger, P., Rinnerthaler, S., Yates, J., Rodan, G.A., Fratzl, P., Klaushofer, K. (2001). Alendronate increases degree and uniformity of mineralization in cancellous bone and decreases the porosity in cortical bone of osteoporotic women. *Bone*, 29(2), 185–91.
40. Jager, I., Fratzl, P. (2000). Mineralized collagen fibrils: A mechanical model with a staggered arrangement of mineral particles. *Biophys. J.*, 79(4), 1737–46.
41. Paris, O., Zizak, I., Lichtenegger, H., Roschger, P., Klaushofer, K., Fratzl, P. (2000). Analysis of the hierarchical structure of biological tissues by scanning X-ray scattering using a micro-beam. *Cell. Mol. Biol. (Noisy-le-Grand)*, 46(5), 993–1004.
42. Rinnerthaler, S., Roschger, P., Jakob, H.F., Nader, A., Klaushofer, K., Fratzl, P. (1999). Scanning small angle X-ray scattering analysis of human bone sections. *Calcif. Tissue Int.*, 64(5), 422–9.
43. Roschger, P., Fratzl, P., Eschberger, J., Klaushofer, K. (1998). Validation of quantitative backscattered electron imaging for the measurement of mineral density distribution in human bone biopsies. *Bone*, 23(4), 319–26.
44. Roschger, P., Fratzl, P., Klaushofer, K., Rodan, G. (1997). Mineralization of cancellous bone after alendronate and sodium fluoride treatment: A quantitative backscattered electron imaging study on minipig ribs. *Bone*, 20(5), 393–7.
45. Fratzl, P., Schreiber, S., Boyde, A. (1996). Characterization of bone mineral crystals in horse radius by small-angle X-ray scattering. *Calcif. Tissue Int.*, 58(5), 341–6.
46. Fratzl, P., Schreiber, S., Roschger, P., Lafage, M.H., Rodan, G., Klaushofer, K. (1996). Effects of sodium fluoride and alendronate on the bone mineral in minipigs: A small-angle X-ray scattering and backscattered electron imaging study. *J. Bone Miner. Res.*, 11(2), 248–53.
47. Fratzl, P., Roschger, P., Eschberger, J., Abendroth, B., Klaushofer, K. (1994). Abnormal bone mineralization after fluoride treatment in osteoporosis: A small-angle x-ray-scattering study. *J. Bone Miner. Res.*, 9(10), 1541–9.
48. Fratzl, P., Fratzl-Zelman, N., Klaushofer, K., Vogl, G., Koller, K. (1991). Nucleation and growth of mineral crystals in bone studied by small-angle X-ray scattering. *Calcif. Tissue Int.*, 48(6), 407–13.
49. Marcott, C., Reeder, R.C., Paschalis, E.P., Tatakis, D.N., Boskey, A.L., Mendelsohn, R. (1998). FT-IR chemical imaging of biomineralized tissues using a mercury-cadmium-telluride focal-plane detector. *Cell. Mol. Biol.*, 44(1), 109–115.
50. Mendelsohn, R., Paschalis, E.P., Boskey, A.L. (1999). Infrared spectroscopy, microscopy, and microscopic imaging of mineralizing tissues.

spectra-structure correlations from human iliac crest biopsies. *J. Biomed. Opt.*, 4(1), 14–21.

51. Mendelsohn, R., Paschalis, E.P., Sherman, P.J., Boskey, A.L. (2000). IR microscopic imaging of pathological states and fracture healing of bone. *Appl. Spectrosc.*, 54, 1183–1191.
52. Paschalis, E.P., Jacenko, O., Olsen, B., Boskey, A.L. (1996). FTIR microspectroscopic analysis identifies alterations in mineral properties in bones from mice transgenic for type X collagen. *Bone*, 19(2), 151–156.
53. Paschalis, E.P., DiCarlo, E., Betts, F., Mendelsohn, R., Boskey, A.L. (1997). FTIR microspectroscopic analysis of normal human cortical and trabecular bone. *Calcif. Tissue Int.*, 61, 480–486.
54. Paschalis, E.P., DiCarlo, E., Betts, F., Mendelsohn, R., Boskey, A.L. (1997). FTIR microspectroscopic analysis of human iliac crest biopsies from untreated osteoporotic bone. *Calcif. Tissue Int.* 61, 487–492.
55. Fratzl, P., Paris, O., Klaushofer, K., Landis, W.J. (1996). Bone mineralization in an osteogenesis imperfecta mouse model studied by small-angle x-ray scattering. *J. Clin. Invest.*, 97(2), 396–402.
56. Bailey, A.J., Wotton, S.F., Sims, T.J., Thompson, P.W. (1992). Post-translational modifications in the collagen of human osteoporotic femoral head. *Biochem. Biophys. Res. Commun.*, 185, 801–805.
57. Bailey, A.J., Wotton, S.F., Sims, T.J., Thompson, P.W. (1993). Biochemical changes in the collagen of human osteoporotic bone matrix. *Connect. Tissue Res.*, 29, 119–132.
58. Knott, L., Whitehead, C.C., Fleming, R.H., Bailey, A.J. (1995). Biochemical changes in the collagenous matrix of osteoporotic avian bone. *Biochem. J.*, 310, 1045–1051.
59. Knott, L., Tarlton, J.F., Bailey, A.J. (1997). Chemistry of collagen crosslinking: Biochemical changes in collagen during the partial mineralization of turkey leg tendon. *Biochem. J.*, 322(Pt 2), 535–42.
60. Knott, L., Bailey, A.J. (1998). Collagen cross-links in mineralizing tissues: A review of their chemistry, function, and clinical relevance. *Bone*, 22(3), 181–7.
61. Masse, P.G., Colombo, V.E., Gerber, F., Howell, D.S., Weiser, H. (1990). Morphological abnormalities in vitamin B_6 deficient tarsometatarsal chick cartilage. *Scanning Microsc.*, 4, 667–674.
62. Masse, P.G., Rimnac, C.M., Yamauchi, M., Coburn, P.S., Rucker, B.R., Howell, S.D., Boskey, A.L. (1996). Pyridoxine deficiency affects biomechanical properties of chick tibial bone. *Bone*, 18, 567–574.
63. Oxlund, H., Barckman, M., Ortoft, G., Andreassen, T.T. (1995). Reduced concentrations of collagen cross-links are associated with reduced strength of bone. *Bone*, 17(Suppl. 4), 365S–371S.
64. Oxlund, H., Mosekilde, L., Ortoff, G. (1987). Alterations in the stability of collagen from human trabecular bone with respect to age. In:

Christiansen, C., Johansen, J.S., Riis, B.J., eds., *Osteoporosis*. Copenhagen, Denmark: Osteopress APS, 309–312.

65. Morris, M.D., Finney, W.F., Rajachar, R.M., Kohn, D.H. (2004). Bone tissue ultrastructural response to elastic deformation probed by Raman spectroscopy. *Faraday Discuss.*, 126, 159–68; discussion 169–83.
66. de Mul, F.F., Hottenhuis, M.H., Bouter, P., Greve, J., Arends, J., Ten Bosch, J.J. (1986). Micro-Raman line broadening in synthetic carbonated hydroxyapatite. *J. Dent. Res.*, 65(3), 437–40.
67. Penel, G., Leroy, G., Rey, C., Bres, E. (1998). MicroRaman spectral study of the PO4 and CO3 vibrational modes in synthetic and biological apatites. *Calcif. Tissue Int.*, 63(6), 475–81.
68. Carden, A., Morris, M.D. 2000. Application of vibrational spectroscopy to the study of mineralized tissues (review). *J. Biomed. Opt.*, 5(3), 259–68.
69. de Carmejane, O., Morris, M.D., Davis, M.K., Stixrude, L., Tecklenburg, M., Rajachar, R.M., Kohan, D.H. (2005). Bone chemical structure response to mechanical stress studied by high pressure Raman spectroscopy. *Calcif. Tissue Int.*, 76(3), 207–13.
70. Schrader, B. (1995). *Infrared and Raman Spectroscopy: Methods and Applications*. Weinheim, Germany: VCH.
71. Levin, I.W., Lewis, E.N. (1990). Fourier transform Raman spectroscopy of biological materials. *Anal. Chem.*, 62(21), 1101A–1111A.
72. Golcuk, K., Mandair, G.S., Callender, A.F., Sahar, N., Kohn, D.H., Morris, M.D. (2006). Is photobleaching necessary for Raman imaging of bone tissue using a green laser? *Biochim. Biophys. Acta*, 1758(7), 868–73.
73. Akkus, O., Polyakova-Akkus, A., Adar, F., Schaffler, M.B. (2003). Aging of microstructural compartments in human compact bone. *J. Bone Miner. Res.*, 18(6), 1012–9.
74. Sahar, N.D., Hong, S.I., Kohn, D.H. (2005). Micro- and nano-structural analyses of damage in bone. *Micron*, 36(7–8), 617–29.
75. van Apeldoorn, A.A., de Boer, J., van Steeg, H., Hoeijmakers, J.H., Otto, C., van Blitterswijk, C.A. (2007). Physicochemical composition of osteoporotic bone in the trichothiodystrophy premature aging mouse determined by confocal Raman microscopy. *J. Gerontol. A Biol. Sci. Med. Sci.*, 62(1), 34–40.
76. Yeni, Y.N., Yerramshetty, J., Akkus, O., Pechey, C., Les, C.M. (2006). Effect of fixation and embedding on Raman spectroscopic analysis of bone tissue. *Calcif. Tissue Int.*, 78(6), 363–71.
77. Bachmann, L., Gomes, A.S., Zezell, D.M. (2005). Collagen absorption bands in heated and rehydrated dentine. *Spectrochim. Acta A Mol. Biomol. Spectrosc.*, 62(4–5), 1045–9.

78. Lees, S. (1981). A mixed packing model for bone collagen. *Calcif. Tissue Int.*, 33(6), 591–602.
79. Lees, S., Heeley, J.D., Cleary, P.F. (1981). Some properties of the organic matrix of a bovine cortical bone sample in various media. *Calcif. Tissue Int.*, 33(1), 83–6.
80. Leroy, G., Penel, G., Leroy, N., Bres, E. (2002). Human tooth enamel: A Raman polarized approach. *Appl. Spectrosc.*, 56, 1030–1034.
81. McCreery, R.L. (2000). *Raman Spectroscopy for Chemical Analysis*. New York: John Wiley & Sons.
82. Carden, A., Rajachar, R.M., Morris, M.D., Kohn, D.H. (2003). Ultrastructural changes accompanying the mechanical deformation of bone tissue: A Raman imaging study. *Calcif. Tissue Int.*, 72(2), 166–75.
83. Timlin, J.A., Carden, A., Morris, M.D. (1999). Chemical microstructure of cortical bone probed by Raman transects. *Appl. Spectrosc.*, 53(11), 1429–35.
84. Rousseau, M.E., Lefevre, T., Beaulieu, L., Asakura, T., Pezolet, M. (2004). Study of protein conformation and orientation in silkworm and spider silk fibers using Raman microspectroscopy. *Biomacromolecules*, 5(6), 2247–57.
85. Tsuboi, M., Benevides, J.M., Bondre, P., Thomas, G.J., Jr. (2005). Structural details of the thermophilic filamentous bacteriophage PH75 determined by polarized Raman microspectroscopy. *Biochemistry*, 44(12), 4861–9.
86. Tsuda, H., Arends, J. (1994). Orientational micro-Raman spectroscopy on hydroxyapatite single crystals and human enamel crystallites. *J. Dent. Res.*, 73(11), 1703–10.
87. Gupta, H.S., Seto, J., Wagermaier, W., Zaslansky, P., Boesecke, P., Fratzl, P. (2006). Cooperative deformation of mineral and collagen in bone at the nanoscale. *Proc. Natl. Acad. Sci. U.S.A.*, 103(47), 17741–6.
88. Kazanci, M., Roschger, P., Paschalis, E.P., Klaushofer, K., Fratzl, P. (2006). Bone osteonal tissues by Raman spectral mapping: Orientation-composition. *J. Struct. Biol.*, 156(3), 489–96.
89. Kazanci, M., Wagner, H.D., Manjubala, N.I., Gupta, H.S., Paschalis, E., Roschger, P., Fratzl, P. (2007). Raman imaging of two orthogonal planes within cortical bone. *Bone*, 41(3), 456–61.
90. Wagermaier, W., Gupta, H.S., Gourrier, A., Burghammer, M., Roschger, P., Fratzl, P. (2006). Spiral twisting of fiber orientation inside bone lamellae. *Biointerphases*, 1(1), 1–5.
91. Landis, W.J., Hodgens, K.J., Arena, J., Song, M.J., McEwen, B.F. (1996). Structural relations between collagen and mineral in bone as determined by high voltage electron microscopic tomography. *Microsc. Res. Tech.*, 33(2), 192–202.
92. Weiner, S., Traub, W., Wagner, H.D. (1999). Lamellar bone: Structure-function relations. *J. Struct. Biol.*, 126(3), 241–55.

93. Hofmann, T., Heyroth, F., Meinhard, H., Franzel, W., Raum, K. (2006). Assessment of composition and anisotropic elastic properties of secondary osteon lamellae. *J. Biomech.*, 39(12), 2282–94.
94. Dehring, K.A., Crane, N.J., Smukler, A.R., McHugh, J.B., Roessler, B.J., Morris, M.D. (2006). Identifying chemical changes in subchondral bone taken from murine knee joints using Raman spectroscopy. *Appl. Spectrosc.*, 60(10), 1134–41.
95. Lane, N.E., Yao, W., Balooch, M., Nalla, R.K., Balooch, G., Habelitz, S., Kinney, J.H., Bonewald, L.F. (2006). Glucocorticoid-treated mice have localized changes in trabecular bone material properties and osteocyte lacunar size that are not observed in placebo-treated or estrogen-deficient mice. *J. Bone Miner. Res.*, 21(3), 466–76.
96. Kubisz, L., Polomska, M. (2007). FT NIR Raman studies on gamma-irradiated bone. *Spectrochim. Acta A Mol. Biomol. Spectrosc.*, 66(3), 616–25.
97. Ramasamy, J.G., Akkus, O. (2007). Local variations in the micromechanical properties of mouse femur: The involvement of collagen fiber orientation and mineralization. *J. Biomech.*, 40(4), 910–8.
98. Schmitt, M., Popp, J. (2006). Raman spectroscopy at the beginning of the twenty-first century. *J. Raman Spectrosc.*, 37, 20–28.
99. Gevorkian, B.Z., Arnotskaia, N.E., Fedorova, E.N. (1984). Study of bone tissue structure using polarized Raman spectra. *Biofizika*, 29(6), 1046–52.
100. Hadrich, A., Lautie, A., Mhiri, T. (2001). Vibrational study and fluorescence bands in the FT-Raman spectra of Ca(10-x)Pb(x)(PO4)6(OH)2 compounds. *Spectrochim. Acta A Mol. Biomol. Spectrosc.*, 57(8), 1673–81.
101. Pasteris, J.D., Wopenka, B., Freeman, J.J., Rogers, K., Valsami-Jones, E., van der Houwen, J.A., Silva, M.J. (2004). Lack of OH in nanocrystalline apatite as a function of degree of atomic order: implications for bone and biomaterials. *Biomaterials*, 25(2), 229–38.
102. Ou-Yang, H., Paschalis, E.P., Boskey, A.L., Mendelsohn, R. (2000). Two-dimensional vibrational correlation spectroscopy of in vitro hydroxyapatite maturation. *Biopolymers*, 57(3), 129–39.
103. Kazanci, M., Fratzl, P., Klaushofer, K., Paschalis, E.P. (2006). Complementary information on in vitro conversion of amorphous (precursor) calcium phosphate to hydroxyapatite from Raman microspectroscopy and wide-angle X-ray scattering. *Calcif. Tissue Int.*, 79(5), 354–9.
104. Crane, N.J., Popescu, V., Morris, M.D., Steenhuis, P., Ignelzi, M.A., Jr. (2006). Raman spectroscopic evidence for octacalcium phosphate and other transient mineral species deposited during intramembranous mineralization. *Bone*, 39(3), 434–42.
105. Weiner, S. (2006). Transient precursor strategy in mineral formation of bone. *Bone*, 39(3), 431–3.

106. Awonusi, A., Morris, M.D., Tecklenburg, M.M. (2007). Carbonate assignment and calibration in the Raman spectrum of apatite. *Calcif. Tissue Int.*, 81(1), 46–52.

107. Xu, J., Gilson, D.F., Butler, I.S., Stangel, I. (1996). Effect of high external pressures on the vibrational spectra of biomedical materials: Calcium hydroxyapatite and calcium fluoroapatite. *J. Biomed. Mater. Res.*, 30(2), 239–44.

108. Timlin, J.A., Carden, A., Morris, M.D., Rajachar, R.M., Kohn, D.H. (2000). Raman spectroscopic imaging markers for fatigue-related microdamage in bovine bone. *Anal. Chem.* 72(10), 2229–36.

109. Bohic, S., Rey, C., Legrand, A., Sfihi, H., Rohanizadeh, R., Martel, C., Barbier, A., Daculsi, G. (2000). Characterization of the trabecular rat bone mineral: Effect of ovariectomy and bisphosphonate treatment. *Bone*, 26(4), 341–8.

110. Freeman, J.J., Wopenka, B., Silva, M.J., Pasteris, J.D. (2001). Raman spectroscopic detection of changes in bioapatite in mouse femora as a function of age and in vitro fluoride treatment. *Calcif. Tissue Int.*, 68(3), 156–62.

111. Tarnowski, C.P., Ignelzi, M.A., Jr., Morris, M.D. (2002). Mineralization of developing mouse calvaria as revealed by Raman microspectroscopy. *J. Bone Miner. Res.*, 17(6), 1118–26.

112. Morris, M.D., Crane, N.J., Gomez, L.E., Ignelzi, M.A., Jr. (2004). Compatibility of staining protocols for bone tissue with Raman imaging. *Calcif. Tissue Int.*, 74(1), 86–94.

113. Akkus, O., Adar, F., Schaffler, M.B. (2004). Age-related changes in physicochemical properties of mineral crystals are related to impaired mechanical function of cortical bone. *Bone*, 34(3), 443–53.

114. Bandekar, J. (1992). Amide modes and protein conformation. *Biochim. Biophys. Acta*, 1120(2), 123–43.

115. Frushour, B.G., Koenig, J.L. (1975). Raman scattering of collagen, gelatin, and elastin. *Biopolymers*, 14(2), 379–91.

116. Dong, R., Yan, X., Pang, X., Liu, S. (2004). Temperature-dependent Raman spectra of collagen and DNA. *Spectrochim. Acta A Mol. Biomol. Spectrosc.*, 60(3), 557–61.

117. Asher, S.A., Ianoul, A., Mix, G., Boyden, M.N., Karnoup, A., Diem, M., Schweitzer-Stenner, R. (2001). Dihedral psi angle dependence of the amide III vibration: A uniquely sensitive UV resonance Raman secondary structural probe. *J. Am. Chem. Soc.*, 123(47), 11775–81.

118. Ager, J.W., Nalla, R.K., Breeden, K.L., Ritchie, R.O. (2005). Deep-ultraviolet Raman spectroscopy study of the effect of aging on human cortical bone. *J. Biomed. Opt.*, 10(3), 034012.

119. Kale, S., Biermann, S., Edwards, C., Tarnowski, C., Morris, M., Long, M.W. (2000). Three-dimensional cellular development is essential for ex vivo formation of human bone. *Nat. Biotechnol.*, 18(9), 954–8.

120. Tarnowski, C.P., Stewart, S., Holder, K., Campbell-Clark, L., Thoma, R.J., Adams, A.K., Moore, M.A. (2003). Effects of treatment protocols and subcutaneous implantation on bovine pericardium: A Raman spectroscopy study. *J. Biomed. Opt.*, 8(2), 179–84.

121. Schulmerich, M.V., Dooley, K.A., Morris, M.D., Vanasse, T.M., Goldstein, S.A. (2006). Transcutaneous fiber optic Raman spectroscopy of bone using annular illumination and a circular array of collection fibers. *J. Biomed. Opt.*, 11(6), 060502.

122. Silva, M.J., Brodt, M.D., Wopenka, B., Thomopoulos, S., Williams, D., Wassen, M.H., Ko, M., Kusano, N., Bank, R.A. (2006). Decreased collagen organization and content are associated with reduced strength of demineralized and intact bone in the SAMP6 mouse. *J. Bone Miner. Res.*, 21(1), 78–88.

123. Choo-Smith, L.P., Edwards, H.G., Endtz, H.P., Kros, J.M., Heule, F., Barr, H., Robinson, J.S., Jr., Bruining, H.A., Puppels, G.J. (2002). Medical applications of Raman spectroscopy: From proof of principle to clinical implementation. *Biopolymers*, 67(1), 1–9.

124. Materny, A., Popp, J., Schmitt, M., Schrötter, H.W., Zheltiko, A.M. (2006). Preface: Wolfgang Kiefer, an appreciation. *J. Raman Spectrosc.*, 37, 1–19.

125. Dippel, B., Mueller, R.T., Pingsmann, A., Schrader, B. (1998). Composition, constitution, and interaction of bone with hydroxyapatite coatings determined by FT Raman microscopy. *Biospectroscopy*, 4(6), 403–12.

126. He, J.P., You, N.T., Chen, Q., Hu, X.N., Zheng, Q.X., Li, S.P., Squire, J. (1999). Raman spectra of porous beta-TCP bioceramics implanted into rabbit femur. *Acta Biochim. Biophys. Sinica*, 31(2), 138–144.

127. Dopner, S., Muller, F., Hildebrandt, P., Muller, R.T. (2002). Integration of metallic endoprotheses in dog femur studied by near-infrared Fourier-transform Raman microscopy. *Biomaterials*, 23(5), 1337–45.

128. Darr, J.A., Guo, Z.X., Raman, V., Bououdina, M., Rehman, I.U. (2004). Metal organic chemical vapour deposition (MOCVD) of bone mineral like carbonated hydroxyapatite coatings. *Chem. Commun. (Camb.)*, 2004(6), 696–7.

129. Lopes, C.B., Pinheiro, A.L., Sathaiah, S., Duarte, J., Cristinamartins, M. (2005). Infrared laser light reduces loading time of dental implants: A Raman spectroscopic study. *Photomed. Laser Surg.*, 23(1), 27–31.

130. Penel, G., Delfosse, C., Descamps, M., Leroy, G. (2005). Composition of bone and apatitic biomaterials as revealed by intravital Raman microspectroscopy. *Bone*, 36(5), 893–901.

131. Tarnowski, C.P., Ignelzi, M.A., Jr., Wang, W., Taboas, J.M., Goldstein, S.A., Morris, M.D. (2004). Earliest mineral and matrix changes in force-induced musculoskeletal disease as revealed by Raman microspectroscopic imaging. *J. Bone Miner. Res.*, 19(1), 64–71.

132. Draper, E.R., Morris, M.D., Camacho, N.P., Matousek, P., Towrie, M., Parker, A.W., Goodship, A.E. (2005). Novel assessment of bone using time-resolved transcutaneous Raman spectroscopy. *J. Bone Miner. Res.*, 20(11), 1968–72.
133. McCreadie, B.R., Morris, M.D., Chen, T.C., Sudhaker Rao, D., Finney, W.F., Widjaja, E., Goldstein, S.A. (2006). Bone tissue compositional differences in women with and without osteoporotic fracture. *Bone*, 39(6), 1190–5.
134. Pillay, I., Lyons, D., German, M.J., Lawson, N.S., Pollock, H.M., Saunders, J., Chowdhury, S., Moran, P., Towler, M.R. (2005). The use of fingernails as a means of assessing bone health: A pilot study. *J. Womens Health (Larchmt)*, 14(4), 339–44.
135. Towler, M.R., Wren, A., Rushe, N., Saunders, J., Cummins, N.M., Jakeman, P.M. (2007). Raman spectroscopy of the human nail: A potential tool for evaluating bone health? *J. Mater. Sci. Mater. Med.*, 18(5), 759–63.
136. Moran, P., Towler, M.R., Chowdhury, S., Saunders, J., German, M.J., Lawson, N.S., Pollock, H.M., Pillay, I., Lyons, D. (2007). Preliminary work on the development of a novel detection method for osteoporosis. *J. Mater. Sci. Mater. Med.*, 18(6), 969–74.

10

RAMAN APPLICATIONS IN CANCER STUDIES

Young-Kun Min, D.Sc.
The University of Tokyo

Satoru Naito, Ph.D.
Tochigi Research Laboratories, Kao Corporation

Hiroya Yamazaki, M.D.
Tokyo Metropolitan Hiroo Hospital

Ehiichi Kohda, M.D.
Ohashi Hospital, Toho University School of Medicine

Hiro-o Hamaguchi, Ph.D.
The University of Tokyo

10.1 INTRODUCTION

Biological soft materials are very complex and almost always contain plenty of water. Usually, they consist of proteins, nucleic acids, lipids, and polysaccharides that have highly ordered and very fragile structures. In order to understand these unique characteristics of biological soft materials, it is necessary to investigate their structures *in vivo* without any treatment. Unfortunately, however, there is no universal analytical method for investigating soft materials *in vivo*. Among the many experimental techniques such as microscopy, rheology,

Raman Spectroscopy for Soft Matter Applications, Edited by Maher S. Amer

calorimetry, and so on, spectroscopic methods provide useful and detailed chemical structural information. Raman spectroscopy is one of the spectroscopic methods that can be suitably applied to biological soft materials [1]. From a viewpoint of *in vivo* capabilities, Raman spectroscopy is superior to other spectroscopic methods, such as nuclear magnetic resonance, infrared absorption, dielectric spectroscopy, and UV-visible spectroscopy. In order to optimize the advantages of Raman spectroscopy, it is essential to use an excitation wavelength as long as possible (see more detailed discussion below). Fortunately, we are equipped in our laboratory with a prototype near-infrared multichannel detector that has the highest sensitivity so far in the deep near-infrared region up to 1.4 μm. In this chapter, we introduce our 1064-nm excited near-infrared Raman system that uses this detector as a state-of-the-art spectroscopic tool for investigating biological soft materials.

Rapid progress in noninvasive clinical diagnoses of cancers has been made with the recent development of medical imaging techniques such as ultrasonography, computed tomography, nuclear medicine, positron emission tomography, and magnetic resonance imaging (MRI). In spite of this progress in noninvasive cancer diagnosis, there still remain a number of occasions in which biopsy has to be chosen. Biopsy is an invasive procedure that can cause serious damage in patients. Furthermore, considerable time and cost have to be spent in taking the biopsy sample, and specialized knowledge of a pathologist is required in the diagnosis. In order to overcome these difficulties with biopsy, introduction of a more advanced noninvasive diagnosis method has been longed for. Raman spectroscopy has already been shown to be ideally suited for noninvasive diagnosis of cancers at the molecular level [2–25].

As was discussed in Chapter 2, Raman spectroscopy is based on Raman scattering, discovered by C.V. Raman in 1928 [26]. Thanks to the technical developments of lasers, optical electronics, and computers, it showed a tremendous development after the 1970s. Raman spectroscopy was born and grew up in the 20th century. The major three advantages of Raman spectroscopy are as follows: (1) The Raman spectrum, referred to as "the molecular fingerprint," sensitively reflects the structure of molecules. It is straightforward to extract information on structures, structural changes, and how they interact with their environments, once Raman spectra are obtained. (2) Since sample pretreatment procedures are unnecessary, it is possible to examine an untreated sample *in vivo* or *in situ*. In fact, Raman spectroscopy is applied to a variety of materials, from molecular systems, such as solids, liquids, solutions, and gases, to biological and medical systems, such as living cells and human tissues. (3) Raman spectroscopy is provided with high time resolution (several picoseconds) and high spatial resolution (several hundred nanometers). It is possible to investigate changing molecular systems in detail by measuring the time- and space-resolved Raman spectra.

10.2 DEEP NEAR-INFRARED EXCITED RAMAN SPECTROSCOPY

In spite of the many advantageous features discussed above, Raman spectroscopy is by no means used more extensively than infrared absorption and nuclear magnetic resonance spectroscopy. The reason is the difficulty caused by the interference from fluorescence. In order to measure Raman scattering, the excitation laser has to illuminate the sample. If the sample itself or the impurities contained in the sample absorb the excitation laser light and emit fluorescence, the much stronger fluorescence totally masks the weak Raman scattering signals. In physical and chemical applications, it is possible to purify samples and therefore, by using highly purified samples, the fluorescence problem can be bypassed. However, in biological, medical, and industrial applications, it is not possible to purify samples because they have to be measured as they are. Therefore, the fluorescence interference becomes more problematic and hinders extensive use of Raman spectroscopy in these applications.

In order to overcome this problem of fluorescence in Raman spectroscopy, many attempts have been made. They can be classified into three categories: (1) the near-infrared excitation method [27–29], (2) the nonlinear Raman method, and (3) the time-resolved method. The near-infrared excitation method will be discussed in detail below. The nonlinear Raman method includes coherent anti-Stokes Raman scattering (CARS) spectroscopy, coherent Raman gain spectroscopy, and inverse Raman spectroscopy [30]. The time-resolved method utilizes the different temporal behaviors of Raman scattering and fluorescence. Time-gated photon counting [31,32], streak-camera [33] and optical Kerr gating [34] have been used to eliminate fluorescence that is emitted with a time delay from the excitation laser pulse. It is fair to say that the near-infrared excitation method is the most versatile among the three.

The near-infrared excitation reduces the possibility of photoexcitation of fluorescent species in the sample and thereby suppresses the fluorescence background to a great extent. It also makes Raman measurements more biofriendly because long-wavelength excitation causes less photodamages to biological samples. Nevertheless, the feasibility of near-infrared excited Raman spectroscopy has not been fully examined until now because of the limitation of the detector sensitivity. With the use of a CCD detector, the possible longest excitation wavelength is 800 nm. We have been exploring the possibility of using the Nd:YAG 1064-nm line for excitation [19,20,22,28,35] and have recently developed a highly sensitive deep near-infrared multichannel Raman system.

10.2.1 The 1064-nm Excited Raman System

The setup of a deep near-infrared multichannel Raman system is shown in Figure 10.1. A Q-switched Nd:YAG laser is used as the near-infrared Raman excitation source. A single polychromator that has two entrance ports on the side and front of the body can be used for the right-angle scattering and the backscattering geometries, respectively. In the right-angle scattering geometry, Raman scattered light is collected by a lens and focused on the entrance slit of the polychromator through a notch filter for cutting Rayleigh scattering. In the backscattering geometry, a He-Ne laser beam can be overlapped with the 1064-nm beam in order to visualize the Raman sampling spot. The 1064-nm laser beam is reflected by a dichroic mirror and then focused on the sample by a lens to a spot size of 100 μm. Raman scattered light is collected by the same lens, passed through the dichroic mirror, and reflected by an Au-coated mirror and finally focused on the entrance slit of the polychromator through a holographic notch filter. A newly developed InP/InGaAsP near-infrared multichannel detector (see below for details) [36] is coupled with a relay lens to the exit slit of the polychromator. In order

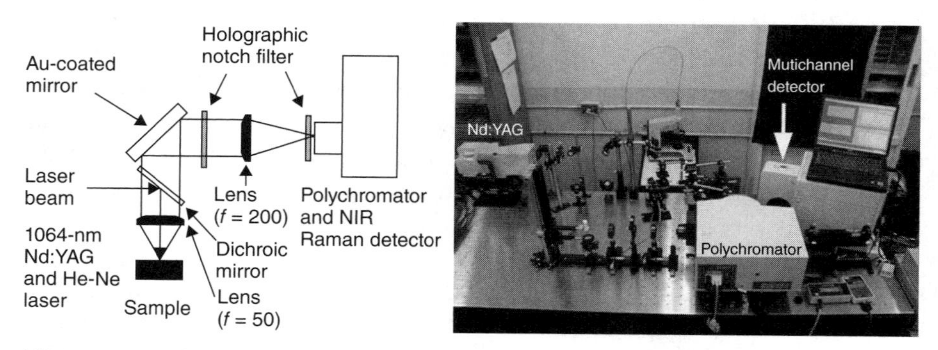

Figure 10.1 Schematic diagram of the backscattering geometry and a photograph of the near-infrared multichannel Raman system.

to eliminate the thermal noise, the multichannel detector is electrically gated (100-ns gated width), synchronizing with the excitation laser pulse. The spectral resolution is ca. $10\,cm^{-1}$.

10.2.2 The Deep Near-Infrared Multichannel Detector

Raman scattering signal is very weak, so an extremely sensitive method such as photon counting is required for its detection. However, detectors so far available in the deep near-infrared region have been much less sensitive than those in the visible/UV region and this fact has hindered the development of deep near-infrared excited Raman spectroscopy. Recently, an image intensifier based on a newly developed InP/InGaAsP photocathode has been developed by Hamamatsu Photonics K.K. to achieve a high sensitivity over the Raman shift range up to $1800\,cm^{-1}$ when excited by 1064 nm [36]. The quantum efficiency curve of the photocathode is shown in Figure 10.2.

10.2.3 Optical Fiber Probe

In order to conduct clinical Raman experiments *in vivo*, it is essential to deliver the excitation laser beam to the inside of the body, say into the lung, through an optical fiber, and to collect Raman scattering from there with another optical fiber. Two possibilities are conceivable for introducing the optical fiber into the human lung. One is to introduce

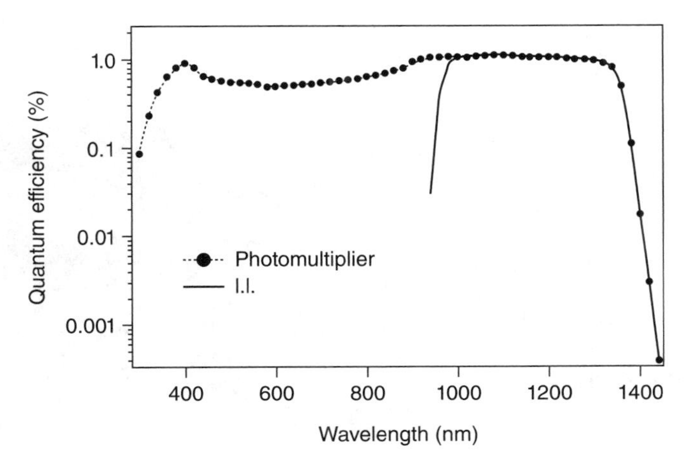

Figure 10.2 Quantum efficiency curve of the new InP/InGaAsP detector (Hamamatsu Photonics K.K.). Rigid line: image intensifier (I.I.); dotted line: photomultiplier.

the fiber probe through one of the holes of an endoscope. The diameter of the fiber probe in this case must be smaller than 2.0 mm. The other is to introduce the fiber probe directly through a thoracoscope. It is often necessary to use a thoracoscope in order to reach the deep part of the lung far from the bronchi. The fiber probe to be used with a thoracoscope can have a considerably larger diameter, up to 10 mm.

Here we show a thick fiber Raman probe that we use for routine measurements of fresh human lung tissues and skins. It is based on a custom-made 10-mm Raman probe for 1064-nm excitation (InPhotonics, Inc., Norwood, MA; Figure 10.3) consisting of a 100-µm excitation fiber and a 300-µm collection fiber. For easier insertion into a human body, the probe tube is made to be 25 cm long. A sapphire ball lens (1-mm focal length) is attached to the end of the probe to measure Raman scattering from a fixed contact point. By assembling a band pass filter, dichroic mirror, edge filter, and lenses, the Raman signals from the fiber itself can be completely blocked. The collection fiber is directly connected to the polychromator with a fiber coupler.

With this thick 10-mm diameter probe, near-infrared Raman spectra of fresh human lung tissues were measured under the conditions of 35-mW laser power and 10-minute exposure time. We have also tested a short probe tube ($L = 5$ cm) assembled with an aspherical lens ($f = 4.5$ mm). We were able to improve the experimental conditions to 22-mW laser power and 3-minute exposure time with the use of this short version of the Raman probe. We are now developing a high-throughput, long, thick fiber probe without using a hollow tube.

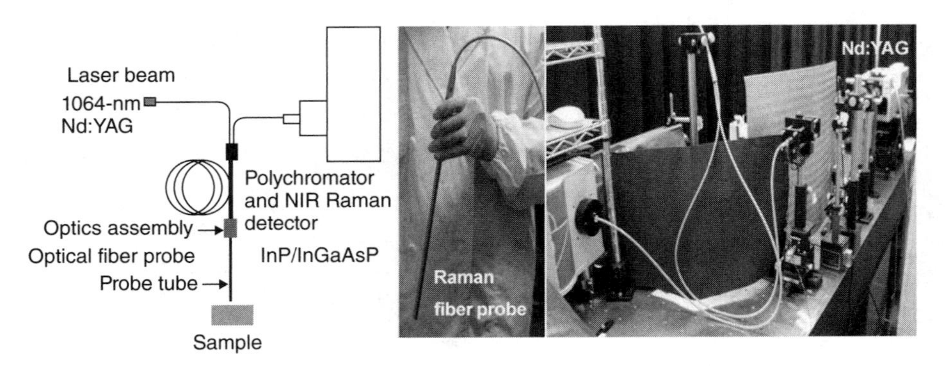

Figure 10.3 Near-infrared (NIR) fiber Raman spectroscopy with the long, thick optical fiber probe.

10.2.4 Effect of the Excitation Wavelength, From Near Infrared to Deep Near Infrared

By using the 1064-nm near-infrared Raman system described above, a variety of biological soft materials have been studied. Among the many biological soft materials investigated, human lung tissues show strong fluorescence if excited with visible/UV light. They also strongly fluoresce even with the near-infrared excitation at 785 nm (Figure 10.4a). The fluorescence entirely masks the much weaker Raman scattering signals at this excitation wavelength. On the other hand, the deep near-infrared excitation at 1064 nm dramatically reduces the fluorescence background and shows several prominent Raman features with high signal-to-noise (S/N) ratios (Figure 10.4b). The background at 1700 cm^{-1} is reduced from 6×10^3 counts/second to 0.04 count/second, five orders of magnitude, ongoing from 785 to 1064 nm. Thus, near-infrared excitation at 785 nm is not good enough for human lung tissues. They require deep near-infrared excitation at 1064 nm.

As another enlightening example, Raman spectra of an aqueous solution of ethidium bromide (EB) [37] with two different excitation wavelengths, 532 and 1064 nm, are compared in Figure 10.5. EB is known as a strong DNA intercalator and is widely used in biochemical research for dyeing DNA. The strong fluorescence from EB makes it impossible to obtain Raman spectra with 532-nm excitation; only a broad and structureless fluorescence background with no recognizable Raman signatures is seen (Figure 10.5a). On the other hand, excitation at 1064 nm decreases the fluorescence background to a great extent, and a high S/N ratio Raman spectrum of EB is obtained (Figure 10.5b).

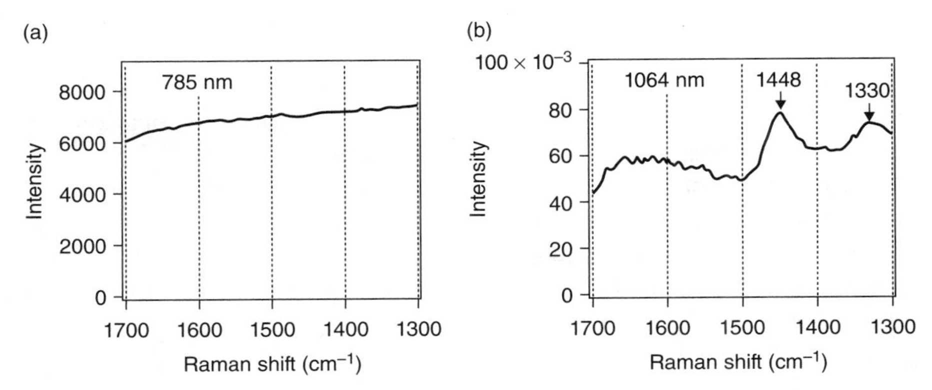

Figure 10.4 Near-infrared Raman spectra of a fresh human lung tissue with cancer. (a) Excited at 785 nm (exposure time, 1 minute; laser power, 46 mW; slit width, 50 μm [5 cm^{-1}]); (b) excited at 1064 nm (exposure time, 5 minutes; laser power, 38 mW; slit width, 150 μm [9 cm^{-1}]).

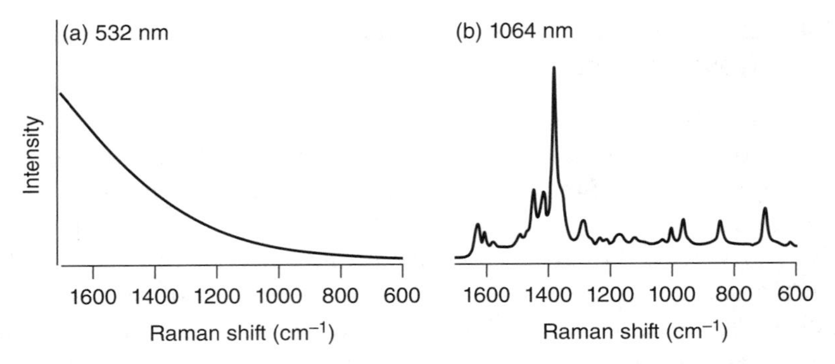

Figure 10.5 Comparison of the Raman spectra of ethidium bromide (EB) in an aqueous solution: (a) 532- and (b) 1064-nm excitation.

The use of a 1064-nm excitation facilitates the measurements of Raman spectra of molecules, such as EB, that generate strong fluorescence with visible excitation. Studies conducted using 1064-nm excitation line for investigating the effect of EB intercalation into DNA showed that the B- to an A-like structural transition occurs with increasing EB concentration [37].

10.3 APPLICATIONS TO HUMAN LUNG CANCER

The mortality from human lung cancer is very high all over the world. In spite of continuous developments of noninvasive cancer diagnosis tools using imaging techniques, such as X-ray computerized tomography, ultrasonography, MRI, and positron emission tomography, early detection and accurate diagnosis of cancers still require more sophisticated tools [25]. Basically, these imaging methods do not lead to decisive disease diagnosis, because they give only limited dimensional information about the disease such as its size, shape, and location. As diseases are induced by biochemical processes, mutation, and/or infection, cancers must necessarily be accompanied by specific molecular composition changes in the tissues. Vibrational Raman spectroscopy is highly sensitive to these molecular composition changes and therefore is potentially suitable for further advanced cancer diagnosis at the molecular level.

10.3.1 Standard Raman Spectra Representing Normal and Cancer Lung Tissues

From the 530 accumulated Raman spectra of 35 patients, standard Raman spectra representing normal and cancerous fresh lung tissues,

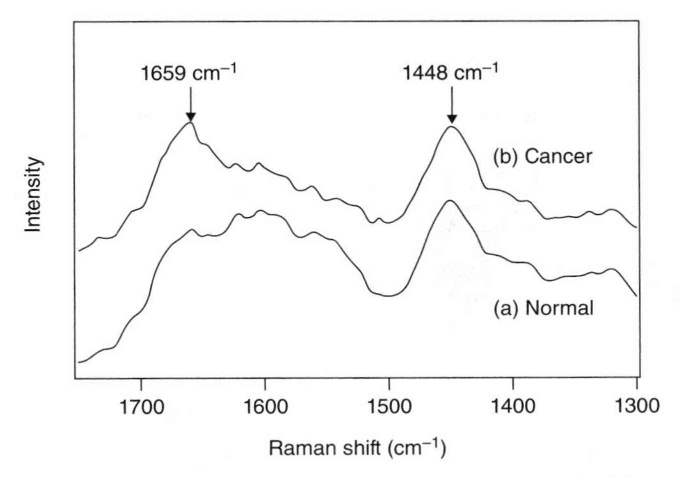

Figure 10.6 Averaged 1064-nm excited Raman spectra obtained from (a) normal and (b) cancerous human lung tissues.

respectively, have been extracted as shown in Figure 10.6. We can make a clear distinction between the normal and cancerous states with fresh tissue samples. The intensity of the 1659-cm^{-1} band (amide I) increases markedly ongoing from the normal state to the cancerous state. Why is the amide I band sensitive to cancer development? The lung has a unique physical structure and chemical composition that differ from other organs. The inside of the lung is filled with alveoli. Owing to the structure of alveoli, normal lung tissues are less dense than the tissues of other organs and contain only a smaller amount of proteins. When cancer develops, the tissues become dense by the increase of fibrous components, mostly collagen-like proteins. This change in the protein concentration is the most likely cause of the increase of the amide I band intensity in the cancerous tissues. The Raman spectra of coagulated blood and the carbon matter engulfed in the lung by scavenger cells were also measured [25]. These spectra and their identification will be indispensable in practical clinical diagnosis for extracting reliable Raman spectroscopic information about the lung tissues themselves.

10.3.2 Noninvasive Selective Measurements

Long-wavelength excitation gives another benefit enabling the acquisition of Raman signals from deeper area of tissues, thanks to the long penetration depth at the near infrared. As a testing sample, fresh lung cancer tissue including pleura was prepared as shown in the photo in Figure 10.7, left. With 22-mW laser power on the sample, Raman spectra

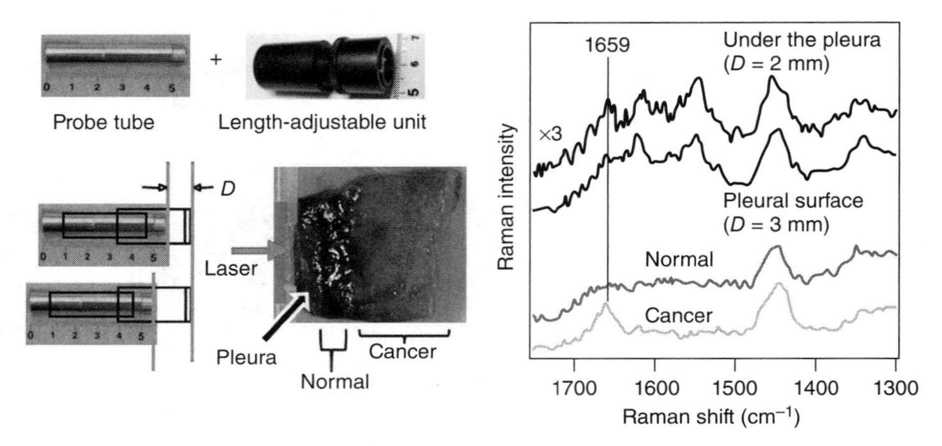

Figure 10.7 Photos of the length-adjustable unit and a fresh lung cancer tissue including pleura (left) and Raman spectra measured by changing the probe position (right).

were measured by 3 minutes acquisition time, using a modified short, thick probe tube that is equipped with a length adjustable unit with a sapphire window. When the Raman probe was placed at the position where the laser beam was focused just on the pleural surface ($D = 3\,\text{mm}$), the Raman spectrum from the surface of the lung was obtained (Figure 10.7, right). If the Raman probe was brought closer ($D = 2\,\text{mm}$) to the pleural surface, a different Raman spectrum having a cancerous character (1659-cm^{-1} amide I band) was observed (Figure 10.7, right). These spectral changes imply that the laser light reaches to the inside cancer tissues through the pleura and excites Raman scattering there. Thus, the possibility of the depth-selective measurement of the inside cancer tissues through outside pleura of the lung was confirmed, simply by adjusting the distance between the end of Raman probe and the surface of the pleura.

10.4 APPLICATIONS TO NEUROBLASTOMA

Raman spectroscopy has already been applied to the diagnosis of hepatocellular carcinoma [10], uterine and cervical cancer [13], breast cancer [8,9], malignant skin lesions [12], and lung cancer [20,22,23,25]. However, it has never been applied to the diagnosis of adrenal tumors because of strong fluorescence emitted from adrenal tissues. Using the 1064-nm laser line, it was possible to bypass the interference of fluorescence and to obtain high S/N Raman spectra of adrenal tissues, which had previously been thought impossible to measure by Raman spectroscopy. A total of 42 Raman spectra were collected from adrenal tissues and neuroblastomas. The spectra were analyzed for diagnosis of neuroblastoma.

10.4.1 Specimens and Condition for Raman Measurements

For Raman investigation, the following procedure was used. Specimens of nine neuroblastomas were extracted during operations. Of those specimens, three were from males and six were from females, with the age ranging from 0 to 4 years. For comparison, specimens of six normal adrenal tissues extracted from autopsies were used. Of those specimens, one was from a male and two were from females, with the age ranging from 77 to 85 years. The specimens were preserved in 10% formaldehyde solutions. A cube of 1 cm was cut from each specimen of neuroblastoma for Raman measurements. The same procedure was followed for each normal adrenal medulla and cortex tissues. The laser power was 52 mW with an exposure time of 27 minutes. All the measurements were carried out under room light and at room temperature. After the experiment, each tissue sample was used to create a hematoxylin–eosin (HE)-stained specimen, and tissue diagnosis was performed by a pathologist to establish the histopathological diagnosis of the sample. No tissue damage by laser irradiation was found in any part of the cancerous or noncancerous samples.

10.4.2 Raman Spectra of Neuroblastomas and Normal Adrenal Tissues

The observed Raman spectra of neuroblastomas and normal adrenal tissues are compared in Figure 10.8. Characteristic peaks are observed at 1254, 1303, 1438, and 1663 cm^{-1}. The 1254- and 1303-cm^{-1} bands are assigned to the amide III bands of proteins. The ~1438-cm^{-1} band is assigned to the CH_2 bending vibrations of lipids and/or proteins and the 1663-cm^{-1} band to the amide I band of proteins [22,27,28,35]. The 1490-cm^{-1} band is assigned to formaldehyde. The other weak bands are assignable to nucleic acids and other molecular species, but their detailed assignment will need further consideration.

The Raman spectra of neuroblastomas (Figure 10.8a) show very similar spectral patterns with one another, but they are distinct from those of normal adrenal tissues (Figure 10.8b). For the neuroblastoma samples, the peak of formaldehyde at 1490 cm^{-1} is characteristically detected with a strong intensity. In addition, fluorescence from the

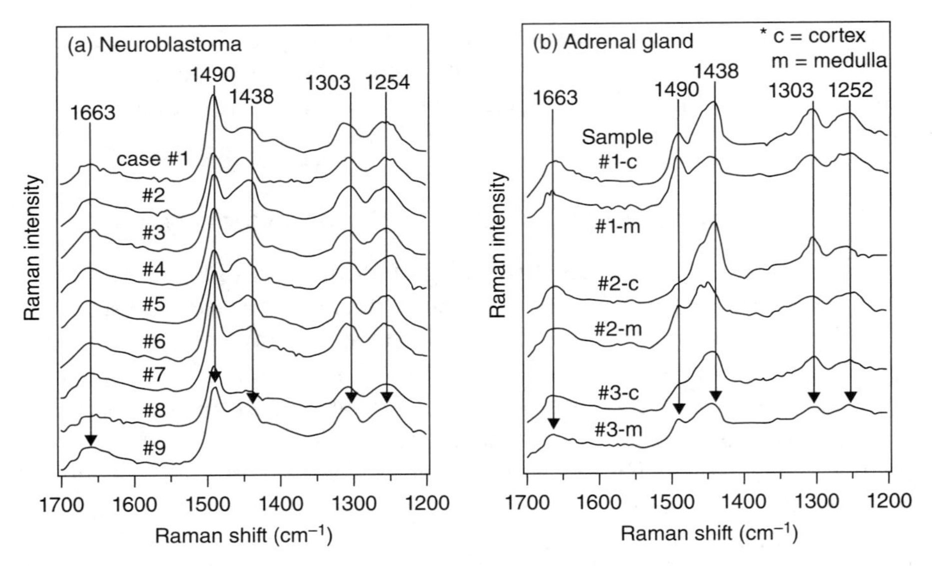

Figure 10.8 Observed Raman spectra of neuroblastomas and normal human adrenal tissues (average power, 52 mW; exposure time, 27 minutes). (a) Raman spectra of nine neuroblastomas. (b) Raman spectra of three normal adrenal glands.

normal adrenal tissues affects their Raman spectra by increasing their backgrounds.

10.4.3 Differentiation of Neuroblastomas From Normal Adrenal Tissues

It is possible to differentiate neuroblastomas from normal adrenal tissues in the spectral region of 1300–1500 cm^{-1}. Characteristic peaks are observed at ~1440 and ~1300 cm^{-1} for both the normal adrenal tissues and nueroblastomas. Analyzing the intensity ratio of the Raman band at 1438 cm^{-1} versus 1303 cm^{-1} shows that the intensity ratio ($\boldsymbol{I}_{1438}/\boldsymbol{I}_{1303}$) is much larger than 1 for each of the normal adrenal tissue. On the other hand, the intensity ratio is comparable or smaller than 1 for each neuroblastoma. Therefore, this intensity ratio is a potential diagnostic parameter for discriminating neuroblastomas from normal adrenal tissues. Such studies show that 1064-nm excited Raman spectroscopy is capable of making a distinction between neuroblastomas and normal adrenal tissues. In spite of the fact that tissue samples used were preserved in 10% formaldehyde solutions, the good correspondence between *in vitro* and *in vivo* measurements has already been established in a number of studies [13,22,23,25]. We believe that similar results as shown here will be obtained from measurement on live patients.

10.5 APPLICATIONS TO SKIN DERMIS

The dermis is responsible for the resilience and elasticity of the skin. The chemical composition of the dermis (collagen, elastin, hyaluronic acid, etc.) relates closely to the skin aging process [38–41]. Very important appendages of the skin, such as hair follicles and sebaceous glands, are also located at the dermis, the lowest layer of the skin. These appendages are involved in the development of acne [42,43]. The *in vivo* analysis of the dermis is highly important for diagnosing these skin functions in terms of chemical composition changes.

Although many analytical methods have been developed for investigating the human skin *in vivo*, most of these methods are for the stratum corneum or for the epidermis. For *in vivo* analysis of the dermis, many physicochemical methods such as ultrasound imaging, optical coherence tomography, MRI, near-infrared diffuse reflectance spectroscopy, and so on are used. However, ultrasound imaging and optical coherence tomography only provide structural information of the dermis. MRI yields only the distribution image of a limited specific nucleus. Near-infrared diffuse reflection gives extremely complicated spectra, and its analytical depth is ambiguous. Although several imaging techniques, such as confocal fluorescence imaging, two-photon fluorescence imaging, and second harmonic generation imaging, are attracting considerable attention as new methods for skin analysis *in vivo*, they show only the distribution image of specific probe molecules. An efficient *in vivo* analysis method for measuring the dermis is yet to be developed.

Raman spectroscopy is a promising method for obtaining chemical information of the dermis *in vivo*. Recently, confocal Raman microspectroscopy has been shown to be powerful for studying the components of the stratum corneum or of the epidermis at a distinct depth [44–46], but not of the dermis. Due to the shorter excitation wavelengths (less than 830 nm), it is not applicable to the dermis because of the short analytical depth (about 200 μm) at these wavelengths.

Thus, 1064-nm near-infrared excited Raman spectroscopy is the most suitable for analyzing the dermis *in vivo* [25]. Incident light with a longer wavelength can penetrate into the dermis of skin owing to low scattering probability. It is also effective in decreasing the fluorescence from the blood in the dermis and that from the melanin in the epidermis. In fact, there are many reports on the 1064-nm excited Raman spectra of skin *in vitro*, as well as a few *in vivo* [47–49]. However, because of the low sensitivity of Fourier transform (FT) Raman spectrometers, high-power irradiation (at least a few hundred milliwatts) was necessary to obtain reliable Raman spectra. Such a high-power irradiation is far from being safe for patients.

10.5.1 Analytical Depth of 1064-nm Excited Raman Spectroscopy

First, to evaluate the quantitative analytical depth of the 1064-nm excited Raman spectroscopy, porcine skin, which was cut into a wedge shape, was prepared. It was placed on a glass plate coated with 4-dimethylamino-4′-niroazobenzene (DANA), which is a Raman-sensitive material, to give strong Raman signals with a weak laser irradiation. Raman spectra were measured by changing the probe position as illustrated in Figure 10.9. The laser power was kept at 30 mW.

The Raman intensities of DANA and skin are obtained from the well-isolated Raman band of DANA at 1585 cm^{-1} and the CH_2 bending band of the skin at 1450 cm^{-1}, respectively (Figure 10.9a). In Figure 10.9b, Raman intensity changes of DANA (i) and the skin (ii) are plotted as functions of the skin thickness. The intensity of the skin

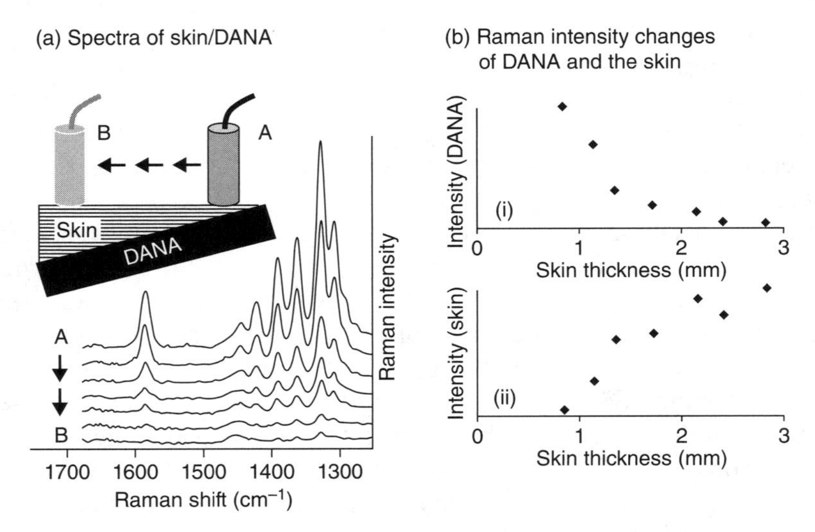

Figure 10.9 Raman spectra of the skin/4-dimethylamino-4′-nitroazobenzene (DANA) sample.

reaches a plateau around 2 mm. Thus, it can be estimated that the maximum analytical depth of the 1064-nm Raman spectroscopy is about 2 mm and that we can directly measure human dermis, the lowest layer of skin, by using this method.

10.5.2 Raman Spectra of the Surface and the Deeper Layers of Porcine Skin

Figure 10.10 shows the Raman spectra of raw porcine skin: (a) an intact porcine skin, (b) the internal layer remaining after surface layer peeling of 450 μm (corresponding to the spectrum of dermis, the thickness of the epidermis being about 200 μm), and (c) the surface layer peeled by a thickness of 450 μm (corresponding to the combination of epidermis and dermis). The peak position of the amide I band is the same in Figure 10.10a,b but is shifted to a lower wavenumber in Figure 10.10c. The difference in the protein secondary structure between epidermis and dermis is likely to cause this difference in the amide I peak position. Keratin is a characteristic protein of the epidermis, whose main

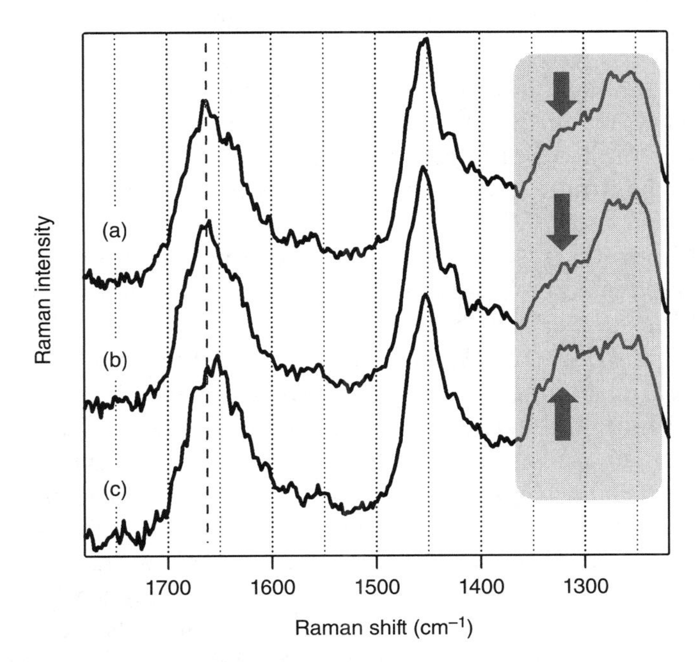

Figure 10.10 Near-infrared excited Raman spectra (1064 nm) of porcine skin at different layers: (a) intact porcine skin, (b) internal layer remaining, and (c) surface layer peeled.

secondary structure is α-helix. On the other hand, collagen, a triple-strand fibrous protein, is a characteristic protein of the dermis.

The spectrum of the surface layer (Figure 10.10c) shows distinct features in the 1200–1350 cm^{-1} region that are absent in Fig. 10.10a,b. The assignment of this variation is not clear at present. We note the possibility that this difference may also have originated from the differences in the protein secondary structure in the epidermis and dermis through the change in the amide III band. It is beyond doubt that the spectrum in Figure 10.10a resembles Figure 10.10b much more than Figure 10.10c, indicating that the 1064-nm Raman spectrum of intact skin corresponds to that of the dermis.

10.5.3 Raman Spectra of Human Skin

Raman spectra of human skin were measured from the cheek, the forehead, the inner side of the forearm, the outer side of the forearm, and the palm of four healthy volunteers (males, age 36–44 years). The recruitment of the volunteers took place according to the recommendations of the Declaration of Helsinki (Edinburgh, Scotland, 2000) and according to the basic International Conference on Harmonisation–Good Clinical Practice (ICH-GCP) guidelines. After wiping the skin with an ethanol/cotton pad, we measured Raman spectrum under the condition of 30-mW laser power and 5-minute accumulation time. The volunteers wore an eye mask during the measurement of their face (cheek and forehead). In order to extract sebum, we applied a sebum-absorbent bentonite paste (bentonite (Wako, Osaka, Japan), 2.0 g; ethanol, 1.5 g; water, 0.5 g) to the forehead (4 × 7 cm) of a man (age 36). After an hour, the resulting powder was retrieved from the forehead and dissolved in ether (3 mL). A sebum residue was left after the evaporation of the supernatant of the extractant.

Figure 10.11 shows the Raman spectra of human skin: the inner side of the forearm (Figure 10.11a), the forehead (Figure 10.11b), the palm (Figure 10.11c), and extracted sebum (Figure 10.11d). The spectra of human skin generally resemble those of porcine skin. The spectrum of the inner side of the forearm is the closest to the intact porcine spectrum among the anatomical sites measured. The spectrum of the forehead shows a shoulder at 1480 cm^{-1} and a peak at 1300 cm^{-1}. These spectral features indicate the existence of sebum (Figure 10.11d). This result is consistent with the fact that sebaceous glands are well developed in the forehead. The spectrum of the palm is close to that of the surface layer of porcine skin. It is probably because the stratum corneum of the palm (about 200 μm) is much thicker than that of the forearm

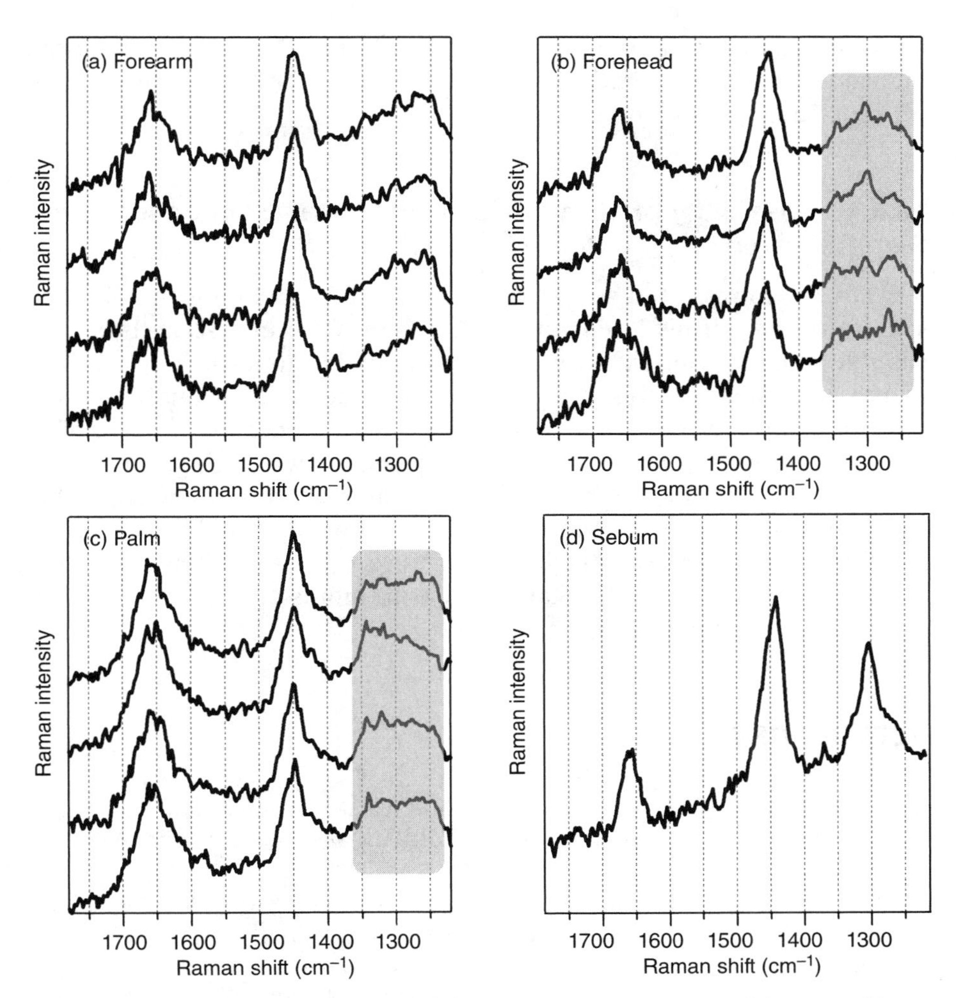

Figure 10.11 1064-nm near-infrared excited Raman spectra of human skin at different anatomical sites and that of human sebum: (a) inner side of the forearm; (b) the forehead; (c) the palm; and (d) extracted sebum.

(about 15 μm). The differences among individual spectra of the palm may reflect the difference in the stratum corneum thickness of the palms among the volunteers.

Both lipids and proteins show strong Raman bands around 1650 and 1450 cm^{-1}. The band area ratio (S [1450 cm^{-1}]/S [1650 cm^{-1}]) of lipids is much larger than that of proteins. Therefore, in the Raman spectra of skin, a larger band area ratio (S [1450 cm^{-1}]/S [1650 cm^{-1}]) indicates a higher content of lipids (sebum). The ratio is smaller for the palm than that for the forehead and the forearm, reflecting the fact that the palm has few sebaceous glands. The band ratio for the forehead shows a larger fluctuation than the other anatomical sites. This large fluctuation indicates that the band ratio depends on whether the Raman probe was set on a sebaceous gland or not.

10.5.4 Possibility of 1064-nm Excited Raman Spectroscopy in Skin Research

We have demonstrated successful attempts for obtaining the Raman spectra of the dermis *in vivo* with a 1064-nm Raman system. It is also capable of measuring skin appendages in the dermis such as sebaceous glands and hair follicles. The method may be able to analyze acne development processes and comedo growth processes *in vivo*. The chemical composition of the dermis may change with aging, skin burn, and skin disease. Deep near-infrared Raman spectroscopy will also be useful in researching the dermis from the viewpoint of chemical composition changes *in vivo*. In the near future, a confocal 1064-nm fiber Raman system that enables Raman measurements at distinct depths of the dermis may be developed.

REFERENCES

1. Hamley, I.W. (2007). *Introductions to Soft Matter: Synthetic and Biological Self-Assembling Materials*, Revised Edition. West Sussex: John Wiley & Sons, Ltd.
2. Liu, C.H., Das, B.B., Sha Glassman, W.L., Tang, G.C., Yoo, K.M., Zhu, H.R., Akins, D.L., Lubicz, S.S., Cleary, J., Prudente, R. (1991). Raman, fluorescence, and time-resolved light scattering as optical diagnostic techniques to separate diseased and normal biomedical media. *Photochem. Photobiol. Biology*, 16, 187.
3. Alfano, R.R., Liu, C.H., Sha, W.L., Zhu, H.R., Akins, D.L., Cleary, J., Prudente, R., Clemer, E. (1991). Human breast tissue studied by IR Fourier transform Raman spectroscopy. *Lasers Life Sci.*, 4, 23.
4. Barry, B.W., Edwards, H.G.M., William, A.C. (1992). Fourier transfer Raman and infrared vibrational study of human skin: Assignment of spectral bands. *J. Raman Spectrosc.*, 23, 641.
5. Williams, A.C., Barry, B.W., Edwards, H.G., Farwell, D.W. (1993). A critical comparison of some Raman spectroscopic techniques for studies of human stratum corneum. *Pharm. Res.*, 10, 1642.
6. Mizuno, A., Kitajima, H., Kawaguchi, K., Muraishi, S., Ozaki, Y. (1994). Near-infrared Fourier transform Raman spectroscopic study of human brain tissues and tumors. *J. Raman Spectrosc.*, 25, 265.

7. Keller, S., Schrader, B., Hoffman, A., Schrader, W., Metz, K., Rehlaender, A., Pahnke, J., Ruwe, M., Budach, W. (1994). Application of near-infrared-Fourier transform Raman spectroscopy in medical research. *J. Raman Spectrosc.*, 25, 663.
8. Frank, C.J., Redd, D.C.B., Gansler, T.S., McGreery, R.L. (1994). Characterization of human breast biopsy specimens with near-IR Raman spectroscopy. *Anal. Chem.*, 66, 319.
9. Frank, C.J., McCreery, R.L., Redd, D.C.B. (1995). Raman spectroscopy of normal and diseased human breast tissues. *Anal. Chem.*, 67, 777.
10. Hawi, S.R., Campbell, W.B., Kajdacsy-Balla, A., et al. (1996). Characterization of normal and malignant human hepatocytes by Raman microspectroscopy. *Cancer Lett.*, 110, 35.
11. Manoharan, R., Wang, Y., Feld, M.S. (1996). Histochemical analysis of biological tissues using Raman spectroscopy. *Spectrochim. Acta A Mol. Biomol. Spectrosc.*, 52, 215.
12. Gniadecka, M., Wulf, H.C., Nielsen, O.F., et al. (1997). Distinctive molecular abnormalities in benign and malignant skin lesions: Studies by Raman spectroscopy. *Photochem. Photobiol.*, 66, 418.
13. Mahadevan-Jansen, A., Mitchell, M.F., Ramanujam, N., Malpica, A., Thomsen, S., Utzinger, U., Robinson, J.S.J. (1998). Development of a fiber optic probe to measure NIR Raman spectra of cervical tissue in vivo. *Photochem. Photobiol.*, 68, 427.
14. Lawson, E.E., Anigbogu, A.N., Williams, A.C., Barry, B.W., Edwards, H.G. (1998). Thermally induced molecular disorder in human stratum corneum lipids compared with a model phospholipid system; FT-Raman spectroscopy. *Spectrochim. Acta A Mol. Biomol. Spectrosc.*, 54, 543.
15. Fendel, S., Schrader, B. (1998). Investigation of skin and skin lesions by NIR-FT-Raman spectroscopy. *Fresenius' Journal of Analytical Chemistry*, 360, 609.
16. Caspers, P.J., Lucassen, G.W., Wolthuis, R., Bruining, H.A., Puppels, G.J. (1998). In vitro and in vivo Raman spectroscopy of human skin. *Biospectroscopy*, 4, S31.
17. Puppels, G.J., van Aken, T., Wolthuis, R., Caspers, P.J., Bakker Schut, T., Bruining, H.A., Romer, T.J., Buschman, H.P.J., Wach, M.L., Robinson, J.S.J. (1998). In vivo tissue characterization by Raman spectroscopy. *Proc. BIOS/SPIE*, 3257 (Infrared Spectroscopy: New Tool in Medicine), 78–83.
18. Gniadecka, M., Faurskov Nielsen, O., Christensen, D.H., Wulf, H.C. (1998). Structure of water, proteins, and lipids in intact human skin, hair, and nail. *J. Invest. Dermatol.*, 110, 393.
19. Kaminaka, S., Yamazaki, H., Ito, T., Kohda, E., Hamaguchi, H. (2001). Near-infrared Raman spectroscopy of human lung tissues: Possibility of molecular-level cancer diagnosis. *J. Raman Spectrosc.*, 32, 139.

20. Kaminaka, S., Ito, T., Yamazaki, H., Kohda, E., Hamaguchi, H. (2002). Near-infrared multichannel Raman spectroscopy toward real-time in vivo cancer diagnosis. *J. Raman Spectrosc.*, 33, 498.
21. Stone, N., Kendell, C.A., Shepherd, N., Crow, P., Barr, H. (2002). Near-infrared Raman spectroscopy for the classification of epithelial pre-cancers and cancers. *J. Raman Spectrosc.*, 33, 564.
22. Yamazaki, H., Kaminaka, S., Kohda, E., Mukai, M., Hamaguchi, H. (2003). The diagnosis of lung cancer using 1064-nm excited near-infrared multichannel Raman spectroscopy. *Radiat. Med.*, 21, 1.
23. Huang, Z., McWilliams, A., Lui, H., et al. (2003). Near-infrared Raman spectroscopy for optical diagnosis of lung cancer. *Int. J. Cancer*, 107, 1047–1052.
24. Crow, P., Stone, N., Kendell, C.A., Ufl, J.S., Farmer, J.A. M., Barr, H., Wright, M.P.J. (2003). The use of Raman spectroscopy to identify and grade prostatic adenocarcinoma in vitro. *Br. J. Cancer*, 89, 106.
25. Min, Y.K., Yamamoto, T., Kohda, E., Ito, T., Hamaguchi, H (2005). 1064 nm near-infrared multichannel Raman spectroscopy of fresh human lung tissues. *J. Raman Spectrosc.*, 36, 73.
26. Raman, C.V., Krishnan, K.S. (1928). A new type of secondary radiation. *Nature*, 121, 501.
27. Min, Y.K., Ito, T., Hamaguchi, H. (2004). Lecture; near-infrared spectroscopy, V. Near-infrared excited Raman spectroscopy. *Bunko Kenkyu*, 53(5), 318.
28. Fujiwara, M., Hamaguchi, H., Tasumi, M. (1986). Measurements of spontaneous Raman scattering with neodymium: YAG 1064-nm laser light. *Appl. Spectrosc.*, 40, 137.
29. Hirschfield, T., Chase, B. (1986). FT-Raman spectroscopy: Development and justification. *Appl. Spectrosc.*, 40(2), 133–7.
30. Hamaguchi, H. (1998). In: Field, R.W., Hirota, E., Maier, J.P., Tsuchiya, S., eds. *Nonlinear Spectroscopy for Molecular Structure Determination*. Oxford: Blackwell Science, 203–222.
31. Duyne, R.P.V., Jeanmaire, R.W., Shriver, D.F. (1974). Mode-locked laser Raman spectroscopy. New technique for the rejection of interfering background luminescence signals. *Anal. Chem.*, 46, 213.
32. Watanabe, J., Kinoshita, S., Kushida, T. (1985). Fluorescence rejection in Raman spectroscopy by a gated single-photon counting method. *Rev. Sci. Instrum.*, 56, 1195.
33. Tahara, T., Hamaguchi, H. (1993). Picosecond Raman spectroscopy using a streak camera. *Appl. Spectrosc.*, 47, 391.
34. Matousek, P., Towrie, M., Ma, C., Kwok, W.M., Phillips, D., Toner, W.T., Parker, A.W. (2001). Fluorescence suppression in resonance Raman spectroscopy using a high-performance picosecond Kerr gate. *J. Raman Spectrosc.*, 32, 983.

35. Engert, C.N., Deckert, V.K., Kiefer, W., Umapathy, S., Hamaguchi, H. (1994). Design and performance characteristics of a near-infrared scanning multichannel Raman spectrometer. *Appl. Spectrosc.*, 48, 933.
36. Niigaki, M., Hirohata, T., Suzuki, T., Kan, H., Hiruma, T. (1997). Field-assisted photoemission from InP/InGaAsP photocathode with p/n junction. *Appl. Phys. Lett.*, 71, 2493.
37. Yuzaki, K., Hamaguchi, H. (2004). Intercalation-induced structural change of DNA as studied by 1064 nm near-infrared multichannel Raman spectroscopy. *J. Raman Spectrosc.*, 35, 1013.
38. Naito, S., Min, Y.K., Sugata, K., Osanai, O., Kitahara, T., Hiruma, H., Hamaguchi, H. (2008). In vivo measurement of human dermis by 1064 nm-excited fiber Raman spectroscopy. *Skin Res. Technol.*, 14, 18.
39. Fligiel, S.E.G., Varani, J., Datta, S.C., Kang, S., Fisher, G. J., Voorhees, J.J. (2003). Collagen degradation in aged/photodamaged skin in vivo and after exposure to matrix metalloproteinase-1 in vitro. *J. Invest. Dermatol.*, 120, 842.
40. Nishimori, Y., Edwards, C., Pearse, A., Matsumoto, K., Kawai, M., Marks, R. (2001). Degenerative alterations of dermal collagen fiber bundles in photodamaged human skin and UV-irradiated hairless mouse skin: Possible effect on decreasing skin mechanical properties and appearance of wrinkles. *J. Invest. Dermatol.*, 117, 1458.
41. Chung, J.H., Seo, J.Y., Lee, M.K., Eun, H.C., Lee, J.H., Kang, S., Fisher, G.J., Voorhees, J.J. (2002). Ultraviolet modulation of human macrophage metalloelastase in human skin in vivo. *J. Invest. Dermatol.*, 119, 507.
42. Thiboutot, D. (2004). Regulation of human sebaceous glands. *J. Invest. Dermatol.*, 123, 1.
43. Plewig, G., Kligman, A.M., Bluhm, C., Hollman, J. (1993). Sebaceous glands. In: Plewig G, Kligman AM, eds. *Acne and Rosacea*, 2nd Edition. Berlin, New York: Springer-Verlag, 33.
44. Chrit, L., Bastien, P., Sockalingum, G.D., Batisse, D., Leroy, F., Manfait, M., Hadjur, C. (2006). An in vivo randomized study of human skin moisturization by a new confocal Raman fiber-optic microprobe: Assessment of a glycerol-based hydration cream. *Skin Pharmacol. Physiol.*, 19, 207.
45. Caspers, P.J., Lucassen, G.W., Carter, E.A., Bruining, H.A., Puppels, G.J. (2001). In vivo confocal Raman microspectroscopy of the skin: Noninvasive determination of molecular concentration profiles. *J. Invest. Dermatol.*, 116, 434.
46. Caspers, P.J., Lucassen, G.W., Puppels, G.J. (2003). Combined in vivo confocal Raman spectroscopy and confocal microscopy of human skin. *Biophys. J.*, 85, 572.
47. Gniadecka, M., Nielsen, O.F., Wessel, S., Heidenheim, M., Christensen, D.H., Wulf, H.C. (1998). Water and protein structure in photoaged and chronically aged skin. *J. Invest. Dermatol.*, 111, 1129.

48. Schallreuter, K.U., Moore, J., Wood, J.M., Beazley, W.D., Gaze, D.C., Tobin, D.J., Marshall, H.S., Panske, A., Panzig, E., Hibberts, N.A. (1999). In vivo and in vitro evidence for hydrogen peroxide (H2O2) accumulation in the epidermis of patients with vitiligo and its successful removal by a UVB-activated pseudocatalase. *J. Invest. Dermatol. Symp. Proc.*, 4, 91.
49. Wohlrab, J., Vollmann, A., Wartewig, S., Marsch, W.C., Neubert, R. (2001). Noninvasive characterization of human stratum corneum of undiseased skin of patients with atopic dermatitis and psoriasis as studied by Fourier transform Raman spectroscopy. *Biopolymers*, 62, 141.

INDEX

Figures and Tables are indicated by *italic page numbers*, and footnotes by suffix 'n'

Raman Spectroscopy for Soft Matter Applications, Edited by Maher S. Amer